CONSTRUCTION DELAY CLAIMS

2007 Cumulative Supplement

This supplement supersedes all previous supplements.

ASPEN
PUBLISHERS

Construction Delay Claims

Third Edition

by Barry B. Bramble and Michael T. Callahan

Construction Delay Claims, Third Edition, provides thorough coverage of delay and impact claims. The authors explain different types of delay, how delays occur, and the defects of delay. Practical information is provided on what to do when delays occur and how to process claims for delay. Information on using the schedule to prove delay and establishing liability for delay is also included. *Construction Delay Claims* discusses how damages are calculated; what types of clauses to include in the construction contract concerning scheduling, delays, and impact; and how project delays arise under the various project delivery methods. In addition, a number of chapters are devoted to impact claims disruption, lost labor productivity, and acceleration. *Construction Delay Claims* is a significant resource for those involved in the management and measurement of time, containing at-a-glance charts and checklists as well as over 1,500 cases.

Highlights of the 2007 Supplement

The following developments and issues are included in the 2007 Cumulative Supplement.

- When a contractor withholds payment and the subcontract does not specify time is of the essence, the subcontractor must only show that it performed within a reasonable time.
- Increasing acceptance of the *Eichleay* formula by state jurisdictions for the recovery of unabsorbed home office overhead.
- Even though minor repairs are needed, duplexes that are occupied by tenants by a contract deadline are considered substantially complete.
- An owner is not required to perform a schedule analysis to allocate delay and liquidated damages among multiple parties when a contractor fails to provide written notice of claims.
- The *Eichleay* formula should not be used to calculate home office overhead recovery for delays prior to contract award.
- A liquidated damages provision is unenforceable where an owner fails to explain how the amount of damages for a bidder's delay has been ascertained or why the amount is a reasonable estimate.

- More restrictive definition and application of the "stand-by" requirement in order to use the Eichleay formula to recover for unabsorbed overhead for government-caused delays.
- The stated requirements for constructive acceleration have remained the same over the last decade. The requirements were re-stated in *Robust Construction, LLC*.
- On-going development of liquidated damages for contractor-caused delays: how liquidated damages are to be assessed and how the daily amount is to be determined in order to survive a challenge.
- When the government grants a time extension for change order work, there is a rebuttable presumption that the time extension is compensable and the contractor may be entitled to extended field overhead costs.
- Effective ways to challenge a claimant's use of the Total Cost Method of calculating damages.
- Some courts use the actual average overhead and profit from the contractor's cost records to determine the amount of recovery for unabsorbed overhead rather than the bid overhead and profit when applying the *Hudson* formula.
- A contractor seeking to recover lost profits in an action for tortious interference must present sufficient evidence from which lost net profits can be calculated.

The Table of Cases has been updated to reflect all the changes to the text.

10/06

For questions concerning this shipment, billing, or other customer service matters, call our Customer Service Department at 1-800-234-1660.

For toll-free ordering, please call 1-800-638-8437.

---

a Wolters Kluwer business

CONSTRUCTION DELAY CLAIMS

Third Edition

2007 Cumulative Supplement

BARRY B. BRAMBLE
MICHAEL T. CALLAHAN

ASPEN
PUBLISHERS

76 Ninth Avenue, New York, NY 10011
www.aspenpublishers.com

a Wolters Kluwer business
www.aspenpublishers.com

Printed in the United States of America

1 2 3 4 5 6 7 8 9 0

ISBN 0-7355-0548-9
ISBN 0-7355-6126-5 (supplement)

About Aspen Publishers

Aspen Publishers, headquartered in New York City, is a leading information provider for attorneys, business professionals, and law students. Written by preeminent authorities, our products consist of analytical and practical information covering both U.S. and international topics. We publish in the full range of formats, including updated manuals, books, periodicals, CDs, and online products.

Our proprietary content is complemented by 2,500 legal databases, containing over 11 million documents, available through our Loislaw division. Aspen Publishers also offers a wide range of topical legal and business databases linked to Loislaw's primary material. Our mission is to provide accurate, timely, and authoritative content in easily accessible formats, supported by unmatched customer care.

To order any Aspen Publishers title, go to *www.aspenpublishers.com* or call 1-800-638-8437.

To reinstate your manual update service, call 1-800-638-8437.

For more information on Loislaw products, go to *www.loislaw.com* or call 1-800-364-2512.

For Customer Care issues, e-mail CustomerCare@aspenpublishers.com; call 1-800-234-1660; or fax 1-800-901-9075.

Aspen Publishers
a Wolters Kluwer business

SUBSCRIPTION NOTICE

This Aspen Publishers product is updated on a periodic basis with supplements to reflect important changes in the subject matter. If you purchased this product directly from Aspen Publishers, we have already recorded your subscription for the update service.

If, however, you purchased this product from a bookstore and wish to receive future updates and revised or related volumes billed separately with a 30-day examination review, please contact our Customer Service Department at 1-800-234-1660, or send your name, company name (if applicable), address, and the title of the product to:

ASPEN PUBLISHERS
7201 McKinney Circle
Frederick, MD 21704

CONTENTS

Sections listed below appear only in the supplement and not in the main volume.

PREFACE

In preparing this year's supplement, we find that the emphasis was exclusively upon developments in case law rather than discussion of trends in how the construction industry deals with delays or delay claims. For example, it is interesting to note that there is a developing divergence between federal contract case law and state case law concerning the recovery of unabsorbed home office overhead. The courts and boards dealing with federal contracts seem to be imposing stricter requirements in order for a contractor to recover home office overhead using a *Eichleay* formula, such as the restrictive conditions relating to the "stand-by" requirement. On the other hand, it seems that more states are accepting the concept of the *Eichleay* formula, and applying it, sometimes even more liberally than the federal courts and boards of contract appeals.

Liquidated damages of owners and total cost claims of contractors are also coming under stricter scrutiny by courts. New grounds for challenging both liquidated damages and total cost claims, and more aggressive pursuit of existing challenges seem to be effective in dealing with these age-old issues. Also, there seem to be more sophisticated analyses of the facts and application of construction case law by trial courts at the state level.

Finally, it appears that courts and administrative hearing boards are refusing to allow claimants to short-cut proof of causation for both delay and loss of productivity claims by using summary conclusions, percentages, and formulas for disruptive events such as change orders. This will likely result in further case law developments in the years to come.

August 2006

Barry B. Bramble
Knowles USA/Hill International
Philadelphia, Pennsylvania
barry.bramble@jrknowles.com

Michael T. Callahan
CCL Construction Consultants, Inc.
Overland Park, Kansas
callahanmt@cclcc.com

[B] Compensable and Noncompensable Delays

Page 1-10, add to note 70:

However, when determining a subcontractor's recovery under a Miller Act bond, a contractor and its payment bond surety may be obligated for another party's delays as long as the contractor was the primary cause of the project's delay. *See* Consol. Elec. & Mech., Inc. v. Biggs Gen. Contracting, Inc., 167 F.3d 432 (1999); see also **§ 15.02**.

[C] Critical and Noncritical Delays

Page 1-14, add after carryover paragraph:

In *Morrison Knudsen Corp. v. Fireman's Fund Insurance Co.,* 175 F.3d 1221 (10th Cir. 1999), Morrison Knudsen Corporation (MK), a federal contractor, terminated its subcontractor, Ground Improvement Techniques (GIT), for an alleged default and sued GIT for damages. GIT counterclaimed for wrongful termination. A jury found the termination wrongful and awarded GIT roughly half the damages claimed.

The United States Department of Energy hired MK in 1983 to manage its Uranium Mill Tailing Remedial Action (UMTA) project, a cleanup of radioactive mill tailings at sites around the country. MK subcontracted with GIT in March 1995 to clean up the Slick Rock, Colorado site. GIT hired several lower-tier subcontractors. The MK-GIT subcontract obligated GIT to complete the project by December 1996. The contractor's default-termination clause incorporated essentially verbatim the standard federal Default clause for fixed-price construction contracts. The Clause allowed MK to terminate GIT if the work had been delayed, not for excusable reasons, but by GIT's lack of diligence. The contract price was roughly $9.3 million. The project did not go well. GIT and its subs encountered delays, difficulties, and increased costs. GIT attributed these to MK's defective specifications, failure to secure permits in a timely fashion, rigid interpretations of specifications and safety requirements, and propensity to reject proposed work plans. During the contractor's performance, GIT requested extra compensation and extensions of time because of delays to, changes in, and increased costs of the work which GIT attributed to MK. GIT's central theory was that its plan to complete the project before the deadline displeased MK, which could not then earn the maximum possible fees from DOE. MK, in GIT's view, thus sought to hinder and delay the work. MK, on the other hand, attributed the delays and increased costs to errors, omissions, and delinquencies by GIT and its subcontractors.

In September 1995, MK terminated GIT for default. The contract allowed MK to do so if GIT was not diligently prosecuting the work to ensure its timely completion. At the same time, MK sued GIT for damages caused by its alleged default. In challenging the termination, GIT claimed that it had been entitled to extensions of time for various "excusable delays" under the subcontract.

MK asserted that the court erred in denying a proposed jury instruction. The proposed instruction specified that, in order to prove a delay "excusable" under the contract, GIT had to show both that something beyond its control delayed part of its work, and that the problem would delay completion of the entire project. The court's instruction #18, "Contractual Excuse of Delays," told the jury that for each claim of excusable delay, GIT had the burden to prove that it met the subcontract requirements for excusable delay. The instruction noted that the subcontract requirements for establishing an excusable delay were explained in instruction #22. Instruction #22, in turn, concerned "GIT's Counterclaim for Extra Work, Delays and Acceleration—Essential Elements." That instruction told the jury what GIT had to prove in order to recover damages on its delay claims. It required GIT to show that something beyond its control or fault, and unauthorized by the contract, had delayed its work; that it had notified MK, and that MK had refused to allow GIT the additional time necessary to complete the work.

MK argued that a contractor was not entitled to an excusable delay unless it proved both that the delay was on the project's critical path and that it would delay completion of the entire project. MK thus asserted that the court erred in refusing to instruct the jury on the element of criticality. GIT responded that MK had not identified any part of the contract or any cases using the terms "critical path" or "criticality" or mandating jury instructions on those terms. Alternatively, GIT claimed that the instructions as a whole were sufficient. MK did not claim that the contract expressly referred to the critical path or required that excusable delays affect the project's overall completion. It argued instead that the latter requirement was a well-established rule of federal-contracting law.

The appellate court found MK's description of federal-contracting law correct. The court referred to a leading treatise that explained that "a contractor is not entitled to relief upon the mere occurrence of an event that qualifies as an excusable delay. The contractor must show that the event caused delay to the overall completion of the contract." A subsection of that treatise entitled "Delay of Overall Completion Required" discussed the case law establishing that a contractor "is not entitled to an excusable delay unless it can prove that the time lost delayed the completion of the job. It is not sufficient to establish that some work was prevented; the work prevented must be work that will delay the overall completion of the job."

MK's proposed instruction comprehensibly conveyed the requirement that a contractor must prove that a delay affected not just an isolated part of a project, but its overall completion, by explaining that GIT had to prove that:

> the delay was a "critical delay." A critical delay is one which would delay not just the particular activity at issue, but the overall completion date of the work. Many activities may be performed on a project at any time without any effect on the completion of the project. A delay in such non-critical activities will not delay the project overall and cannot constitute an excusable delay. Only delays to activities on the critical path—activities with no leeway in the schedule—may give rise to excusable delay.

problem. Three options were discussed at the meeting: (1) issuance of a change order to allow additional payment to Anderson, (2) termination of the contract without financial liability to either party, and (3) execution of the work by Anderson according to Anderson's interpretation of the contract.

On August 24, 1993, Anderson wrote:

> To prepare the islands for placement of dredged materials will cost approximately $27,000.00. If the S.I.D. is prepared to issue a Change Order to that effect Anderson will begin as agreed. However, if the S.I.D. is unwilling to issue the Change Order, then there appears to be only two other alternatives.
>
> First, they could rebid the dredging portion of the contract, and include the areas left unaddressed such as the island preparation. In this case Anderson would be willing to relinquish all rights under this contract without any further expense to the S.I.D., provided that the performance and payment bonds are returned. . . .

The second alternative required all parties to prepare for litigation and rely upon the judicial system for determination. Should this become necessary, Anderson was also prepared to exercise this alternative. The SID did not terminate the contract, but behaved as if there was a contract still in place after the August 24, 1993, letter was received. Anderson never commenced work on the dredging project, and SID did not demand that work be commenced. On December 20, 1996, Anderson brought suit against the SID alleging that (1) the SID refused to go forward with the contract or to sign a proposed change order to the contract about the responsibility for the preparation of the islands; (2) the SID abandoned and breached the contract; and (3) Anderson incurred expenses in preparing to begin work on the contract, including costs incurred for insurance and performance bonds. Anderson's petition alleged a cause of action for breach of contract seeking lost profits, and expenses for preparation for work on the project. The SID denied the allegations and alleged that Anderson had breached and abandoned the contract.

The SID moved for summary judgment. The district court granted the motion and dismissed Anderson's petition, and Anderson appealed. The Nebraska Court of Appeals determined that there were genuine issues of material fact preventing summary judgment on the issue of Anderson's reliance on the contract and reversed, and remanded. On remand, the SID moved for partial summary judgment on the breach of contract cause of action, arguing that the Court of Appeals' decision determined that there were issues of fact only on the "detrimental reliance" cause of action. The district court granted the motion. After a bench trial, the court determined that Anderson had repudiated the contract when it sent the August 24, 1993, letter demanding a change order or that the project be rebid and threatening litigation. The court determined that the unilateral repudiation of the contract precluded any equitable claims for recovery or for breach of contract. The court dismissed the petition. Anderson appealed.

Anderson was neither sent a "notice to proceed" nor was it sent a notice that the contract had been terminated. Anderson also provided evidence that a notice to proceed was different from a change order. The SID, however, claimed that the change order fulfilled the notice-to-proceed requirement in the contract because it set a date fixing the date on which work would begin under the contract. The SID failed to generate a notice to proceed for Anderson to begin work. Anderson contended that the district court erred by failing to find that the SID was in breach of the contract by failing to issue a notice to proceed and by delaying Anderson's performance of the contract. When Anderson repudiated the contract, the SID was excused from performing the condition of issuing a notice to proceed and was excused from performing any of its contractual duties. The judgment was affirmed.

Page 2-13, add at end of first full paragraph:

If notice to proceed has not been given and the construction contract has been terminated for convenience, the contractor may be entitled to recover for unabsorbed overhead, but not according to the *Eichleay* formula. In *Nicon, Inc. v. United States*, 51 Fed. Cl. 324 (2001), the government did not issue the notice to proceed on a contract for the repair of dormitories because of a bid protest. The government later decided to terminate the contract for convenience. Nicon submitted a claim for unabsorbed home office overhead calculated using the *Eichleay* formula (see **§ 12.06[A]**). In considering the matter, the U.S. Court of Federal Claims held that *Eichleay* damages only apply when contract performance has been delayed and, without notice to proceed, there was no contract performance. In further considering the matter, the U.S. Court of Appeals for the Federal Circuit in *Nicon, Inc. v. United States*, 331 F.3d 878 (Fed. Cir. 2003), agreed that recovery under the "rigid" *Eichleay* formula is not appropriate where the contract was terminated for convenience. Rather, the contractor should use a method to allocate its fixed home office expenses during the suspension period between the suspended contract and its other contracts.

The lack of a designated period for notice to proceed may result in disputes over when the contract time is to begin and end. For example, in *Daniel E. Terreri & Sons, Inc. v. Mahoning County Board of Commissioners*, 786 N.E.2d 921 (Ohio Ct. App. 2003), the contractor was awarded a contract for the renovation of a building, but the owner did not give notice to proceed. The contract required the renovations to be complete within 120 days after the notice to proceed was issued, but did not provide a time frame in which the owner had to issue the notice to proceed. Sixteen months after bid opening, the contractor demanded that the owner issue a notice to proceed and demanded a 15 percent increase in the contract price. Later, the contractor wrote the owner that the contractor considered the contract terminated and sued for overhead and loss of profit. In considering the matter, the Court of Appeals of Ohio determined that there was no contractual requirement for the owner to issue a notice to proceed in a given time frame, and that the contractor had by its actions repudiated the contract.

Page 2-14, add after first full paragraph:

Although an industry custom, a preconstruction meeting is not a prerequisite to starting work unless required by the contract. In *Program & Construction Management Group, Inc. v. Davis*, 246 F.3d 1363 (Fed. Cir. 2001), the contractor argued that the project was delayed as a result of the owner's late preconstruction meeting. The General Services Administration contracted to upgrade the heating, ventilation, and air conditioning of the cafeteria, kitchen, and other facilities in a government building. The contractor submitted a delay claim to the contracting officer. The contractor argued that the government did not schedule a preconstruction meeting shortly after issuing the notice to proceed, but instead waited 46 days before holding one. It asserted that it could not proceed with the work until the preconstruction meeting was held. Nothing in the contract, however, required a preconstruction meeting. Regardless of whether there was an industry custom to hold a meeting shortly after issuance of the notice to proceed, a preconstruction meeting was not a prerequisite to the start of work. If the contractor believed a preconstruction meeting was necessary, it should have undertaken to initiate the process by contacting the contracting officer. The time between the issuance of the notice to proceed and the convening of a preconstruction conference did not establish compensable delay to the contractor. The General Services Administration was not responsible for the contractor's erroneous impression that a preconstruction meeting was required to be held before the contractor could start work at the site.

Page 2-15, add at end of subsection:

When a contract does not include a start date, the parties may agree to a project start. Consider *Missouri Department of Transportation v. Safeco Insurance Company of America*, 2002 Mo. App. LEXIS 1761 (2002) (opinion withdrawn). Robertson Contractors, Inc. was the general contractor, and PR Developers, Inc. (PRD) was the subcontractor for the base rock and paving for the construction of a bridge, embankments, ramps, and roadways at the interchange of an interstate highway for the Missouri Highway and Transportation Commission (MHTC). The contract provided that Robertson would have 130 working days to complete the Project, starting from March 15, 1999. The subcontract stated that time was of the essence, but did not specify a start date. After the subcontract was signed, Robertson told PRD to prepare to begin its work on March 15, 1999. PRD verbally agreed to this starting date, expecting to be done within 80 working days. PRD had indicated that it would like to start work on the west side of the Project. PRD was not able to begin its work until the end of May 1999, and had to begin work on the east side of the Project, a delay of approximately 50 working days. PRD filed suit against Robertson for breach of contract. The jury returned verdicts in favor of PRD on its claims, and awarded damages for delay of $1,581,192.50 against Robertson and $74,810.16 for non-delay damages.

On appeal, Robertson contended that there was not sufficient competent evidence that Robertson made the specific promises to have the job site ready for PRD to proceed by March 15, 1999. PRD repeatedly testified that Robertson wanted PRD ready to start work on the Project on March 15, 1999. After PRD signed the subcontract, PRD was told to begin its work on the Project on March 15. PRD agreed to be ready and acted in reliance upon that agreement.

Robertson contended that this testimony violated the parol evidence rule. The subcontract was not a completely integrated document, and that PRD's testimony concerning the agreement to have PRD start work, or be prepared to start work, on March 15, 1999, did not vary or contradict any terms of the subcontract, which was silent as to the start date for PRD. If a written agreement was not completely integrated, the parol evidence rule had no application, and parol evidence was competent to resolve the meaning. PRD and Robertson did not make the agree ment prior to or contemporaneous with PRD signing the subcontract. Accordingly, the testimony concerning the agreement did not violate the parol evidence rule.

Robertson also contended that there was no consideration for the start date agreement, and that it therefore failed. PRD testified that by agreeing to be ready to start on March 15, 1999, PRD, a single project company, was unable to bid on any other projects and undertake work while it was committed to working as a subcontractor for Robertson. By agreeing to be ready to start on March 15, PRD effectively committed equipment and labor to the Project, which resulted in those resources being unavailable for other work in March, April, and May 1999. PRD also stated that PRD had estimated that its work would take 80 work days, and had expected to be able to work on other projects beginning in August 1999. PRD could have started work on the Project much later without causing any difficulties to Robertson, which had 130 days to complete the Project for MHTC. The delays that the Project experienced resulted in PRD not being able to allocate its equipment and employees to other projects until late October 1999. PRD's agreement to commit its equipment and personnel to starting on March 15, earlier than it needed to do to perform the subcontract work, was sufficient consideration to support the Agreement.

[B] Contract Completion

Page 2-16, add to note 49:

Lange v. City of Batesville, 832 So. 2d (2002) (When no performance time is stated a reasonable time to complete construction is implied. Panola County developed plans for construction of an arena. The drawings included a five-lane road leading from the project and intersecting with a highway. The Whitakers and the Langes' property and the arena property shared a common boundary. The county agreed to transfer to the owners an 11-acre parcel of property to the east of the road, but only if a road was built on the Whitaker-Lange property. The county later determined that it did not have the financial resources to continue with the project. The county further stated that the "project must be ready to be occupied by owner by the 22 of August." However, the "22" was crossed out, "30" was written in above, and the change was initialed by both the subcontractor and the contractor. A letter from the contractor was admitted that indicated that performance by the subcontractor as late as November of 1995 was acceptable. The subcontractor installed the wrought-iron fence in the first week of October 1995, and submitted an invoice for the work. The contractor tendered a check for the contract amount minus liquidated damages. The subcontractor sued the contractor to recover the remaining amount. The trial court rendered judgment in favor of the subcontractor and the contractor appealed.

The contractor argued that the trial court erred in admitting parol evidence regarding discussions on August 22, 1995, between the contractor and the subcontractor that it would take six to eight weeks for the wrought-iron gates to arrive once they were ordered from its supplier. The contractor argued that the admission of this evidence precluded its argument that the subcontractor materially breached its subcontract by not completing construction of the wrought-iron fence by August 30, 1995. Based on the entire record, the trial court could have concluded that the subcontractor did not materially breach its contract by installing the wrought-iron fence in October 1995. The contractor had not shown that the alleged error was reversible error.

In *Keeney Construction v. James Talcott Construction Co., Inc.*, 45 P.3d 19 (Mont. 2002), the subcontract for site work did not provide for a specific completion date or reference a schedule. The terms of the subcontract required that the subcontractor would maintain progress as directed by the general contractor, and that "time was of the essence." However, the general contractor's contract with the owner provided for a stated completion deadline. During the course of construction, the general contractor negotiated a time extension with the owner. The site work subcontractor filed a delay claim against the general contractor, based upon the impact of the time extension granted to the general contractor by the owner. The general contractor's defense was that it had the right to direct the subcontractor's work without regard to the schedule in effect between the owner and the general contractor. In litigation, the general contractor's position was upheld, precluding the subcontractor's delay claim for damages; the court reasoned that there was no commitment concerning the subcontract performance schedule.

§ 2.04 SUBSTANTIAL COMPLETION

Page 2-18, add after first full sentence of carryover paragraph:

The judicial doctrine of substantial performance involves something less than perfection and this often leads to disputes. In essence, whether the contractor has achieved substantial completion is a factual issue that is often resolved in the contractor's favor if the owner has beneficial use of the project, even if serious defects exist. However, if important safety features are not installed, such as missing safety rails, handrails, or a fire alarm system, a court may find that the project is not complete. Bumb, *Substantial Completion*, Constr. Law. 6 (Winter 2003).

Page 2-18, add after note 61:

An owner may waive its right to enforce a substantial completion deadline in the construction contract by issuing numerous change orders and allowing the deadline to pass without setting a new one. For example, in *RDP Royal Palm Hotel, L.P. v. Clark Constr. Group, Inc.*, 2006 U.S. App. LEXIS 3815 (11th Cir. Fla. Feb. 17, 2006), RDP Royal Palm Hotel, L.P. (owner) entered a contract with Clark Construction Group, Inc. (contractor) for the construction of a resort. The contractor agreed to construct the resort for an amount not to exceed a specified Guaranteed Maximum Price (GMP). The contract also included a substantial completion deadline of February 28, 2000, and provided liquidated damages for failure to meet that date. The owner was permitted to require additional work by issuing a change order or construction change directive. Hundreds of change orders were issued by the owner or its architect. The contractor requested numerous extensions of the substantial completion deadline and increases to the GMP, none of which were resolved. The deadline of February 28, 2000, passed and construction continued. Over two years later the contractor ceased work because the owner discontinued payments for work completed under the change orders. The owner hired another contractor to complete the resort. The owner sued the contractor citing the two-year delay in the resort's completion date. The contractor filed a counterclaim. The district court entered a judgment denying all of the owner's claims against the contractor and granting the contractor's claims against the owner. The owner appealed.

The owner argued that the district court erred in finding that the owner had waived its claim for liquidated damages and for not holding the contractor liable for the delays caused by its subcontractors through an apportionment of responsibility for the total delays. Some delays were caused by the subcontractors, but those delays did not substantially postpone the eventual completion of the resort. The substantial delays which severely impacted the resort's construction were all attributable to the owner. The owner did not established a reasonable basis for an apportionment nor show that apportionment was required. The owner allowed the substantial completion date of February 28, 2000, to pass without setting a new deadline and continued issuing change orders and construction change directives requiring the contractor to perform additional work. The owner's conduct in issuing hundreds of change orders after expiration of the substantial completion date, constituted waiver of the "time is of the essence" provision of the contract. The owner failed to set a new substantial completion date, thus it failed to reserve its right to enforce the liquidated damages provision for any date after February 28, 2000. The appellate court, therefore, affirmed the district court's judgment.

[A] Definition of Substantial Completion

Page 2-18, add to note 66:

The EJCDC form contracts state that the work is substantially complete when "the Work or a specified part thereof has progressed to the point where, in the opinion of Engineer as evidenced by Engineer's definitive certificate of Substantial Completion, it is sufficiently complete, in accordance with the Contract Documents, so that the Work (or specified part) can be utilized for the purposes for which it is intended; or if no such certificate is issued, when the Work is complete and ready for final payment as evidenced by Engineer's written recommendation of final payment in accordance with paragraph 14.13." EJCDC Standard Form No. 1910-8 (1990 ed.), General Conditions of Construction Contract, Art. 1.38.

Page 2-18, add after second sentence of second full paragraph:

Federal leases have also adopted the concept of substantial completion. Substantially complete and substantial completion means that the work, common and other areas of the building, and all other things necessary for the government's access to the premises and occupancy, possession, use, and enjoyment thereof, as provided in this lease, have been completed or obtained, excepting only such minor matters as do not interfere with or materially diminish such access, occupancy, possession, use, or enjoyment. Federal Acquisition Regulations (FAR) § 552.2.

Page 2-18, add at end of second full paragraph:

Federal contracts often contain specific definitions of substantial completion. For example, in *Kinetic Builders, Inc. v. Peters*, 226 F.3d 1307 (Fed. Cir. 2000), the definition of completion for a fire alarm and smoke detection system in a building alteration contract provided:

> After all equipment for this system has been installed and made operational, and at a time directed by the Contracting Officer, the contractor shall conduct tests to demonstrate that the installation and system operation is in accordance with plans and specifications.

The project was late, and the contractor and the government disputed the reasons for delay and the date of substantial completion. The contractor filed suit to recover its delay costs and the return of liquidated damages withheld by the government. The contractor claimed that the system was installed, fully operational, and capable of passing the required system tests on a date the contractor claimed was the date of substantial completion. However the actual testing of the system did not occur until 22 days later. The court ruled that the date of substantial completion required compliance with the specific provisions concerning testing. It was reasonable for the government to expect the system to be tested before it occupied the building and that the successful completion of the required tests was part of the contract definition of substantial completion.

Page 2-19, add after second full paragraph:

In *Kinetic Builders, Inc. v. Secretary of Air Force*, 226 F.3d 1307 (Fed. Cir. 2000), the Air Force awarded Kinetic the contract for alteration of a building. The government took beneficial occupancy of the building on March 1, 1996, 37 days after the scheduled completion date. The Air Force withheld $4,775.70 in liquidated damages for late completion and credit for unperformed work. The contractor submitted multiple claims to the contracting officer (CO) for additional costs incurred as a result of government-caused delays. The CO determined that the contractor should be permitted 20 days of delay due to a sewer backup. The contractor appealed to the Board. The Board determined that the contractor was entitled only to 17 days of excusable delay because of the Air Force's delay in handling the sewer backup, because substantial completion did not occur until February 26. The contractor appealed the Board's decision to the Federal Circuit court.

The contractor argued that testing of the fire alarm system had no bearing on substantial completion of the contract. The contractor contended that it was prepared to begin carpet installation on February 1, and that the only reason this installation was not finished was the Air Force's delay in dealing with the sewage backup. The contractor asserted that if it had not been for the Air Force's delay in dealing with the sewage backup, it could have completed carpet installation and installation of the fire alarm system, which was capable of being successfully tested by February 4. Moreover, because the Air Force took beneficial occupancy on March 1 and the Air Force delayed dealing with the sewage backup for 17 days, the contractor contended that it was entitled to a total time extension on the contract of 26 days. According to the Federal Circuit court, in addition to bargaining for operational installation of the fire alarm system, the Air Force had bargained for testing of that system to demonstrate that the system was operational. Fire alarm testing was an important requirement to the Air Force before the building could be used for its intended purpose. The determination of when the work on the fire alarm system was substantially complete required more than evidence that the fire alarm system was installed, operational, or capable of being successfully tested. The fire alarm testing did not actually occur until February 26, 1996. The Board's finding was not arbitrary, capricious, or unsupported by substantial evidence.

Page 2-19, add at end of section:

For example, even though minor repairs are needed, duplexes that are occupied by tenants by a contract deadline are considered substantially complete. In *All Seasons Constr., Inc. v. Mansfield Hous. Auth.*, 2006 La. App. LEXIS 69 (Jan. 25, 2006), Mansfield Housing Authority (housing authority) and All Seasons Construction (contractor) entered into a contract requiring the contractor to re-roof and install siding on 28 duplexes. The contract required the contractor to achieve substantial completion by a certain date and provided for liquidated damages for each day beyond the completion date. Before the completion date, the contractor informed the construction inspector of the housing authority's architect that the project was complete, other than punch list items that had not yet been received, and requested a certificate of substantial completion. After the completion date had passed, the construction inspector declared that the project was not substantially complete. After the punch lists were completed, the certificate for substantial completion was issued. In its final payment to the contractor, the housing authority deducted liquidated damages for 111 days. The contractor filed suit against the housing authority to recover the amount deducted for liquidated damages. The trial court determined that the work was substantially complete by the contract deadline and rendered judgment in favor of the contractor. The housing authority appealed.

The housing authority argued that the trial court erred in concluding that substantial completion occurred on the contract deadline rather than when the certificate for substantial completion was finally issued. The housing authority asserted that the trial court did not consider the construction inspector's field reports showing problems and lack of progress during the construction period. The housing authority further argued that the punch lists showed that significant work remained on the project after the contract completion date. According to the housing authority, the months it took the contractor to complete the punch lists showed that substantial completion was not achieved by the contract deadline. The contractor argued that the issuance of the punch lists with dollar amounts listed for the items to be addressed indicated that substantial completion had occurred. The contractor described the items on the punch lists as minor and maintained that the duplexes were fit for their intended purpose by the deadline. The record established that the majority of the work was performed on time. The contractor had roofed the buildings and installed the siding by the contract deadline. The punch lists showed that there were deficiencies that needed correction, but the existence of defects did not necessarily preclude a finding of substantial completion. Most items listed were cosmetic in nature, including caulking, removing debris, and repairing gaps. Although some repairs were needed, the buildings were able to be used by the housing authority and were occupied by tenants during the construction process. Because the project was substantially complete by the contract deadline, liquidated damages could not be deducted from the final payment. The trial court's findings were reasonable based upon the record, and the judgment was affirmed.

[B] Consequences of Substantial Completion

Page 2-20, add after note 72:

A project that is not substantially complete does not entitle a contractor to receive its unpaid contract retainage. In *RLI Ins. Co. v. MLK Ave. Redevelopment Corp.*, 2005 Ala. LEXIS 97 (June 17, 2005), MLK Avenue Redevelopment Corporation (developer) purchased land to develop into a residential subdivision. R. P. Colquett (contractor) was awarded a construction contract, which stated that the developer would retain 10 percent of the amount until final completion and acceptance of the contractor's work. RLI Insurance (surety) issued a performance bond naming the contractor as principal and the developer as owner. The developer approved a change order to remedy an unsuitable soil condition. The contractor performed excavation and backfill work on the site. The developer's soils engineer found that the contractor's work was in compliance with the project requirements. A city engineer requested that additional work be performed after inspecting the project. After the contractor performed some of the requested work, the subdivision plat was recorded. Problems later developed with the contractor's improvements. The city engineer wrote the developer, stating that specific items had to be corrected to bring the subdivision up to city standards. The contractor refused to perform further work. The surety refused to have the work completed or to pay the developer the cost of having the work completed. The developer sued the surety, seeking specific performance under the performance bond. The trial court granted judgment to the developer. The surety appealed.

The surety argued that the trial court erred in failing to reduce the damages awarded to the developer by the amount of the unpaid contract retainage and/or the amount claimed for extra work performed. The surety argued that the contractor substantially performed its work under the contract and was thus able to enforce the developer's promise to pay. The surety asserted that it was entitled to an offset equal to the retained amount. The surety also argued that the trial court erred in refusing to allow it an offset for extra work performed by the contractor for which the developer did not pay. The developer argued that since final completion and acceptance never occurred, the contractor was not entitled to the retained amounts. There was credible evidence to support a finding that there was not substantial performance. The city engineer's letter to the developer indicated that a significant amount of work had to be performed to make the project comply with the contract and city standards. The contractor failed to complete the project within the time provided in the contract and failed to perform a substantial amount of extra work called for in a change order. The contractor's claim was entirely offset by the amount of liquidated damages it owed to the developer for its failure to complete the project on time and by the amount of extra work not performed. The trial court's finding that there was not substantial performance was not plainly and palpably wrong, and the judgment was affirmed.

Page 2-20, add to note 76:

Subsequent owner repairs after substantial completion, which constitute only a small portion of the project, will not preclude the builder from recovering on the theory of substantial performance. In Teramo & Co., Inc. v. O'Brien-Sheipe Funeral Homes, Inc., 725 N.Y.S.2d 87 (App. Div. 2001), O'Brien-Sheipe Funeral Homes, Inc., hired Teramo & Co., Inc. to construct an addition to its funeral home. The parties' contract did not specify a date for completion of the proposed work and time was not of the essence. The project was substantially completed and a certificate of completion was issued. Although the owner made periodic payments to the builder during the course of construction, the owner failed to make full payment of the builder's final bill, leaving a balance due of $17,950. The builder commenced an action to recover the outstanding balance. The owner asserted counterclaims for business allegedly lost due to delays in construction, and for the cost of hiring other contractors to correct problems caused by poor workmanship. The trial court dismissed the builder's complaint, finding that the builder's performance was unsatisfactory in both time and skill. The court also awarded the owner damages based on the profits lost due to construction delay. The builder appealed. There was no clause in the parties' contract that made time of the essence, and the owner had continued to make periodic payments to the builder. Some delay in construction was attributable to the owner's request that no work be performed at the premises while funeral services were in progress. Although the evidence presented in support of the owner's counterclaims established that the owner had expended $6,180 to repair certain items that had not been properly completed in a workmanlike manner, this constituted only a small portion of the project, and did not preclude the builder from recovering on the theory of substantial performance. The appellate court found that the builder should be permitted to recover the outstanding balance due on the contract price, but that the recovery should be reduced by $6,180 for the cost of repairing work found to be unacceptable.

§ 2.06 INTERIM MILESTONES

Page 2-26, add at end of section:

In *Safeco Credit v. United States*, 44 Fed. Cl. 406, 1999 U.S. Claims LEXIS 178 (Ct. Cl. July 27, 1999), the parties disagreed about the effect of modifications on the contract completion dates and, in turn, on Safeco's possible liability for failure to complete the project on time. The court found that the plain meaning of the time extension clause entitled the government to assess liquidated damages for delays beyond the date that each phase was specifically extended. The dispute arose from a 1985 contract for berthing improvements to a naval dock. Safeco Credit was the surety to the contractor and succeeded to its responsibilities and interests. The contract was divided into seven construction phases identified as 1A, 1B, 2A, 2B, 3A, 3B, and 3C, and a mobilization phase. The mobilization phase and construction phases 1A, 1B, 2A, and 2B ran consecutively. With the exception of phase 1B, which was to begin before the completion of phase 1A, each phase of 1A, 1B, 2A, and 2B had to be completed before work on the next phase could commence. Phases 3A, 3B, and 3C ran consecutively, but phase 3A began concurrently with phase 1A. Each construction phase was for construction on one berth. The Navy added or changed work on the project with 53 modifications to the contract. Many of the modifications extended the time period for completion of phases 1A, 1B, 2A, and 2B—those phases of the project which (with the exception of phase 1B, which was to begin before the completion of phase 1A) were to run consecutively.

The government calculated damages based on the dates of completion set forth in the contract and the modifications. If a phase of the project was not completed by the date in the contract, the government assessed liquidated damages for each day of delay beyond the contract completion date for that phase of the project. Safeco interpreted the modifications to extend both the project completion and subsequent commencement dates. If a phase of the project was extended, in Safeco's view, that extension altered the commencement date for the subsequent phase so that it would begin after the completion date of the previous phase. Safeco's interpretation would greatly reduce the amount of damages that could be assessed under the contract. In support of its reading of the contract, Safeco argued that the project was not driven by any particular requirement for completion which drove a project deadline.

To determine the correct method for computing liquidated damages, the court looked to the plain language of all of the relevant contract provisions. The contract expressly provided for time extensions:

> Notwithstanding any other provisions of this contract it is mutually understood that the time extensions for changes in the work will depend upon the extent, if any, by which the changes cause delay in the completion of the various elements of construction. The change order granting the time extension may provide that the contract completion date will be extended only for those specific elements so delayed and that the remaining contract completion dates for all other portions of the work will not be altered and may further provide for an equitable readjustment of liquidated damages pursuant to the new completion schedule.

The court found that the plain language of the contract provision was inconsistent with Safeco's interpretation of the contract. A change in the completion dates for one phase of the contract "may provide" for extension of completion dates for only some portions of the contract work without affecting the commencement and completion dates of other portions of the contract work. For a phase to be affected, the language of a contract modification must explicitly state the intent to change the completion date of that phase. The terms of the contract modifications made it clear that a change in the completion date for one phase of the contract did not lengthen the period of time in which a subsequent phase was to be completed.

Each modification in which a change was made had specified the phases affected and the new completion date(s). For example, Modification P00010 described the change order work to be completed and stated that "the contract amount is herefor changed to read from $9,885,812.00 to $9,896,018.00, a net increase of $10,206.00 and the contract completion date is extended three calendar days up to and including 87MAY08 for Phase 1B, 87OCT08 for Phase 2A, and 88MAR10 for Phase 2B." That unambiguous language provided for extended completion dates for each phase specified. The court found that the plain meaning of the contract language was consistent with the government's view that the government was entitled to the assessment of liquidated damages for delays beyond the dates that each phase was specifically extended in accordance with the contract terms. The court denied Safeco's recovery of any of the liquidated damages withheld by the government.

§ 2.08 NOTICE PROVISIONS

Page 2-29, add new footnote after second sentence in first paragraph:

[121.1] *See* E.C. Ernst, Inc. v. General Motors Corp., 482 F.2d 1047, 1055 (5th Cir. 1973) ("The purpose of a notice provision in a contract is to alert the other party that the claimant has a grievance against it.").

Page 2-29, add to beginning of second paragraph:

Generally courts will enforce notice requirements for a delay claim, even if the result is that the claimant is unable to recover on its claim. *United States v. Centex Constr. Co., Inc.,* 638 F. Supp. 411 (W.D. Va. 1985); *Westates Constr. Co. v. City of Cheyenne*, 775 P.2d 502 (Wyo. 1989); *County Comm'rs of Caroline County v. J. Roland Dashiell & Sons, Inc.*, 747 A.2d 600 (Md. 2000); *Byron's Constr. Co. v. State Highway Dep't,* 448 N.W.2d 630 (N.D. 1989).

Page 2-29, add after note 127 in third full paragraph:

Similar contractual notice provisions concerning delay claims and extensions of time are legal issues also considered by Commonwealth jurisdictions. Typical government contracts of the Commonwealth nations require notice of delay-causing events within a specified number of days, followed by a more detailed description of the cause, effect cost impact and relief claimed within a specified number of days after the initial notice. The contracts also make compliance with these provisions a condition precedent to recovery for claims. Such provisions are generally upheld with certain exceptions. One exception is if the contract does not have a mechanism for time extension for owner-caused delay. Another exception is where the contract provides that an owner's agent, construction manager, or superintendent can make an independent determination as to the time extension due to the contractor. The owner's agent may be required to grant an extension of time on its own volition in the interests of both the owner and the contractor. Another exception is when the owner or architect waived or dispensed with the notice requirement. Bircham, "Conditions Precedent to Extension of Time Claims," *In-House Lawyer* 43 (July/Aug. 2003).

[A] Time of Notice

Page 2-31, add after carryover paragraph:

In *Commissioners of Caroline County v. J. Roland Dashiell & Sons, Inc.*, 358 Md. 83, 747 A.2d 600 (2000), as a first step in the process of renovating and adding to the County's correctional facility, the County entered into a Standard Form of Agreement Between Owner and Architect with Greenhorne, an architect, for improvements to the Caroline County Detention Center. Pursuant to it, Greenhorne was responsible for designing the renovation and for providing supervision and monitoring services during the actual construction of the building addition. The project consisted of an addition to the existing detention center, and renovation of the original facility. The County contracted with Dashiell for construction of the proposed renovation. For a lump sum, the contractor agreed to furnish all labor, equipment, materials, and services, and to perform all of the work necessary to renovate and expand the detention center by a date no later than 425 calendar days after the date of commencement. Section 3.2 of the Dashiell contract specifically provided that liquidated damages of $500 per calendar day would be assessed if Dashiell failed to complete the project within the 425 calendar-day period. The parties included American Institute of Architects, General Conditions of the Contract for Construction, Document A201 (1987). If the contractor wished to make a claim for an increase in the contract sum, written notice was required before proceeding to execute the work. If the contractor believed additional cost was involved, claims were to be filed in accordance with the procedure established therein.

Almost immediately after receiving a notice to proceed, the contractor began to encounter construction delays for which it requested an extension. By letter, a 60-day extension of time was granted. No further completion date extensions were granted. Ultimately, construction of the project was not completed by the required date.

The contractor claimed that an additional extension of time, due to delays, was needed. By letter, the contractor's executive vice-president informed the architect that it was working to develop a claim for lost time due to weather. Subsequent letters tendered "formal notice that J. Roland Dashiell & Sons, Inc. will be preparing a claim against Caroline County as provided under the General Conditions of the Contract for Construction." The contractor later submitted a claim in accordance with Section 4.3 to extend the contract completion date by 522 days and to increase the contract price by $1,061,038.00 for delays allegedly due to architectural and engineer design deficiencies, weather delays, and concealed or unknown conditions. The court found that neither of these letters constituted a proper claim in compliance with Section 4.3 of the General Conditions of the Contract for Construction. Instead, these letters merely indicated that the respondent intended to file claims in the future.

Late notice does not bar all claims, but only limits damages to those incurred after notice. For example, in *Associated Mechanical Contractors, Inc. v. Martin K. Eby Construction Co.*, 271 F.3d 1309 (11th Cir. 2001), Associated and Eby signed a $3,150,000 subcontract under which Associated was to perform the mechanical, heating, ventilation, air-conditioning, and plumbing work at a correctional center. The subcontract specified that Associated's work would begin on or before May 2, 1990, and would be completed by July 19, 1991. The subcontract required that claims for delay damages must be filed in writing with the contractor within 10 days from the commencement of the alleged damage and a full accounting filed within 10 days after the extent of damage was known or be considered void. The subcontract also provided that Associated agreed to make all claims for damages against the prime contractor in the same manner provided in the prime contract for claims by the prime contractor against the owner. That article provided: "No claim of the contractor for damage shall be valid unless written notice thereof shall have been received by the owner by registered mail within 15 days after occurrence of the event on which the claim is based."

From the beginning, the prison project was plagued by delays. The footings for the buildings were not finished on time. Eby acknowledged the delay at project meetings, but announced at an August 7 meeting that it would make up the lost time by revising the construction schedule. By letter of August 11, 1990, Associated complained to Eby that Associated's excavation and backfilling work was being held up improperly. The project fell further behind schedule over the next month, mainly due to problems with the masonry work, which affected Associated's work schedule because Associated's work had to be imbedded in the masonry. By September 5, 1990, Associated wrote Eby complaining that it could not be responsible for making sure the masonry contractor's work was up to specifications. On September 13, 1990, Associated wrote Eby complaining of delay from Eby's failure to transmit Associated's "submittal data" to the project architect or engineer for approval. According to the letter, Associated had just learned of the problem, although it had been going on since April.

Associated sent letters on October 1, 1990 and November 8, 1990, notifying Eby that Associated was suffering delay damage from the project being 29 days behind schedule. At the November 13 project meeting, Eby announced that the masonry subcontractor had defaulted. Associated wrote Eby again on November 26, 1990, saying that without any bricklayers on site, Associated would have to lay off workers on November 30 because there was nothing for them to do. Associated said it would incur expenses from this delay and would expect compensation. Again, on February 15, 1991, Associated wrote Eby complaining that the masonry problem and other changes had delayed its own work, and that Eby's attempt to make up lost time by accelerating the work was going to cost Associated money. From early 1991 until the project was finally finished, a constant stream of correspondence continued between Associated and Eby about the delays and problems on the project. Associated repeatedly described in its letters expenses it incurred from the rescheduling of its work, which sometimes left it with idle hands and sometimes caused it to have to pay overtime or to be short of workers. In November 1992, Associated presented Eby with a Request for Equitable Adjustment, asking for $737,343.96 and a 462-day time extension. When Eby denied Associated's request, Associated brought suit. Eby filed a summary judgment motion, contending that Associated had failed to give timely notice as required by the subcontract and by the section incorporating notice requirements of the prime contract. The trial court held Associated did not provide Eby with notice of an intent to seek damages on account of those delays until August of 1990, at the very earliest—well beyond the 10-day period provided for in the subcontract for notice of delay claims.

Associated admitted that it knew of certain delays as early as May and June 1990. It argued that the delays did not constitute damages because Associated and Eby both hoped that the early schedule slippage might be recouped. Eby responded that it was patently impossible for the damages to be eliminated, because once the equipment had sat idle and the overhead had accrued, damages had commenced. The subcontract required Associated to give notice when it incurred damage, without waiting to see whether mitigation efforts would be successful. But even assuming that Associated failed to give timely notice for the delays suffered from May through July 1990, and that damages for such delays were not recoverable, those damages were a small part of the total damages Associated claimed for delay and disruption. Associated could not have suffered 462 days of delay between May and July 1990. Many of the problems occurred after the May–July 1990 period. The timeliness of claims based not on the postponed mobilization, but on discrete events such as the default of the masonry subcontractor, must be evaluated separately from those based on the late start. Because the district court did not appear to have taken into account the damages that were caused by events after July, remanded for the district court to determine which claims were barred by failure to give timely notice and which were preserved.

[B] Notice Procedures

Page 2-32, add to end of first paragraph:

For example, a prime contractor has a contractual duty to provide timely notice of other prime contractor's delays as required by the contract. *Biemann & Rowell Co. v. Donohoe Cos., Inc.*, 147 N.C. App. 239, 556 S.E.2d 1 (2001). Biemann & Rowell Co. and Donohoe Cos., Inc. entered into separate contracts with the State of North Carolina to build a multi-million dollar neuropsychiatric hospital under multiple-prime contracts. Each separate contractor was directly liable to the state and to the other contractors. Donohoe was assigned the role of project expediter. The contractor was responsible for maintaining the progress schedule, making monthly adjustments, updates, and corrections, as well as keeping all contractors and the architect fully informed. Delays occurred throughout the project and the contractors were frequently forced to complete work out of the anticipated sequence. To accelerate completion of the project, it was the understanding of the contractors, architect, and schedule coordinator that Donohoe would "dry-in" the building by installing a moisture seal. Donohoe never installed the building seal. The architect indicated that Donohoe was failing to complete work according to the project schedule and to fulfill its duties as project expediter. Biemann and Rowell maintained that the contractor's failure to install the building seal, and its failure to supervise and properly schedule its subcontractors, caused delays to Biemann and Rowell's work. The trial court rendered judgment in favor of Donohoe. Biemann and Rowell appealed.

Biemann and Rowell contended that the trial court erred in finding an express contractual obligation to provide notice of Donohoe's delay to the architect, and that Donohoe's actual knowledge of its potential liability to co-prime contractors was sufficient notice. As arbiter of disputes between the prime contractors, it was necessary for the architect to be notified when one contractor caused delay to another. The general conditions of the project contract provided that a contractor who was delayed by another contractor was to request an extension of time in writing to the architect and owner within 20 days following the cause of the delay. Biemann and Rowell had a contractual duty to provide notice of delay caused by Donohoe. Discussions at weekly foreman's meetings and monthly progress meetings with the architect and owner did not constitute sufficient notice. While the meetings may have provided constructive notice, Biemann and Rowell never gave written or verbal notice of potential claims at the meetings, nor did Biemann and Rowell ever give notice that it was suffering economic harm. Failure to provide proper notice was a breach of Biemann and Rowell's duty to mitigate damages, and prejudiced Donohoe by not providing an opportunity to cure. Biemann and Rowell failed to provide Donohoe with timely notice of its claims. Affirmed.

Page 2-32, add after second full paragraph:

The contractor should make an attempt to comply with requirements to provide delay damages shortly after the occurrence of a delay or it may lose its rights to recover on a delay claim. For example, in *Heckler Electric Co., Inc. v. City of New York*, 715 N.Y.S.2d 619, 186 Misc. 77 (2000), the public construction contract contained a two-part notice and documentation requirement. First, the contractor had to provide written notice to the City within 10 days of any act giving rise to a claim. Then before the 15th day of the following month, the contractor had to provide an itemized statement of the delay damages. In response to a suit filed by the contractor alleging delay, the City brought a motion for summary judgment based upon the contractor's failure to comply with the two-part notice requirement. The contractor unsuccessfully argued that it was impossible to itemize the delay costs in the time required because the delay costs were unknown at the time when the second part of the notice was required. The court reasoned that precise quantification was not required but an estimate of the cost would comply with the itemization requirement. Later the contractor would be able to amend its estimate when it determined the actual damages. Other New York courts have also enforced similar requirements for itemized statements of claim damages, especially when the contractor failed to make an attempt to comply. *A.H.A. General Constr., Inc. v. New York City Hous. Auth.*, 92 N.Y.2d 90, 699 N.E.2d 368 (1998); *F. Garofalo Electric Co. v. New York Univ.*, 270 A.D.2d 76, 705 N.Y.S.2d 327 (2000).

[C] Actual Notice

Page 2-33, replace subsection title with:

[C] Actual Knowledge

Page 2-33, add to note 146:

Allgood Electric Co. v. Martin K. Eby Constr. Co., 959 F. Supp. 1573 (M.D. Ga. 1997).

Page 2-33, add to note 147:

See also Metric Constructors, Inc. v. National Aeronautics & Space Admin., 169 F.3d 747 (Fed. Cir. 1999) (court accepted critical path schedule with no activities for replacing lamps as evidence that trade practice did not require contractor to replace all light bulbs when project finally completed).

Page 2-34, add after note 151 in first carryover paragraph:

The written notice requirement may also be met by other project documentation. For example, in *Welding, Inc. v. Bland County Service Authority*, 541 S.E.2d 909 (Va. 2001), the contractor maintained that the project architect's meeting minutes documented the contractor's intention to file a claim. At issue was whether the contractor complied with the statutory requirement that public works contractors submit written notice of their intention to file a claim at the time of its occurrence. Because of the lack of formal written notice from the contractor, the trial court granted summary judgment in favor of the public owner, ruling that the architect's meeting minutes were not sufficient to meet the statutory notice requirement. Reversing the summary judgment, the Supreme Court of Virginia held that as a matter of law, the meeting minutes could not be ruled out as a way to comply with the statutory notice requirement. Whether the contractor complied with the statutory notice requirement from the architect's meeting minutes was a determination to be made at trial.

Page 2-34, replace first sentence in first full paragraph with:

In countering an argument that actual knowledge should satisfy the notice provision, it may be asserted that although the owner may have been aware of the event, the owner did not realize that it would be the basis of a claim for additional costs.

Page 2-34, add after note 152 in first full paragraph:

Similarly, a general contractor may argue that the contractor's actual knowledge of a subcontractor's delay should not excuse the subcontractor from providing written notice of a delay to the general contractor. For example, in *Associated Mechanical Contractors, Inc. v. Martin K. Eby Constr. Co., Inc.*, 271 F.3d 1309 (11th Cir. 2001), the subcontract contained a provision requiring the subcontractor to give written notice of delay within 10 days of its commencement, followed by a "full accounting" of the delay impact and costs "within 10 days after the extent of the damage is known or the cause of damage ceases, whichever is sooner." The contract between the owner and the general contractor also contained written notice requirements for delay, and the subcontract incorporated the provisions of the prime contract.

The subcontractor incurred "stop and start" delays, resequencing requirements by the general contractor and other delays, but did not provide written notice of the delays to the general contractor. The subcontractor argued that the general contractor had actual knowledge of the delays since it ordered the resequencing of the work. The subcontractor further argued that the two aspects of the notice provision (written notice and full accounting) in effect did not require notice to the contractor until it was possible to quantify the delay damages. However, the court held that written notice by the subcontractor was required despite actual knowledge of the delay by the general contractor. The court reasoned that the general contractor's actual knowledge would not enable it to satisfy its own written notice obligation under the contract with the owner. The court opined that the function of the notice provision in the subcontract was to permit the general contractor to include the subcontractor's claims in its notices to the owner.

[F] Waiver

Page 2-36, add to note 162:

Service Steel Erectors Co. v. SCE, Inc., 573 F. Supp. 177 (W.D. Va. 1983); E.C. Ernst, Inc. v. Koppers Co., Inc., 626 F.2d 324 (3d Cir. 1980); Transpower Constructors v. Grand River Dam Auth., 905 F.2d 1413 (10th Cir. 1990)

Page 2-36, add after note 168 in third full paragraph:

In another case, the court held that the construction manager lacked authority to waive the contractual requirement for written notice of claims. New York University retained an agency construction manager for a medical school building construction project. The University held all of the multiple trade contracts. The terms of the electrical trade contract required prompt written notice followed by written itemization and documentation of costs as conditions precedent to recovery of additional compensation. The electrical contractor admitted that it hadn't strictly complied with the notice procedures but asserted that the construction manager was fully aware of the circumstances giving rise to the claim, receiving written reports of associated costs. The University, however, refused to pay the claimed costs, arguing that the construction manager had no authority to waive the written notice requirements on the University's behalf. When the matter reached the appellate court, it noted that neither the electrical prime contract nor the construction management contract gave the construction manager the authority to change or revise the terms of the electrical trade contract. Rather the electrical trade contract stated that modifications could only be made by written amendment signed by the University. *F. Garofalo Electric Co. Inc. v. New York Univ.*, 270 A.D.2d 76, 705 N.Y.S.2d 327 (2000).

Waiver of notice of delay claims may be more liberally construed on privately financed construction than on public construction. In *Brooklyn Navy Yard Cogeneration Partners v. PMNC*, 716 N.Y.S.2d 64 (2d Dep't 2000), the contractor submitted claims for unpaid extra work and delay on a privately financed project.

The owner argued that the claims were barred by the contractor's failure to comply with the notice requirements. In seeking to defend against the owner's summary judgment motion, the contractor successfully argued that New York law had a more liberal legal standard for waiver of notice on private construction contracts. The court agreed.

§2.09 SCHEDULING PROVISIONS

Page 2-37, replace note 172 with:

For a more complete discussion of other scheduling clauses, *see* Callahan & Hohns, Construction Schedules at 47–86 (2d ed. 1998).

Page 2-38, replace note 175 with:

Callahan & Hohns, Construction Schedules at §3.3 (2d ed. 1998).

Page 2-38, add to note 179:

A clause requiring the contractor to phase its work with the owner may not give the owner the ability to direct the work in a particular manner. In DRC Corp. v. General Servs. Admin., GSBCA No. 14919-COM, 99-1 B.C.A. (CCH) ¶30,649 (1999), the contract for the renovation of portions of two buildings contained a provision that stated "the work shall be closely staged, phased, and coordinated with the Contracting Officer's Techinical Representative [COTR] in order to avoid interruption of ongoing Government operations." The contractor allegedly based its bid on a plan dividing the work into two phases to avoid disruption, but the COTR required a three-phase plan that delayed the contractor's timely completion of the contract. When the Board of Contract Appeals decided the merits of the contractor's delay claim, it ruled that the contractor's two-phase plan met the contract requirements, and the government's directive to use a three-phase plan was unreasonable.

Page 2-39, add after carryover paragraph:

Proper scheduling provisions are necessary for the management of construction projects, as well as providing tools and a schedule baseline against which delays and claims can be measured. The schedules developed during construction may be used as proof in court to prove and disprove delays. Construction schedules are often manipulated to justify certain positions, as discussed in **§11.06.** Schedules may also be manipulated during the course of construction with the purpose of justifying or refuting time extensions, minimizing the effect of a party's own delays, maximizing the effect of other parties' delays, and increasing recovery for change orders and claims. There are many ways to protect against schedule manipulation. Two examples are contractual defenses written into the scheduling specifications and schedule software tools to identify errors. Zack, *Schedule "Games" People Play, and Some Suggested "Remedies,"* 8 J. Mgmt Engineering 2, 138 (Apr. 1992). Many owners find it useful to require the contractor to provide "predecessor-successor" reports with all schedule submissions. Other owners require the contractor to use certain scheduling software and to submit all schedules and updates on computer disks, allowing the owner to analyze the schedules.

Scheduling of owner-furnished material or information may also be manipulated in the schedule. For example, the delivery activities for owner-furnished equipment may be inserted into the schedule on or near the critical path to show that if the owner misses the delivery date, the overall completion of the project will be delayed. To deal with this situation, some owners have provided in the contract that the delivery date for owner-furnished equipment should be within a window of "not earlier than" and "not later than" dates.

Another scheduling issue of concern to owners is early completion schedules. See **§ 11.05[A]**. Specifications may require the contractor to indicate its intention for an early completion when submitting its bid, or require the contractor to "escrow" its bid documents reflecting an early completion. Other contracts provide that there will be no time extensions for excusable delays unless the delays drive project completion beyond the contractually required completion date. Another approach is to require a change order or contract amendment for an early project completion.

"Preferential logic" and "sequestering of float" are other scheduling dilemmas that owners may encounter. Preferential logic allows the scheduler to develop schedules that create maximum opportunities for the schedule to reflect owner delays and interferences if events occur, or scheduling sequences are changed that are the owner's responsibility. Sequestering float is similar in that the schedules are created or updated with the intention of allowing little or no float or flexibility for potential delay-causing events. Preferential logic and sequestering of float techniques are difficult to define, hard to detect, and even harder to prevent, generally requiring a thorough review of the schedule. However, there may be ways to limit the scheduler's ability to employ preferential logic, by requiring that the activities be resource-loaded with labor crews and equipment, that the subcontractors accept the schedule, and that the contractor submit the schedule for the owner's review and comment.

[A] Noncompliance with Scheduling Provisions

Page 2-40, add after carryover paragraph:

Some contract provisions govern contractors' schedules even though they are not found in the scheduling provisions of the contract. For example, in *P.R. Burke v. United States,* 47 Fed. Cl. 340 (2000), aff'd, 277 F.3d 1346 (Fed. Cir. 2002), the government awarded a contract to Burke for improvement to a sewage treatment plant at Camp Pendleton, California. The contract required the plant to stay in operation during the work, and the scheduling provisions did not specify a sequence or schedule. Burke submitted a schedule indicating that all demolition would be done in the first phase of its work and that the secondary treatment operation of the plant would be temporarily suspended due to the bypass of the trickling filter system while it worked on a replacement filter system. The government refused to approve the submitted schedule, and Burke resequenced the work. Burke later claimed that it was entitled to a logical sequence of the work, and that there was an implied warranty that the government's design would accommodate a logically sequenced schedule. The courts reviewing the matter agreed that the schedule did not meet the contract requirements, and the Court of Appeals ruled that requirement for continuous operation created a performance-type scheduling and sequencing requirement, but that there was no implied right to a "logical sequence" in the scheduling of the work.

Page 2-40, add the following new subsection:

[B] Owner Approval of Schedules

In many instances, the contractor will prepare and submit the schedule to the architect or owner "for their information." Such is the requirement under the American Institute of Architect contracts. *See* AIA Doc. A201-1997, ¶3.10.1. Some contracts go further to review, comment upon, and even approve the contractor's schedule. This may raise a number of legal issues relating to the owner's liability for its involvement in the scheduling process.

Why would an owner want to approve a contractor's schedule? It may be that the owner may want to reject a schedule that does not comply with the contract requirements. The owner may object if the contractor does not use the start and completion dates set in the contract. The owner may decide to review the schedule for compliance with seasonal and weather restrictions, phasing, milestone provisions, and to coordinate with other contractors performing work for the owner. The owner may want to prevent contractor manipulation of the schedule relating to owner activities or float, or other scheduling elements that may unnecessarily put the owner at risk for claims of delay or disruption. On a practical level, it may be in the owner's best interest to evaluate the initial project schedule to prevent or lessen the impacts of potential changes during the course of construction. Smith & Cook, *Practical Uses of CPM Scheduling*, Constr. Briefings No. 2002-2 (Feb. 2002) at 3. The owner's approval of an early completion schedule may not satisfy the contractor's burden of proving it would have met that early completion date but for the owner's actions. *Wickham Contracting Co., Inc. v. Fischer*, 12 F.3d 1574 (Fed. Cir. 1994) (even if GSA had approved the early completion date, such agreement did not prove that the contractor would have completed the project by the projected early completion date).

§ 2.11 CHANGES CLAUSE

Page 2-44, add after note 216 in second full paragraph:

Contractors not only may be delayed by the performance of the change order work, but by delays leading up to the changed work, such as the owner's delay in making a decision whether to proceed with a change order. For example, in *Insulation Specialties, Inc.*, ASBCA No. 52090, 2003-2 B.C.A. (CCH) ¶ 32,361 (2003), the contractor was awarded a contract for the rehabilitation and painting of three water towers to be completed within 360 days. During the course of performance of the contract work, the government issued a change order for the addition of a corrosion coating system for the exterior of the tower and agreed to a time extension of 212 days for performing the additional work. However, the contractor alleged that the government further delayed the project by its failure to make a timely decision to use the corrosion coating system and to issue the change order. The specific terms of the change order did not purport to deal with the government delay in making the change order decision, and the contractor was allowed to recover under the "suspensions" clause for the impact of the delay attributable to the late decision making.

Page 2-44, add after first sentence in third (carryover) paragraph:

The constructive change theory of recovery may allow a contractor to obtain limited relief when the terms of the changes clause do not literally apply, and the contractor may be precluded from recovering under other contractual and legal theories such as differing site conditions, suspension of work, undisclosed superior knowledge, and constructive acceleration. For example, in *The Sherman R. Smoot Corp.*, ASBCA No. 52173, 2003-1 B.C.A. (CCH) ¶ 32,212 (2003), the contractor was to renovate buildings where lead-based paint was present as indicated by the contract documents and a pre-bid site inspection. However, the deteriorated condition of the lead-based paint resulted in a continuous generation of lead dust even without it being disturbed by construction activities. This continuous release of lead dust was not noted in the contract documents and increased the cost of the remediation. It also resulted in delays and disruptions in the work, and the contractor sought relief before the board of contract appeals. The contractor's claim included a count for "suspension of work" that was denied because the board was not able to apportion the additional costs caused by the lead dust from construction activities and the lead dust released in the absence of construction activities. The contractor's claim based upon undisclosed superior knowledge was denied because there was no evidence that the government knew that lead dust would be released without any construction activity. The contractor's differing site conditions claim was denied because there was no affirmative representation in the contract concerning the continuous release of lead dust. However, the contractor did recover its increased remediation costs under the constructive change count as the government directed more thorough cleaning and remediation beyond what was supported by the contract specifications.

Page 2-45, add after note 219 in first carryover paragraph:

The legal theory of constructive changes is not universally accepted by all jurisdictions, although another legal theory may achieve the same results. For example, in *Sentinel Industrial Contracting Corp. v. Kimmins Industrial Service Corp.*, 743 So. 2d 954 (Miss. 1999), the subcontractor-claimant was able to recover for its costs in performing extra work without a written change order under an estoppel-type theory, as the jurisdiction did not accept the theory of constructive changes. The general contractor awarded the subcontractor a subcontract to disable a portion of an ammonia refinery. The subcontract provided that written change orders were required for any adjustments to the terms or price. The general contract was incorporated by reference into the subcontract, and contained a provision requiring the contractor to comply with directives and instructions even without a written change order. The general contractor directed changes to the sequence and schedule of the subcontract work as well as requiring performance of work not in the subcontract scope. The subcontractor performed as directed and submitted change order requests to the general contractor who refused to approve them.

The subcontractor sued the general contractor to recover for the extra work costs and at trial was awarded $966,375. On appeal, the general contractor argued the lack of written change orders prevented any recovery by the subcontractor. The subcontractor argued that it had to comply with the general contractor's directives or face a possible default termination or breach of contract. Although the legal concept of constructive changes was not recognized under Mississippi law, its highest court allowed recovery for extra work without a written change order under the theory that one may not impose extra-contractual work while denying change order requests.

Page 2-46, add at end of section:

The theory of cumulative impact of change orders is argued when a claimant signs a change order accepting payment of the direct costs and executing a release for all costs associated with the change but subsequently seeks to recover other delay, disruption, and impact costs. When arguing this in the face of executed change order releases, the claimant may allege that it was not the isolated affect of the individual change order but the overwhelming affect of the numerous quantity of change orders that resulted in impact costs down the line. A cumulative impact claim may also be made even where the contractor has not executed change orders or change order releases. The claimant may allege that the issuance of numerous change orders results in impact costs that are greater than the impact cost caused by an individual change order when viewed independently. Furthermore, numerous change orders may constitute a constructive change above and beyond the individual change orders that may have been executed. See **§ 5.05[H]**.

Multiple change orders do not entitle a contractor to additional payment without a breach of a fundamental obligation under the contract. In *Dugan & Meyers Construction Co. v. State of Ohio Department of Administrative Services*,

2005 Ohio 3810, 2005 Ohio App. LEXIS 3503 (July 28, 2005), the State of Ohio Department of Administrative Services (ODAS) engaged Dugan & Meyers Construction Co., Inc. (D&M) to act as the prime contractor for construction of three buildings at the Fisher College of Business at The Ohio State University (OSU). The Resource Center Building was to be completed by May 13, 1999, and the Undergraduate and Executive Education Buildings by June 12, 1999. The contract included a provision that time was of the essence and another provision for the assessment of liquidated damages in the amount of $3,000 per day. ODAS engaged an associate architect (associate) to respond to requests for information (RFIs) sent by the contractors and administer any design changes necessitated by any conflicts or discrepancies in the plans or specifications. ODAS retained Gilbane Building Company (Gilbane) to act as the construction manager. D&M was required to publish monthly schedule updates and to propose an affirmative recovery plan in the event that it appeared that contract completion dates would not be met. Beginning in June 1998, work was delayed significantly. Between June 1998 and the end of September 1998, 176 RFIs had been submitted, 48 field work orders (FWOs), and 15 architect supplemental instructions (ASIs) had been issued. Each RFI contained a space to indicate any impact the RFI was expected to have on the construction schedule or the cost of the project. D&M reported no cost impact on any RFI that it issued, and attributed only four days of delay to any of its RFIs. In early April 1999, OSU, the associate, and Gilbane indicated to D&M that it was most critical to complete the Undergraduate Building and several floors of the Resource Center by the contract completion date. Progress continued to be slow in May and June 1999, with the continued submission and issuance of RFIs, ASIs, and FWOs, but with no requests for extensions of time. The Undergraduate Building and Resource Center were completed in time for fall quarter classes, but the Executive Education Building was completed six months after the contract completion date. OSU assessed liquidated damages against D&M for part of the delay in the amount of $325,000. D&M filed a claim to recover its contract balance, reversal of the liquidated damages, and extra amounts attributed to delay damages. OSU rejected D&M's claim. The contract required any claim include information regarding "activities on the Construction Schedule affected by the claim or new activities created by any delay and the relationship with existing activities," and "anticipated duration of any delay." D&M instituted an action for breach of contract. OADS asserted counterclaims seeking liquidated damages. D&M recovered $602,745 as the balance of its contract; a complete reversal of the liquidated damages; and $730,760 in "cumulative impact" damages, plus interest on those damages.

On appeal, OADS argued that Ohio law did not recognize claims for "cumulative impact" damages. The trial court was persuaded to award delay damages because of the "sheer numerosity" of the RFIs. However, the record failed to demonstrate that the substantive concerns addressed in any one RFI or group of RFIs rendered the owner-furnished plans unbuildable or otherwise wholly inadequate to accomplish the purpose of the contract. There was not one identifiable defect or set of defects in the plans and specifications that proved to be so unassailable that the contractor's mere performance of the contract served to undermine the basic purpose thereof. OADS' assignment of error was sustained. The trial court's judgment was affirmed in part, reversed in part, and remanded with instructions.

§2.12 DIFFERING SITE CONDITIONS CLAUSE

Page 2-47, add to note 245:

Ronald Adams Contractor, Inc. v. Mississippi Transp. Comm'n, 770 So. 2d 649 (Miss. 2000) (failure to provide written notice would not bar the claim because the state's on-site engineer had actual knowledge of the soil conditions, investigated the conditions, and directed the undercutting work. Under these circumstances, written notice by the contractor would have been a "vain and useless act"); Dan Nelson Constr., Inc. v. Noland & Dickson, 608 N.W.2d 267 (N.D. 2000) (actual knowledge of a potential differing site condition claim would not estop a public owner from raising the notice defense absent a showing that the failure to provide written notice was induced by the actions of the public owner's representatives).

One purpose of the differing site conditions clause is to take at least some of the "gamble" on subsurface or other unknown conditions; Bidders need not weigh the cost and effort of performing their own site investigations (i.e., soil borings) against the risk of encountering an adverse unknown condition, or provide a contingency pricing factor in the bid to cover the risk. They will not have a pricing windfall if no problem conditions are found or a disaster should an unknown condition be encountered. Owners benefit from more accurate bidding without paying the price for contingencies for conditions that may not be encountered. Owners pay for the effort to deal with unknown conditions only when they arise. Foster Construction C.A. and William Brothers v. United States, 435 F.2d 873, 887 (Ct. Cl. 1970).

Page 2-47, add at end of second full paragraph:

In *Ronald Adams Contractor, Inc. v. Mississippi Transportation Commission*, 770 So. 2d 649 (Miss. 2000), the contractor was awarded a state contract to place the base for a highway. The scope required the contractor to cut and fill to a specified elevation. During the performance, the state's on-site engineer directed the contractor to undercut below grade and backfill with borrow material. The contractor alleged that the directive had a disruptive effect on its labor productivity and costs of performance. However, the only additional compensation given to the contractor for the cut and fill work directed by the state's on-site engineer was payment for unclassified excavation at the contract unit price rates. Although the contractor failed to give prompt written notice as required by the contract, it submitted a differing site condition claim. The claim was rejected by the state for various reasons, one being the failure to comply with the notice provisions of the differing site conditions clause. Summary judgment was granted to the state. In reviewing the notice issue, the Supreme Court of Mississippi held that the failure to provide written notice would not bar the claim because the state's on-site engineer had actual knowledge of the soil conditions, investigated the conditions and directed the undercutting work. Under these circumstances, written notice by the contractor would have been a "vain and useless act."

Actual knowledge alone may not be satisfactory to meet the notice requirement of a differing site conditions clause. *Ronald Adam Contracting* involved a directive by the owner's agent who had actual knowledge of the conditions and also investigated the conditions, serving some of the purposes of the notice requirements. Mere knowledge of the condition may not be satisfactory to overcome the written notice requirements. For example, in *Dan Nelson Construction, Inc. v. Noland & Dickson*, 608 N.W.2d 267 (N.D. 2000), the claimant argued that the public owner either waived or would be estopped from asserting the failure to comply with the written notice requirement because of the actual knowledge of its on-site representatives. The court rejected this argument, reasoning that the actual knowledge of a potential differing site condition claim would not estop a public owner from raising the notice defense absent a showing that the failure to provide written notice was induced by the actions of the public owner's representatives.

Page 2-48, add after carryover paragraph:

Some industry commentators have indicated that there may be a third type of differing site conditions claim—unanticipated and unknown hazardous material encountered at the project site. The contractual basis for this special category of differing site conditions claim is found in clauses that assign the risk of encountering hazardous material. For example, the American Institute of Architects general conditions provide that if the contractor discovers hazardous material at the project site during construction, it is to stop work in the area and notify the owner of the presence of suspected hazardous material. The owner is then to retain a licensed laboratory to determine whether a hazardous material is present, abate the material, and verify that it has been rendered harmless. The contractor may then submit a request for an adjustment of the contract time and cost. AIA Document A201-1997, and ¶ 10.3.

Even without such a provision, a contractor encountering hazardous material at the project site during the course of construction may be able to recover under the differing site conditions clause, if it can establish the elements of recovery for a Type I or Type II condition. The hazardous material may be shown to be unanticipated, concealed, and not readily discovered during a pre-bid site visit because the hazardous environmental conditions were present in limited areas. However, it may be difficult for a contractor to recover on a Type II differing site conditions claim for hazardous material involving the removal of PCBs in an old building in the vicinity of an old transformer since these types of conditions may not be "unusual" in the context of the type of the project site. Beard, *Contractor Claims Based on Encountering Unanticipated Hazardous Materials on Site*, 20 Con. Law. 24, 26 (July 2000).

Page 2-48, add at end of section:

The existence of a differing site conditions claim puts the design professional in a position of possible exposure to claims of professional malpractice from the owner or to be joined as a defendant in any litigation brought by the contractor. Typically, the design professional prepares many aspects of the contract documents or other information relating to the contractor's claim of differing site conditions. In the instance of subsurface geological conditions, the design professional working for the owner may be a geotechnical engineer. In the instance of differing site conditions concerning pre-existing buildings and structures, the project architect may be involved. Such design professionals may investigate the subsurface or existing building conditions as part of their pre-design or design services to the owner, prepare the contract documents, and may be called upon to review the contractor's claims as part of their contract administration responsibilities. As a result, they may be drawn into the litigation. Stein & Popvsky, *Design Professional Liability for Differing Site Conditions and Risk-Sharing Philosophy*, Con. Law. 13 (Apr. 2000).

[A] Type I Conditions

Page 2-49, add after note 259 in first full paragraph:

This may include unanticipated hazardous material. For example, in *Shumate Constructors, Inc.*, VABCA No. 2772, 90-3 B.C.A. (CCH) ¶22,946 (1990), the contract scope included demolition, site work and asbestos abatement. The contract drawings contained a note that the drawings were not definitive nor did they limit the amount of the asbestos to be abated. The notes further stated that the contractor was required to make its own survey of the asbestos and remove all asbestos encountered. The contractor found asbestos in two different areas that generated disputes. The contract drawings indicated that limited amounts of asbestos were located on certain pipes, but the contractor found additional quantities. The contractor also found asbestos in a boiler where the contract drawings did not indicate that any asbestos would be found. As the contractor was not paid for its abatement costs for either location, the contractor filed a claim for differing site conditions, and the dispute reached the Board of Contract Appeals.

In considering the matter, the board made a distinction between the two situations. The contract drawings indicated that asbestos would be found on the pipes, and the contract disclaimer did not affirmatively represent that all asbestos was shown. The contractor was therefore responsible for removing all of the asbestos on the pipes. However, the contract drawings did not indicate that any asbestos was present in the boiler. The Board held that the contractor could infer that there would be no asbestos in the boiler, and that this constituted a Type 1 differing site condition.

Page 2-49, add to note 261:

For example, in Youngdale & Sons Constr. Co. v. United States, 27 Fed. Cl. 516 (1993), *as amended by* 31 Fed. Cl. 167 (1994), the contractor admitted that he did not review the boring logs made part of the contract documents. The court found that there was no reliance by the contractor and that the failure to review the logs was grossly unreasonable and negligent.

Page 2-49, add after note 261 in second paragraph:

The standard for comparison in a Type I differing site conditions claim is not just what is contained in the four corners of the contract documents. For example, in *Hunt Willett, Inc. v. United States,* 351 F.2d 980, 986 (Fed. Cir. 1964), the court stated that the "contractor cannot rest 'content' with the materials furnished to him; he must also refer to other materials which are available and about which he is told by the contract documents." In *Hunt & Willett*, the reference documents were the owner's soil reports, not specifically a part of the contract documents.

Page 2-50, add after note 265 in carryover paragraph:

Withholding of information by the owner is often characterized as "superior knowledge" and often works against an owner defending against a differing site conditions claim. However, the contractor's past experiences may not work against a contractor pursuing a differing site conditions claim. For example, in *Massman Constr. Co. v. Missouri Highways and Transportation Comm'n*, 31 S.W.2d 109 (Mo. App. 2000), the contractor was awarded a contract to construct a bridge and discovered that an existing rock revetment that was not shown on the contract drawings (or otherwise made known to the bidders by the owner) would interfere with one of the bridge piers to be constructed. The contractor placed the revetment four years earlier under a contract with the federal government, and the state argued that, from the perspective of the contractor, it was not "unforeseen." However, the court did not charge the contractor with "superior knowledge," reasoning that it would be inappropriate to charge the contractor with either the exact location of the revetment or the fact that it would interfere with the construction of the bridge.

[B] Type II Conditions

Page 2-51, add after note 275 in carryover paragraph:

For both Type I and Type II differing site conditions claims, adverse weather conditions generally do not constitute a differing site condition. For example, in *Luhr Brothers, Inc.*, ASBCA No. 52887, 01-1 B.C.A. (CCH) ¶ 31,443 (2001), the contractor was to replace dikes in the channel of the Mississippi River and claimed that persistently undulating waves impeded the pile driving. According to the claimant, the waves were not the result of adverse weather conditions but due to the unusual contours and alignment of the channel bottom. The Board considered expert testimony on the issue, but found that the waves were produced by distant storms in the Gulf of Mexico and so denied the claim.

Page 2-51, add at end of second full paragraph:

In a Type II claim, the contractor alleges to have encountered something materially different than the "known and usual." This is a more difficult test because of the variety of conditions that may be found below the surface of the earth or hidden below the surface of existing structures. *Charles T. Parker Constr. Co., Inc. v. United States*, 433 F.2d 771 (Ct. Cl. 1970).

[C] Site Inspection

Page 2-52, add after carryover sentence of carryover paragraph:

Nor is a site inspection clause an absolute disclaimer of responsibility if the actual conditions differ from those represented by the information provided to the bidders in the contract documents. For example, in *SMC Corporation v. New Jersey Water Supply Authority,* 334 N.J. Super. 429, 759 A.2d 1223 (App. Div. 2000), the contract documents contained a site investigation clause whereby the contractor acknowledged that it investigated the site and became familiar with "reasonably ascertainable" conditions and all other matters upon which information was "reasonably obtainable." The contractor was performing the reconstruction of a culvert using cofferdams that required dewatering. The contractor encountered a "scour hole" in the dewatering area not indicated on the contract drawings that allegedly increased its dewatering costs. The contractor properly submitted notices and a differing site conditions claim, and when it was rejected, filed suit. Upon motion of the owner, the trial court granted summary judgment based upon its understanding that the site investigation clause placed the risk of unforeseen site conditions upon the contractor. On appeal, the court found that the clauses were not an absolute disclaimer, but required a fact-finding as to whether the scour hole was reasonably ascertainable from a reasonable pre-bid site investigation.

Page 2-52, add after note 284 in first full paragraph:

The contractor must be careful in drawing inferences from information in the bid package, site information, and specifications. For example, the contractor in *J.B. Construction v. Social Security Administration*, GSBCA No. 14260-SSA, 00-1 B.C.A. (CCH) ¶ 30,686 (1999), the specifications for the replacement of carpet tiles required that the new replacement tiles be installed with easily releasable adhesive. Therefore, the contractor argued that it was reasonable to assume that the old tiles to be removed also should have had the easily releasable adhesive. However, the tiles to be replaced had a general purpose adhesive that required more time and cost to remove which was the basis for the contractor's Type II differing site conditions claim. The board rejected the contractor's argument that the easily releasable adhesive on the tiles to be removed was implied by the specifications. Rather it was an assumption made by the contactor that could have been dispelled by a reasonable pre-bid site inspection.

Page 2-53, add at end of carryover paragraph:

Restrictions on pre-bid inspections in essence requires the bidders to rely upon the subsurface and other information on existing conditions provided by the owner. This increases the risk of delay claims for the owner and reduces the likelihood of them enforcing other owner remedies such as default termination when it involves the restrictions on the pre-bid site investigation. For example, in *Marshall Associated Contractors, Inc.* and *Columbia Excavating, Inc., A Joint Venture*, IBCA No. 1901, 3433-3435, 01-1 B.C.A. (CCH) ¶ 31,248 (2001), the Bureau of Reclamation awarded a contract to the joint venture to produce sand and course aggregate to be used in the construction of a dam. The material for the contract was to be obtained from a government-owned borrow site. The Bureau did not allow the bidders to perform any pre-bid inspections of the borrow site. The bidders were provided geological information, but the Bureau did not provide a particular test pit analysis that was done. In performing the contract, the joint venture found it to be more difficult, time consuming and costly than it anticipated to produce sand that would meet the specifications, and fell behind schedule. The Bureau terminated the contract for default due to the joint venture's inability to produce the required quantity of sand by the completion date. Allegedly, if the joint venture had been able to perform a pre-bid site inspection, it would have been able to determine the actual conditions and would have bid differently, but it was prevented from doing so, and was not given the particular test pit analysis that would have indicated some of the problems encountered. The termination for default was set aside by the Board and converted to a termination for convenience.

[F] Disclaimers

Page 2-54, add to note 299:

Morris, Inc. v. South Dakota Dep't of Transp., 598 N.W.2d 520 (S.D. 1999). The disclaimer clause provided that "the information covering the pit for the project is given to you for informational purposes only. The Department of Transportation does not guarantee the quality or quantity of the material listed in the above information. Interested contractors should investigate the area before considering it for bidding purposes." Unknown to the contractor, the test boring logs were over

10 years old, and dealt with an area where the aggregate was removed by others. Further, the Department had not even visited the pit location in over 10 years, but included the test borings in the bidding documents. In submitting its bid, the contractor relied on the test boring logs in the state-leased pit, and during the course of performance discovered the pit could not produce the required aggregate for the highway project. The contractor filed suit to recover its delay costs. The state supreme court reasoned that the Department understood that bidder would likely rely on the test borings, and had an affirmative obligation to provide more accurate information. Further, the court reasoned that the test borings were included in the bid documents to avoid the cost of borings and to obtain lower bids. It deemed the borings to be a material misrepresentation of the conditions, making the disclaimer unenforceable.

Page 2-54, add after note 299 in first full paragraph:

Disclaimers may not be effective in situations where the owner can be demonstrated to have "superior knowledge"—that is, the owner has information in its possession that is relevant to the site conditions but was not made available to the bidders. The elements of superior knowledge may include (1) the owner possesses information that it knew or should have known would be material to the bidders concerning the site conditions; (2) the contractor was not aware of the information or was not responsible to obtain it; (3) the owner knew of the bidder's lack of knowledge concerning the site information; (4) the owner failed to disclose the site information to the bidders; and (5) the contractor suffered damages as a result of the lack of information. *City of Indianapolis v. Twin Lakes Enterprises, Inc.*, 568 N.E.2d 1073 (Ind. App. 1981); *Horton Industrial, Inc. v. Village of Moweaqua*, 492 N.E.2d 220 (Ill. App. 1986); *P.T. & L. Constr. Co., Inc. v. State of New Jersey*, 531 A.2d 1330, 108 N.J. 539 (1987); *R.J. Wilder Contracting Co., Inc. v. Ohio Turnpike Comm'n*, 913 F. Supp. 1031 (N.D. Ohio 1996).

Page 2-56, add after note 305 in first paragraph:

Courts may enforce a disclaimer of soil borings or other information provided by the owner where it finds that the disclaimer was unambiguous, and includes such contract terms as "the data is provided for informational purposes only," "not to be relied upon," "not part of the contract documents," "no implied warranties," "not complete information," or "only approximations." *Air Cooling & Energy, Inc. v. Midwestern Constr. Co.*, 602 S.W.2d 926 (Mo. App. 1980); *Frontier Foundations, Inc. v. Layton Constr. Co., Inc.*, 818 P.2d 1040 (Utah App. 1991); *Mooney's Inc. v. South Dakota Dep't of Transp.*, 482 N.W.2d 43 (S.D. 1992); *Jahnke v. Georgia Dep't of Transp.*, 322 S.E.2d 505 (Ga. App. 1984); *Green Constr. Co. v. Kansas City Power & Light Co.*, 1 F.3d 1005 (10th Cir. 1993); *Bilotta Constr. Corp. v. Village of Mamaroneck*, 604 N.Y.S.2d 966, 199 A.D.2d 230 (1993); *H&S Corp. v. United States Fidelity and Guaranty Co.*, 667 So. 2d 393 (Fla. App. 1995); *Empire Paving, Inc. v. City of Milford*, 747 A.2d 1063 (Conn. App. 2000).

Page 2-56, add after third full paragraph:

Characterizing all subsurface excavation material as "unclassified" may also assist the owner in avoiding liability for delay claims involving differing site conditions. In *Basin Paving Co. v. Mike M. Johnson, Inc.*, 27 P.3d 609 (Wash. App. 2001), the contract for the construction of water and wastewater lines contained three provisions that the court found to be relevant in denying a contractor's differing site conditions claim. The first provision stated that all excavation was to be "unclassified and payment for rock excavation is not authorized." The court found this to be important because the owner made no representations concerning the proportion of rock and other material that might be encountered. Rock would not be a differing site condition as all subsurface conditions were labeled as unclassified. Second, the contract advised the bidders that subsurface rock conditions would be encountered, and further placed the burden of predicting its presence on the contractor. Third, the owner disclaimed responsibility for the test boring logs that later proved to be inaccurate. The contract stated that the information was provided without any owner responsibility for its accuracy or "for any conclusions that the Contractor might draw therefrom." Even though the contract contained a differing site conditions clause, the court held that the contractor was not entitled to additional compensation.

Page 2-57, add at end of section:

Another approach to allocating responsibility for subsurface conditions is to provide the contractor with a fairly detailed baseline of expected conditions to be used as a comparison to the actual conditions encountered. This is the approach recommended by the U.S. National Committee on Tunneling Technology (USNCTT). It recommended providing in the bid documents "all factual subsurface data, professional interpretation thereof and design consideration thereby raised." USNCTT, *Better Contracting For Underground Construction*, 1 (National Academy of Sciences 1974). In furtherance of this approach, another industry group recommended that for projects with significant underground construction, the designer should prepare a geotechnical design summary report (GDSR) to be referenced in the contract. The GDSR would describe and define the anticipated subsurface conditions and ground behavior in precise terms, and set forth a statement by the designer of the anticipated impact of the conditions upon design and construction. Underground Technology Research Council, American Society of Civil Engineers, and the American Institute of Mining, Metallurgical and Petroleum Engineers, *Avoiding and Resolving Disputes During Construction* 1, 25–26 (1991). Coupled with a differing site conditions clause, this approach provides a detailed description of anticipated conditions allowing the bidders to more accurately prepare bids as well as providing a clearer standard of comparison in the event that the actual conditions encountered are materially different than described in the contract documents.

§2.13 SUSPENSION OF WORK CLAUSE

[A] Owner's Responsibility for Suspension

Page 2-59, add at end of second full paragraph:

The contract may contain provisions requiring the contractor to suspend work at the owner's direction due to environmental conditions. In *Housatonic Valley Construction Co., Inc.*, AGBC No. 1999-181-1, 00-1 B.C.A. (CCH) ¶30,869 (2000), the U.S. Forest Service awarded a contract to install a sand filter sewage system in a national forest in Oregon. The contract contained provisions requiring the work to comply with the regulations of the Oregon Department of Environmental Quality, allowing the government to suspend work under the standard Suspensions Clause, and allowing the government to suspend work for environmental protection or the contractor's failure to comply with the specifications. The last provision did not allow the contractor to receive a price adjustment for suspensions under its terms. The government ordered the contractor to stop work on the installation of the leach field, and later denied the contractor's claim for remobilization costs and unabsorbed overhead for the period of suspension. Even though the stop work order did not specifically refer to the suspensions clause for environmental protection, the Board of Contract Appeals did not allow the contractor to recover its suspension-related costs because the installation in high groundwater would violate the state environmental regulations.

§2.14 TERMINATION CLAUSES

[A] Termination for Default

Page 2-62, add after note 358:

There is a difference between termination for failure to make progress and termination for failure to achieve timely completion under federal contracts. If the termination is based on failure to make progress, the government must give the contractor a 10-day cure notice and an opportunity to cure. Further, the government must be able to demonstrate that there is no reasonable likelihood that the contractor can complete the work by the required contract completion date. If a valid completion date has already passed, grounds for termination are more properly based upon the failure to achieve timely completion. *Abcon Associates, Inc. v. United States*, 44 Fed. Cl. 625 (1999).

Page 2-62, add after note 360:

The claimant may also allege that the original completion date was waived by additional work and other delays, that no attempt to enforce the original completion deadline was made, and that he was encouraged to continue working. *Quinn Bros., Inc. v. Whitehouse*, 737 A.2d 1127 (N.H. 1999).

Page 2-63, add after note 367 in first full paragraph:

For example, in *AFV Enterprises*, PSBCA No. 2691, 2001 PSBCA LEXIS 8, 01-1 B.C.A. (CCH) ¶ 31, 388 (2001), the U.S. Postal Service (USPS) awarded a contract to AFV to construct a building and lease the facility back to the USPS. AFV was unable to get a building permit, but kept pursuing the matter with the local building officials. Thirty days after the required contract completion, AFV still did not have the building permit, and the USPS notified AFV that the contract was being terminated for default. AFV appealed the default termination, alleging that the USPS had waived the completion date by not acting until 30 days after the required contract completion date. In examining the actions and communications between the parties, the board found that AFV did not continue to pursue contract performance, and the 30-day period was not unreasonably long so as to lead AFV to believe the USPS would not enforce the contract deadline.

Page 2-64, add after note 371:

Granting a time extension subsequent to a contract delay also waives the owner's right to use the delay to justify terminating the contract for failing to progress. Consider *R.W. Granger & Sons Inc. v. City School District of Albany*, 296 A.D.2d 636, 744 N.Y.S.2d 567 (2002). The City School District awarded R.W. Granger & Sons the prime contract for a renovation and addition project at the Albany School of Humanities. The contract encompassed three phases. Phase II dealt primarily with the construction of an addition to the school and was to be completed by February 7, 1995. However, numerous delays extended the scheduled date of completion and, in May 1995, the parties agreed that Phase II was to be substantially completed by June 23, 1995, with total completion by September 5, 1995. While Phase II was completed on June 23, 1995, defendant nevertheless terminated the contract on June 28, 1995 because of plaintiff's failure to timely progress its work. As a consequence, plaintiff commenced its action alleging that it was wrongfully terminated. Following a nonjury trial, the Supreme Court determined that plaintiff's termination was without cause and awarded plaintiff $1,412,745 in stipulated damages, plus interest. Defendant appealed.

Granger's expert engineer testified that based upon a "critical path" analysis, the project was delayed for a total of 133 days. Of the 133 days, the expert attributed all but 15 days to defendant and its agents by reason of their failure to coordinate the work of the various prime contractors and to prepare adequate construction schedules. These failures, in turn, resulted in the mechanical, electrical and plumbing work interfering with plaintiff's work, thereby causing the complained-of delay. The expert attributed only 15 days of the delay to plaintiff, which, in his opinion, was neither substantial nor material to completion of the project. There was contrary testimony on the part of defendant's expert that the trial court resolved. With regard to the 15 days of delay attributable to plaintiff, defendant urged that such was sufficient to justify termination of the contract, which provided that plaintiff's "failure to comply with any of the terms" of the contract would constitute a basis for termination. The 15-day delay attributable to plaintiff was from December 13, 1994 to January 10, 1995. Subsequent to the contractor's delay, the parties mutually agreed to an extension of the completion date of the contract and plaintiff continued working thereon. Under the circumstances, defendant waived any complaint as to that 15-day delay. Moreover, to permit such an insignificant delay to constitute a basis for termination would amount to a forfeiture, which the courts in New York were loath to enforce. The judgment was affirmed.

Page 2-64, add after note 372:

and for abandonment indicated by the removal of equipment from the project site, (*Great Western Constr. Co. v. Martin Corp.,* 13 P.3d 1016 (Or. App. 2000)).

Page 2-65, add at end of carryover paragraph:

The owner's representative responsible for making the decision to terminate the contract for default should be careful in assessing the facts involved, not totally relying upon the information and assessment provided by its own staff. For example, in *The Ryan Company*, ASBCA No. 48151, 00-2 B.C.A. (CCH) ¶ 31,094 (2000), the owner's representative responsible for making the decision whether to terminate the contract for default (the contracting officer), relied upon the project engineer's assessment of the situation. When reporting to the contracting officer, the project engineer withheld information and misrepresented the severity of the contractor's performance problems. Relying upon the project engineer's information without performing an independent fact investigation, the contracting officer terminated the contract for the replacement of electrical substation switchgear. When the default termination was appealed to the Armed Sevices Board of Contract Appeals, the Board ruled that the termination was an abuse of the contracting officer's discretion. It found that the contracting officer owed the contractor an assessment of all the relevant circumstances.

The level of care that owners may have to use in terminating a contract for default is further illustrated by *CJP Contractors v. United States*, 45 Fed. Cl. 343 (Fed. Claim. 1999). In *CJP Contractors*, the General Services Administration (GSA) awarded a contract to CJP to replace the heating system in a storage depot. A two-phase performance period was required—170 days to install new gas piping and 310 days to replace the existing oil-fired heaters and replace them with gas-fired heaters. The contracting officer issued a stop-work order to CJP to prevent it from removing the existing oil heaters until the new gas-fired units were delivered to the site. Even after the stop-work order was lifted 47 days later when the new units were delivered, GSA was concerned whether CJP's workforce would be adequate to complete the installation of the new gas-fired heaters prior to the start of the heating season. GSA representatives prepared a rough estimate of the required labor hours to complete the work. Based upon this rough estimate and the time left to complete the work, the GSA contracting officer concluded that there was no reasonable likelihood that CJP would complete the installation in a timely manner. Therefore, CJP's contract was terminated for default.

In considering the termination on appeal, the U.S. Court of Federal Claims critically examined the actions of the contracting officer. The contract did not require CJP to have the new heating units on-site before removing the old heaters. Thus, the court did not find merit in the stop-work order, and granted CJP a 47-day time extension. When the contracting officer evaluated CJP's performance and workforce, CJP was measured against the original contract duration without the time extension to which it was entitled and granted by the court. Further, the completion estimate prepared by the GSA representatives was also found to be flawed in that the estimate was prepared in only a few minutes, based solely upon the limited experience of the GSA representatives in such work, and without reference to other sources, such as estimating guides or other corroboration. The contracting officer relied upon the flawed estimate without question or critical examination. The court determined that the contracting officer demonstrated "unreasonably poor judgment" and based its decision on "materially erroneous information."

Page 2-65, add after note 381 in first full paragraph:

However, the estimate to complete the work must be based upon a reliable estimate of the labor, materials, and equipment required to complete the tasks and must also take into consideration any time extensions that may be due to the contractor, even if not granted. *JP Contractors v. United States*, 45 Fed. Cl. 343 (1999).

Page 2-65, add at end of first full paragraph:

To supplement the evaluation, the owner may retain the services of an independent scheduling expert to assess the status of the project, analyze the schedule with all owner-caused delays and other excusable delays, and determine whether the contractor would be able to finish in a timely manner after extending the schedule for all excusable delays. *Thomas & Sons, Inc.*, ASBCA No. 51874, 00-2 B.C.A. (CCH) ¶ 31,166 (2000).

Page 2-65, add after first sentence of second full paragraph:

Under federal contract law, a contractor's failure to make satisfactory progress gives the government the basis to request assurance that the contract work will be completed in a timely manner and corrective measures will be taken to "cure" the default, even though the circumstances at the time may not justify termination or default. The contractor's failure or refusal to provide such assurances to the government may be grounds to terminate the contract for default. *Danzig v. AEC Corp.*, 224 F.3d 1333 (Fed. Cir. 2000).

Page 2-66, add after note 387 in carryover paragraph:

Under the American Institute of Architect's contracts, an architect's certification of default must cite a material breach by the contractor related to the contractual grounds for default. In *Ingrassia Construction Co., Inc. v. Vernon Township Board of Education*, 345 N.J. Super. 130, 784 A.2d 73 (App. Div. 2001), the owner awarded a contract to Ingrassia for renovations and expansion of a high school building. The contract documents included the AIA General Conditions, Document A201, requiring an architect's certification that there exists grounds for default termination before the owner could terminate the contract for default. The contract schedule contained several milestones that were not met. The owner requested that the architect certify that the contractor failed to achieve the progress to meet the schedule milestones due to its failure to provide an adequate labor force. The architect was reluctant to do so, stating that it did not have an on-site representative present on a daily basis and could not make the requested certification from its own independent observations. However, the architect prepared a certification that stated that the contractor failed to achieve the schedule milestones. Based upon this certification, the owner terminated the contract for cause. The contractor filed suit claiming wrongful termination. Upon the contractor's motion for summary, the trial court found that although the architect's certification reflected that the milestones were missed, it did not indicate that the contractor was at fault. The certification was deficient in this respect, and the contractor was entitled to recover damages for wrongful termination. Upon appeal, it was held that although the certification was defective, this did not automatically establish that the termination was unjustified. The owner still could prove at trial that the contract was in default. Reliance upon the designer's certificate of default may not be sufficient for the owner to base its defense against a claim of wrongful termination. For example, in *Driver Pipeline Co., Inc. v. Mustang Pipeline Co., Inc.*, 69 S.W.3d 779 (Tex. App. 2002), the owner (Mustang) awarded a contract to Driver to construct a 100-mile section of gas pipeline by a stated completion date. Heavy rains occurred in the area, and Mustang granted a time extension to a contractor installing an adjoining section of pipeline, but did not grant a time extension to Driver. Further, Mustang's consulting engineer later issued a certificate of default stating that Driver was not staffing the project with an adequate work force or with adequate equipment for the contract work, and that adequate grounds for default termination existed. Relying upon the engineer's certificate, Mustang terminated Driver's contract for default, and Driver brought suit for wrongful termination. Upon consideration by the Texas Court of Appeals, it was decided that Mustang's reliance upon the engineer's certificate of default was not justified. The court noted that the engineer never visited the site, never observed the construction work, did not have accurate information as to the number of workers that Driver had on the project, and did not know what type of equipment was required for Driver's contract work.

Page 2-67, add after note 391 in carryover paragraph:

An important procedural requirement on terminating a contract for default is the owner's notice to the contractor. The notice should provide specifics as to the deficiencies in the contractor's performance. Generalized statements, such as the contractor is "jeopardizing" its contract, may be held to be inadequate notice because of the contractor's inability to cure its alleged default. *Blaine Economic Development Authority v. Royal Electric Co., Inc.*, 520 N.W.2d 473 (Minn. App. 1994). In some situations, the owner's inadequate vague notice of default can be clarified by subsequent discussions that specify the deficiencies. *Danrenke Corp.*, VABCA No. 3601, 93-1 B.C.A. (CCH) ¶ 25,365 (1992). The key is to provide the contractor with enough specific information so it can exercise its right to cure the default. Also, the reasons given on the notice of default should be consistent with the reasons given in the actual default termination. For example, in *Allstate Contractors, Inc. v. Marriott Corp.*, 273 Ill. App. 3d 960, 652 N.E.2d 1113 (1995), the notice of default dealt with noncompliant concrete placement at a specific location, whereas the default termination was based upon deficient concrete placement in a different location. The court found this to be the basis for liability for wrongful termination.

Page 2-67, add at end of subsection:

An owner need not perform a delay analysis in order to collect additional time costs to complete after a default termination. Consider *In Re Stone & Webster, Inc.*, 279 B.R. 748 (Dist. Del. 2002). Maine Yankee Atomic Power Company hired Stone & Webster Engineering Corporation (SWEC) to decommission Maine Yankee's Wiscasset nuclear power generating facility. The Maine Yankee project involved licensing, engineering, health, physics, and radioactive waste cleanup. Pursuant to the Decommissioning Agreement, Maine Yankee issued a notice to SWEC stating that it was terminating the Decommissioning Agreement based upon SWEC's insolvency and because SWEC had not adequately performed under the contract. Soon thereafter, SWEC filed voluntary petitions for bankruptcy relief. Maine Yankee filed proofs of claim in the bankruptcy cases against SWEC. The Debtors objected to Maine Yankee's claims, arguing that Maine Yankee did not properly terminate the Decommissioning Agreement for either insolvency or failure to perform and did not have a right to damages for terminating the agreement on account of SWEC's insolvency.

Debtors argued that Maine Yankee failed to do a proper "delay analysis" of the cause of the six-month delay from the completion date reflected in the Decommissioning Agreement to Maine Yankee's current projected completion date of October 2004. Debtors contended that Maine Yankee could not recover damages associated with delays unless they proved that those delays were chargeable to SWEC and would not have been avoided by SWEC. Debtors argued that because Maine Yankee failed to prove that the delay was chargeable to SWEC, Maine Yankee's damages must be reduced by at least $2.17 million, which was the cost for supervisory labor associated with the six-month delay. The court found that Maine Yankee mitigated its damages by weighing other contractors' bids against its own self-performance bid. Maine Yankee was owed damages based on the cost of completion for the SWEC scope of work. The fact that Maine Yankee now projected that it would not complete that scope of work until six months after the date of completion in the Decommissioning Agreement did not mean that any "delay" costs adjustments were properly compensable to SWEC. So long as the damages that Maine Yankee were recovering were based on Maine Yankee's cost to complete the SWEC scope of work, it was proper for Maine Yankee to recover that entire amount.

After default termination of a contract, the owner may have liability to the subcontractors of the defaulted contractor if it requests them to continue to deliver or install materials after the effective date of the termination. This is particularly true if the owner takes possession of the project site and all materials and equipment on site in order to complete the work. In sorting out the owner's responsibility to the subcontractors and suppliers, courts may find the owner liable to them under informal written agreements, oral agreements or theories of unjust enrichment. *Encore Constr. Corp. v. S.C. Bodner Constr., Inc.*, 765 N.E.2d 223 (Ind. App. 2002).

[B] Termination for Convenience

Page 2-68, add after note 401 in first full paragraph:

The percentage of the profit the contractor is entitled to is based upon the profit percentage the contractor would have earned on the entire contract had it not been terminated for convenience. For example, in *Charles G. Williams Construction, Inc.*, ASBCA No. 49775, 00-2 B.C.A. (CCH) ¶ 31,047 (2000), the contract was terminated for convenience after the contractor completed Phase 1 of the improvements to a medical center. The contractor claimed a 10 percent profit markup on the costs of the work prior to termination, based upon the fact that the 10 percent profit factor was used for earlier change order work approved by the government. The Board, however, ruled that the appropriate measure of profit is the profit margin the contractor would have earned if the contract were not terminated. A government audit reflected that if the Phase 2 work was not terminated by the government, the contractor would have earned only 3.61 percent profit. The Board applied the 3.61 percent profit factor to the cost of the work completed in calculating the contractor's recovery for the termination for convenience.

Page 2-68, add after note 402 in first full paragraph:

In a situation where a termination for default is converted to a termination for the convenience of the owner, a contractor may not recover damages for home office overhead for the period between the default termination and the resumption of the work as a subcontractor to the takeover surety. *Walsky Constr. Co.*, ASBCA No. 52772, 01-2 B.C.A. (CCH) ¶ 31,557 (2001).

If notice to proceed has not been given and the construction contract has been terminated for convenience, the contractor may be entitled to recover for unabsorbed overhead, but not according to the *Eichleay* formula. The contractor should use a method to allocate its fixed home office expenses during the suspension period between the suspended contract and its other contracts. *Nicon, Inc. v. United States*, 331 F.3d 878 (Fed. Cir. 2003).

Page 2-68, add after first full paragraph:

Also, if the contract is terminated for the convenience of the owner because the work is impossible to perform, the contractor may be able to recover performance costs in excess of the contract price, even if the contractor was in a loss position at the time of termination. Although this may not be the case otherwise, the impossibility of performance results in a situation where the usual limitations on the amount and type of damages may not apply. *D.E.W., Inc. and D.E. Wurzbach, a Joint Venture*, ASBCA No. 50796, 00-2 B.C.A. (CCH) ¶ 31,104 (2000).

Page 2-68, add after the first full paragraph:

In the event of termination for the convenience of the owner, the contractor may be able to recover certain costs other than for work completed. For example, the recovery may include certain start-up and preparatory costs incurred prior to contract award. In *Barsh Company*, PSBCA No. 4481, 2000-2, B.C.A. (CCH) ¶ 30,917 (2000), the Postal Service notified Barsh that it would be awarded a contract for the construction of security improvements at a federal building. The government also scheduled a pre-construction conference prior to the award that Barsh to believe that the project was on a tight schedule. Barsh prepared subcontracts, purchase orders, bonds, and job cost files. Subsequently, the contract was awarded to Barsh, but three weeks later the contract was terminated for the convenience of the government. Barsh's termination settlement proposal included costs for these pre-award activities, but the Postal Service opposed recovery of any pre-award costs. In considering Barsh's appeal, the board noted that these costs were incurred after Barsh learned that it would receive the contract and reasonably believed there was urgency to the project. The Board found that these costs were incurred in order to meet the tight contract schedule. However, Barsh's costs incurred prior to submitting its bid proposal were not recoverable. The bid proposal costs that were eliminated from its recovery included costs for estimating, negotiating subcontract prices, and its proposal preparation costs.

Page 2-69, add after note 408 in first full paragraph:

Other jurisdictions may limit the owner's ability to use the termination for convenience provisions. For example, in *RAM Engineering & Construction,*

Inc. v. University of Louisville, 2003 WL 22971158 (Ky. 2003), the University of Louisville put out for bid the site development package for a stadium project. However, the bids received exceeded the University's budget. After the University negotiated with the three lowest bidders, it declared RAM to be the lowest responsible bidder, and issued a notice to proceed to RAM. Another bidder, MAC, who tendered the lowest bid before the negotiations, filed a bid protest. The University then declared the award to RAM to be null and void, and put the matter out to bid for a second time. RAM again bid the work but its price was approximately $600,000 lower than its previous bid. The University again awarded the contract to RAM; however, RAM filed a protest because the contract was awarded to it in a reduced amount from the original award. A trial court upheld the University's denial of the protest on the grounds that the previous contract was terminated for convenience.

The matter was appealed to the Kentucky Supreme Court which reasoned that a public entity has the duty to undertake everything necessary to carry out a contract despite the presence of a termination for convenience clause. Although the government should be able to terminate a contract for convenience when it is in the government's best interest, it is reasonable to expect that the government's interest will only change if the circumstances surrounding the contract substantially change. That is, for the government to terminate a contract for convenience, there must be a "substantial change in circumstances." In the matter of RAM's first contract, there was no substantial change in circumstances that justified a termination for convenience. For example, even though there was delay to the project due to MAC's bid protest, a provision in the first RAM contract anticipated possible legal delays without indicating that termination for convenience would result.

Page 2-69, add after note 409:

If the federal government elects to terminate the contract for convenience, the contractor has one year to submit a termination for convenience settlement proposal. If it fails to do so, the contracting officer may make a unilateral determination, and the contractor may waive its right to appeal the matter. *Earth Burners, Inc. v. United States*, 43 Fed. Cl. 481 (1999).

§2.15 LIQUIDATED DAMAGES

Page 2-70, add after note 417 in carryover paragraph:

To determine the validity of liquidated damages, courts may consider whether the delay damages would be difficult to prove *at the time the parties entered into the contract*, not at the time of the breach. *Phillips v. B.M. Hogan Co.*, 54 S.W.2d 39 (Ark. Ct. App. 1980). The *prospective* perspective is important in allowing an owner to pre-establish liquidated damages, not a *retrospective* perspective.

Page 2-71, add after note 421 in carryover paragraph:

Liquidated damages determined by the architect without explanation for their computation are an unenforceable penalty. In *Fuschino v. Smith*, 2001 Ohio App. LEXIS 21 (2001), the contract contained a liquidated damage clause requiring the contractor to pay $50 for each day after the completion date that a deck was not completed. The construction was not completed and the owner terminated the contract, completed the deck, and then sued the contractor. The contractor counterclaimed for extras. At a bench trial, judgment was entered for the contractor. The trial court found that the liquidated damage clause constituted a penalty because it bore no reasonable relationship to the actual damages suffered by the owner, and was therefore unenforceable.

The owner appealed and argued that the liquidated damages clause was created to compensate for the loss of use and enjoyment of the deck for each day that its construction exceeded the assigned completion date. It was clear that the owner wanted to use the deck during the summer months, and had even planned a large party for which extra space was needed, which the deck would have provided. Damages for missing these activities would ordinarily be difficult to determine. However, the reviewing court found that the amount of the liquidated damages clause had been set by the architect without evidence to explain the basis for the amount. The only testimony regarding the amount of liquidated damages was provided by the contractor's son and indicated that the normal amount provided for liquidated damages for similar projects was less than one-half of the amount provided for in the contract. The owner's claim for liquidated damages bore no reasonable relationship to the actual damages. At most, the damages from the loss of use of the deck were nominal. The trial court did not err by deeming the liquidated damages provision to be unenforceable.

[A] Apportioning Delay

Page 2-73, add after note 433 in first full paragraph:

For example, under the first rule, if the owner admits that it delayed the project, it will not be able to assess any liquidated damages. Under the second rule, in order for the owner to assess liquidated damages, it must be able to apportion responsibility for delays between the owner and contractor with sufficient credibility that the court will be able to clearly apportion the responsibility for delay. *PCL Construction Services, Inc. v. United States*, 53 Fed. Cl. 479 (Fed. Cl. 2002).

Page 2-73, add at end of first paragraph:

The liquidated damage clause may prevent the apportionment of delay. For example, in *Brashear v. Richardson Construction, Inc.*, 10 P.3d 115 (Wyo. 2000), the court held that the subcontractor could not have delayed completing the project "by April 7, 1998" as stated in the liquidated damage clause, if it did not begin its work until April 13, 1998. The subcontractor and the general contractor in *Brashear* contracted for remodeling of a church. The contract provided:

> In addition to Contractor's other remedies set forth herein, in the event that contractor is delayed in completing the project by April 7, 1998 due to actions or inactions of Subcontractor, Subcontractor shall be liable to Contractor for Liquidated Damages of Three Hundred Twenty[-]Five Dollars ($325.00) per calendar day of such delay.

Initially, the subcontractor was scheduled to begin work on February 1, 1998. The start date was delayed by weather and change orders unrelated to the subcontractor. The subcontractor did not begin work until April 13, 1998. At that time, sheetrockers were still not finished with their work, and they and other trades limited the subcontractor's efforts. In late June, the architect noticed certain problems with the subcontractor's installation of the sisal wall covering. The owner gave written notice of rejection of the sisal installation and a demand to cure. The subcontractor demanded payment for the work completed, but the general contractor withheld payment pending correction of the noted problems with the sisal installation. The subcontractor wanted to try to correct the problems without removing the sisal, but the architect insisted that removal and reinstallation consistent with the manufacturer's instructions were required. After the end of June, the subcontractor performed no significant work. On August 10, 1998, the general contractor terminated the contract and removed and reinstalled the sisal. The project was accepted by the owner as complete on October 1, 1998. The owner assessed liquidated damages to the general contractor pursuant to the general contract for more than 118 days of delay in the project completion. The general contractor allocated 54 days of delay to this subcontractor based upon what the general contractor considered to be a "reasonable" basis, with the remaining days of delay being allocated to the general contractor and other subcontractors. The record contained no evidence concerning the basis for that allocation. No request for extension of time was made by the subcontractor or the general contractor of one another. The record reflected no change order or orders effecting an extension of the April 7, 1998, deadline that triggered liquidated damages in the contract between the general contractor and the subcontractor.

The contract provided for liquidated damages only if the subcontractor delayed completion of the project beyond April 7, 1998. The record contained no evidence that the project was delayed past April 7, 1998, by the actions of the subcontractor. The district court erred when it awarded liquidated damages beyond those contemplated in the contract. If the general contractor had wanted a liquidated damages provision that would apply after the date specified in the contract, it could have provided one, because it drafted the contract. The contract provided for change orders and extensions of time. However, neither party used those remedies to alter the explicit terms of the contract when the April 7, 1998, date passed. Consequently, only the explicit terms of the written contract applied.

The general contractor argued that the subcontractor's testimony showed that the subcontractor understood the contract applied even though work was commenced later than contemplated and that liquidated damages applied whenever a delay was caused. However, the subcontractor's testimony in no way undermined the conclusion that the express terms of the contract did not authorize liquidated damages when the project was delayed long after April 7, 1998, by persons other than the subcontractor. The general contractor argued that, by commencing work after the April 7, 1998, date, the subcontractor had waived its right to claim that liquidated damages would apply only pursuant to the terms of the contract. In fact, the opposite was true. The general contractor waived its right to rely on the liquidated damages provision when it allowed the subcontractor to commence work after April 7, 1998—the contractually specified liquidated damages date—without requiring a change order or amendment to the contract. The award of liquidated damages was based upon the allocation of all the days of delay after the contract was terminated until the project was accepted as complete by the owner. No provision in the contract provided for such an allocation of damages. The district court inappropriately developed its own formula to calculate liquidated damages to be assessed against the subcontractor. The court reversed that portion of the district court's judgment that awarded the general contractor liquidated damages in the amount of $17,550, and remanded.

Page 2-75, add after note 440:

An owner is not required to perform a schedule analysis to allocate delay and liquidated damages among multiple parties when a contractor fails to provide written notice of claims. For example, in *Dugan & Meyers Construction Co. v. State of Ohio Department of Administrative Services*, 2005 Ohio 3810, 2005 Ohio App. LEXIS 3503 (July 28, 2005), Dugan & Meyers Construction Co., Inc. (D&M) acted as the prime contractor for construction of three buildings at the Fisher College of Business for the State of Ohio Department of Administrative Services (ODAS) at Ohio State University (OSU). The Resource Center Building was to be completed by May 13, 1999, and the Undergraduate and Executive Education Buildings by June 12, 1999. The contract included a provision that time was of the essence and another provision for the assessment of liquidated damages in the amount of $3,000 per day. ODAS engaged an associate architect (associate) to respond to requests for information (RFIs) sent by the contractors and administer any design changes necessitated by any conflicts or discrepancies in the plans or specifications. ODAS retained Gilbane Building Company (Gilbane) to act as the construction manager. D&M was required to publish monthly schedule updates and to propose an affirmative recovery plan in the event that it appeared that contract completion dates would not be met. Work was delayed significantly. The Undergraduate Building and Resource Center were completed in time for fall quarter classes, but the Executive Education Building was completed six months after the contract completion date. OSU assessed liquidated damages against D&M for part of the delay in the amount of $325,000. OSU apportioned the remainder of the delay to other contractors, and did not apportion any of the delay to itself. OSU apportioned the delay equitably based upon design team members' assessment of the relative responsibility for the delay attributable to D&M and to other contractors. However, the parties' agreement did not require OSU to engage in such "equitable" apportionment; rather, it was contractually permitted to assess liquidated damages against D&M in the amount of $3,000 per day. D&M filed a claim to recover its contract balance, reversal of the liquidated damages, and extra amounts attributed to delay damages. OSU rejected D&M's claim. D&M instituted an action for breach of contract. OADS asserted counterclaims seeking liquidated damages. D&M recovered $602,745 as the balance of its contract; a complete reversal of the liquidated damages; and $730,760 in "cumulative impact" damages, plus interest on those damages.

ODAS argued that D&M failed to make written requests for extensions of time. D&M did not dispute that it failed to request extensions of time. Rather, it argued that submitting requests for extensions of time would have been futile because OSU had made it clear that no extensions of time would be granted. According to D&M, it should not be required to do a vain act, even if such an act was required by the unambiguous terms of a contract. At most, all parties were aware that if completion of the project (and particularly the Undergraduate Building and the Resource Center) was delayed beyond the start of the fall quarter, OSU did not have alternative arrangements in place for holding many of the classes that the College of Business had planned to conduct. The fact that OSU had likely not prepared itself for a delay extending beyond mid-September of 1999 did not prove that OSU would have refused time extensions upon the proper request under the contract. D&M had not demonstrated that it was entitled to disregard its obligations under that part of the contract, and to claim later, in its complaint, that OSU unreasonably withheld liquidated damages for delay in completing the project. D&M was obligated to request an extension of time for completion of the contract, within ten days following its formation of a belief that a particular condition would entitle it to such an extension. Had D&M done so, it could have relied upon the contract, which provided time was to be extended for such reasonable time which the associate determined. Having failed to comply with its obligations to engage in such communication, D&M could not complain how OSU assessed liquidated damages for delay. This was true notwithstanding the fact that OSU's "equitable" allocation of liquidated damages was based upon the design team's "gut feeling" as to the contractors' relative responsibility for the delays. This did not overweigh the fact that the amount of liquidated damages to be assessed against D&M, as well as the mechanism D&M, was required to follow to preserve its right to challenge the imposition of such damages; the amount was prescribed by the plain language of the parties' agreement. A dissenting opinion stated that the trier of fact appropriately addressed the issue of liquidated damages finding that OSU did not objectively evaluate the project delays in assessing $325,500 in liquidated damages based on its "gut feeling." The trial court judgment was affirmed in part, reversed in part, and remanded with instructions.

[B] Termination and Abandonment

Page 2-76, add after note 450 in first paragraph:

In *Fuschino v. Smith*, 2001 Ohio App. LEXIS 21 (2001), the owner contracted for the construction of a deck. The contract provided that "time is of the essence" and specified a completion date. It contained a liquidated damage and termination clause. The construction was not completed and the owner terminated the contract. The owner completed the project with the contractor's labor, then sued for both liquidated damages and the cost to complete. At a bench trial, judgment was entered in the contractor's favor. The trial court found that the owner "legally may not collect" both liquidated damages for delay and additional cost to complete the work.

The owner was entitled to damages for the cost of completion of the deck in accordance with Article 20.2 of the contract, which stated that the measure of damages was the difference between additional costs and the unpaid balance of the contract price. The appellate court thought that the trial court's finding ignored the fact that the owner had stated two separate claims for damages. The first claim, for actual damages, arose from the owner's having had to hire another to finish the project. The second claim, for liquidated damages, was based upon the loss of use and enjoyment of the deck during the interval between the time that it was supposed to have been completed and the time when it was actually completed. The owner was not barred from seeking both actual and liquidated damages, because the two are based upon separate losses.

Page 2-76, add to note 452:

Another case indicates that unless the owner actually reprocures the completion of the work, it may not be able to assess and collect liquidated damages after the contract was terminated for default. In Standard Coating Service, Inc., ASBCA No. 49201, 48611, 00-1 B.C.A. (CCH) ¶ 30,725 (2000), the contractor was awarded a fixed-price contract to paint four different sections of fuel piping and the structural steel trestles that supported the piping. After completing three sections of the piping and trestles, a dispute arose between the government and the contractor concerning time extensions and additional compensation. The contractor demobilized from the project site without performing any of the work on the fourth section, and its contract was terminated for default. Although the government issued a solicitation for the reprocurement of the painting of the fourth section, the solicitation was cancelled. Nevertheless, the government assessed liquidated damages against the defaulted contractor for the period from the time of termination to the estimated completion date of the work that was to be reprocured. The contractor appealed the assessment of liquidated damages for this time period. The Board ruled that liquidated damages could not be recovered for the period of the attempted but cancelled reprocurement. The government could not collect liquidated damages for work that it decided that it did not want.

[C] Assessment of Liquidated Damages

Page 2-77, add to note 457:

However, in the event of default termination of the contract, unless the owner actually reprocures the completion of the work, it may not be able to assess and collect liquidated damages from a contractor after the contract was terminated for default. Standard Coating Service, Inc., ASBCA No. 49201, 48611, 00-1 B.C.A. (CCH) ¶30,725 (2000).

Page 2-77, add after note 457 in first full paragraph:

However, in the case where the liquidated damages are tied to passing of performance tests, the owner must undertake to perform the tests in a timely manner in order to recover liquidated damages. For example, in *Yorkshire Water Services Ltd. v. Taylor Woodrow Construction Northern Ltd, BiWater Treatment Ltd. and Elga Ltd.* [2004] EWHC 1660 (TCC), 2004 B.L.R. 409, Yorkshire Water contracted with Taylor Woodrow for improvements to a sewage treatment plant. The contract provided for liquidated damages not for late completion of the construction but for the failure to pass performance tests. Yorkshire Water claimed that certain equipment installed by Taylor Woodrow's subcontractor did not operate at the contractually required capacity, and that it was entitled to costs of remedial work, costs for the plant not meeting the requirements in the contract specifications, and liquidated damages for delayed completion of the work. However, at the time that legal action was commenced and at the time of the last amendment of the legal proceedings, Yorkshire Water had not carried out the performance tests. The court noted that the contractual obligation to carry out the performance tests rested upon Yorkshire Water, and that allowing Yorkshire Water to recover the liquidated damages without carrying out the tests would produce the "bizarre result" of allowing it to recover liquidated damages for its own breach for failing to perform the tests.

Page 2-78, replace note 467 with:

195 W. Va. 714, 466 S.E.2d 782 (1995).

Page 2-78, add after first full paragraph:

In *Casson Construction Co.*, GSBCA 4884, 78-1 B.C.A. (CCH) ¶13,032 (1978), the contractor, though finishing on time, failed to complete a phase of its work that delayed a follow-on contractor. The follow-on contractor was ordered to accelerate. It did so by using overtime and double shifts for several weeks. Casson claimed that its liability was limited to the liquidated damage amount of $240 a day contained in its contract; the real cost of acceleration was $644 a day. The Board pointed to another provision in the contract, which stated that the contractor would indemnify the public agency for acceleration payments made to other contractors. The court held that the latter clause controlled and that the same breach could invoke different clauses. The Board would not apply the liquidated damages clause to milestone date delays, only to delay in completing the entire contract performance. The Board noted that the standard language used in the contract was adopted long before the advent of fast-track construction. In *M. Eisenberg & Bros., Inc. v. White Plains Housing Authority*, 55 A.D.2d 599, 389 N.Y.S.2d 390 (1976), the court held that the liquidated damage award did not preclude recovery of additional damages under indemnification provisions in the contract. However, in *Burns v. Hanover Insurance Co.*, 454 A.2d 325 (D.C. 1982), the court held that damages in addition to those provided for under the liquidated damage clause were not recoverable, even though the clause expressly stated that the liquidated sums were for "packing and storing of furnishings" and "temporary accommodations." The court concluded that to recover more than liquidated damages, the contract must state expressly that the clause is limited to specified types of damages and that other damages are recoverable.

In *Northern Petrochemical Co. v. Thorsen & Thorshov*, 297 Minn. 118, 211 N.W.2d 159 (1973), the court declined to enforce a liquidated damage clause for unanticipated, unreasonable delay. In *Northern Petrochemical Co.*, after completion, serious structural defects were discovered for which large-scale redesign and reconstruction were required. Correction of the defects took eight months. The court affirmed an award based on lost profits because of delayed occupancy and excess operating costs while awaiting occupancy. The court rejected application of the liquidated damages clause, noting that it was intended to apply to normal delay, not to an extraordinary eight-month delay due to redesign and reconstruction. The court's conclusion would limit liquidation to normal, expected delays.

Page 2-79, add after note 475 in third full paragraph:

However, a contractor is liable for liquidated damages for those delays that it cannot establish are attributable to the owner's conduct. In *A.G. Cullen Constr., Inc. v. State Sys. of Higher Educ.*, 2006 Pa. Commw. LEXIS 128 (Mar. 15, 2006), A.G. Cullen Construction, Inc. (contractor) was hired by the Commonwealth of Pennsylvania (State) to renovate a school administration building by a scheduled completion date specified in the contract. The contractor experienced significant delays after discovering that wood window frames, which were to be replaced, were painted with hazardous lead-based paint. The State was aware of the lead paint; however, there was no mention of it in the project specifications due to an oversight. The contractor had to suspend work while its employees were trained to safely perform the lead abatement work. The contractor missed the scheduled completion date by 172 days. The State substantially reduced its payments to the contractor pursuant to a liquidated damages clause in the contract which assessed $500 per day for the full 172 days. The contractor filed a claim with the Board of Claims (Board) seeking damages for breach of contract related to the reduced payments. The Board reduced the liquidated damages for the period the contractor had to suspend work to implement its abatement program and based on its finding of an earlier substantial completion date. The Board held that the State was entitled to retain liquidated damages for the remaining delays. The contractor appealed.

The contractor contended that, in order to recover liquidated damages, the State was required to prove the contractor was solely responsible for the project delays. The contractor argued that the State was at least partially responsible for the delays. The Board determined that the State's assessment of liquidated damages did not account for project delays that were out of the control of the contractor or were due to the acts or omissions of the State. The Board reduced the assessment of liquidated damages by 31 days, the period during which the contractor suspended work to implement its lead paint abatement program. In addition, the Board determined the project was substantially complete on January 11, 2002 rather than February 12, 2002, and reduced the assessment of liquidated damages by an additional 32 days. Based on these reductions, the Board determined the State was entitled to retain liquidated damages for a period of 109 days. The contractor was experienced and agreed to a liquidated damage clause on a project where delay was a risk and damages were difficult to prove. The contractor did not establish any additional periods of delay that were attributable to the State's conduct and was thus liable for the remaining 109 days of liquidated damages.

Page 2-79, add after note 477 in third full paragraph:

However, the contract must be clear that the liquidated damages provisions apply to the distinct phasing requirements or milestones. For example, in *Abcon Assocs., Inc. v. United States*, 49 Fed. Cl. 678 (2001), the contract for additions to a mail distribution center provided for three different phases of completion, and required the overall project completion in 510 calendar days. Another provision of the contract set liquidated damages of $750 per day "if the contractor fails to complete the work . . . within the time specified in this contract. . . ." The government withheld liquidated damages based upon the missed milestone date rather than the overall completion date. Subsequently, the contract was terminated for default for failure to make progress. The contractor appealed the default termination arguing that the government wrongfully withheld liquidated damages based upon the failure to meet the milestone associated with the phase, rather than assessing liquidated damages for the overall completion. The U.S. Court of Federal Claims held that the liquidated damages were improperly assessed because the liquidated damages clause was not explicitly tied to the interim phase milestones, only the overall completion date.

However, where there is substantial evidence that an owner granted a contractor's request for additional time to complete a project, the contractor is entitled to recovery of withheld liquidated damages. In *Swink v. Lasiter Constr., Inc.*, 2006 Ark. App. LEXIS 171 (Feb. 22, 2006), the owner entered into a contract with Lasiter Construction (contractor) for the construction of a project. The contract provided that the contractor would complete the project within 120 days of being ordered to start construction. The contract also provided that if the contractor failed to timely complete the project, it would pay the owner liquidated damages of $500 per day to be withheld from the final amount due the contractor. In addition, the contract provided a method for the contractor to request additional time to complete the project. The contractor took more than eight months to substantially complete the project. The contractor filed suit against the owner, seeking to collect the balance due under the contract. The complaint also sought extended overhead costs generated by delays allegedly caused by the owner or his engineer. The owner filed a counterclaim, alleging that the contractor had breached the contract by not completing work within 120 days and by not requesting extra time. A jury verdict awarded the contractor a judgment for the money withheld by the owner as liquidated damages for delays in the project. The owner appealed.

The owner asserted that there was no substantial evidence that the owner granted a request by the contractor for additional time to complete the project. The owner also argued that the award of liquidated damages withheld should be remitted because the award was irrational or based upon conjecture. Both parties agreed that the jury's award corresponded to 121 days of liquidated damages, but the owner argued that there was no way to calculate how the jury arrived at the 121-day figure. The contractor contended that he sent three letters to the project engineer that served as requests for additional time to complete the project. The letters identified problems that caused delays, including not having a full set of plans at the start of the project, a waterline not being noted on the plans, and groundwater seeping into the work. The owner's testimony indicated that he did grant an extension based on one of the letters, which contained a schedule calling for an additional 29 work days to complete the project. The owner's agreement to extend the deadline served as substantial evidence to support the jury's verdict, as the jury was not asked how many days the owner agreed to extend the deadline. One plausible explanation for how the jury arrived at the figure for the liquidated damages award was that the jury extended the 120-day deadline 29 days in accordance with the contractor's letter, resulting in a new deadline that was 121 days from the date the City of Little Rock accepted the work. The jury's award of withheld liquidated damages was supported by substantial evidence.

[D] Waiver of Liquidated Damages

Page 2-81, add at end of first paragraph:

For example, in *Textor Construction, Inc. v. Forsyth R-III School District*, 60 S.W.3d 692 (Mo. App. 2001), it was determined that the owner had waived liquidated damages based upon assurances to the contractor from the owner's superintendent, the architect, and the architect's representative that liquidated damages would not be assessed.

Page 2-82, add to note 487:

Brashear v. Richardson Constr., Inc., 10 P.3d 1115 (Wyo. 2000) (general contractor waived the right to collect liquidated damages when it allowed the subcontractor to commence work after the contractually specified liquidated damages date without establishing a new completion date, or requiring a change order or amendment to the subcontract).

[E] Establishing Amount and Terms

Page 2-83, add after note 490 in carryover paragraph:

For example, in *Brinich v. Jencka*, 2000 Pa. Super. 209, 757 A.2d 388 (2000), the Jenckas, the homeowners, signed an agreement with Brinich, the contractor, which provided for the work to be "substantially completed" by December 31; that date was subsequently changed to 270 days from the date of their construction loan agreement. Despite the revised completion date, the homeowners contended that Brinich had assured them that the house would be completed by December 31, 1993. Brinich encountered numerous problems that impeded timely construction. First, the excavator encountered an underground spring that continually filled the foundation with water. To correct the problem, Brinich had to extend the foundation and install a sump pump. In the meantime, the blocklayer with whom Brinich had subcontracted for the job began work on another project; therefore, Brinich was forced to find a replacement at a higher price. In January 1994, construction was delayed by extremely cold weather. In addition to these problems, Brinich testified that the Jenckas continually requested additions and/or modifications that not only involved additional costs, but also delayed construction. Construction continued into the spring of 1994.

The parties realized that the house would not be completed on time and executed an agreement extending the time for completion of the home until July 21, 1994. Attached to the agreement was a list of uncompleted work. The agreement also contained a liquidated damages clause in the event that construction was not completed by the extension deadline. The homeowners later refused to allow Brinich on the property after they obtained an occupancy permit and moved into the house in early October. In addition, the Jenckas refused to authorize disbursement of the final draw.

Brinich commenced suit against the Jenckas, alleging breach of written and oral contracts, unjust enrichment, tortious interference with contractual relations, and defamation. The jury returned a verdict in favor of Brinich for $87,635.21 for breach of written and oral contract.

The Jenckas contended that the trial court erred in failing to grant JNOV or a new trial on their counterclaim for liquidated damages. Because it was undisputed that Brinich failed to complete construction prior to the Jenckas' moving in, the Jenckas argued that Brinich was responsible for liquidated damages. The Jenckas provided very little evidence of any actual costs they incurred due to the delay in the construction of their home. Mr. Jencka testified that the liquidated damages provision was suggested by others to "put some teeth" in the agreement, and to encourage Brinich to finish the project on time. He claimed that the amount represented "costs that we were experiencing out of pocket." However, testimony concerning these out-of-pocket expenses was not limited to the delay period. The only expenses relevant to determining the legitimacy of the liquidated damages clause were storage and living expenses of approximately $550 per month, or less than $20 per day. Even if Jenckas' mileage expenses were included, the total still accounted for less than $30 per day. Because the Jenckas failed to present sufficient evidence justifying the $250-per-day liquidated damages figure, the court concluded that it was a penalty and thus unenforceable. The judgment was affirmed.

A liquidated damages provision is unenforceable where an owner fails to explain how the amount of damages for a bidder's delay has been ascertained or why the amount is a reasonable estimate. In *Eatherly Constr. Co. v. HTI Mem. Hosp.*, 2005 Tenn. App. LEXIS 575 (Sept. 12, 2005), HTI Memorial Hospital Corporation (owner) planned to build a new acute care hospital. The owner awarded a contract to Eatherly Construction Company (bidder) for the construction of a pumping station and installation of a water line under a highway. The bidder filed suit against the owner to recover the cost of installing some iron pipe that was omitted from its bid. The owner denied the bidder's claims and filed a counterclaim to recover liquidated damages of $500 per day for each day the construction was delayed beyond the agreed completion date. The trial court dismissed the bidder's bidding error claim. The trial court also dismissed the owner's counterclaim for liquidated damages due to construction delays. The trial court held that the liquidated damages provision in the contract was a penalty and thus was unenforceable. Both parties appealed.

The bidder challenged the validity of the liquidated damages provision and asserted that its work was completed in a timely manner because it was granted additional time to complete the work. The owner contended that the bidder did not complete its work until 63 days after the expected completion date. The owner contended that $500 per day was reasonable because the parties agreed to that amount in the contract. Yet the owner offered no guidance regarding how the amount was ascertained or its reasonableness. The owner did not explain how the contract was negotiated or how the amount of damages was calculated. The parties' contractual agreement as to the amount of liquidated damages was not sufficient to show that the amount was reasonable. There was no proof in the record to support a finding that the liquidated sum was a reasonable estimate of the potential damages to be incurred by the owner. Because the evidence offered by the owner was inadequate to establish that the liquidated sum agreed upon was a reasonable estimate, the liquidated damages provision was a penalty and unenforceable. The owner failed to demonstrate that the evidence supported a finding of fact different than that of the trial court's finding of fact. The trial court's judgment was affirmed.

Page 2-83, add at end of carryover paragraph:

Liquidated damages of $5,000 a week were found not to be a penalty for prolonged inconvenience, discomfort, invasion of privacy, and forced renegotiation with another contractor to complete the job. *The Gables, Gables/Kovens, Inc. v. Choate*, 792 So. 2d 520 (Fla. Dist. Ct. App. 3d Dist. 2001). Choate signed a purchase and sale agreement with The Gables, a developer for a luxury condominium unit. The contract provided that the unit would be completed by May 1, 1996, and that if the unit was not completed on time, the owner would receive a credit of $5,000 per month until closing. By December 1996, the unit was still incomplete and the developer was in bad financial straits. An improvement contract was incorporated into the closing agreement, providing that if the unit was not finished within 30 days the developer would pay the owner $5,000 per week should completion of all the work be delayed beyond 30 days. The interior contractor also agreed to pay the owner $5,000 for each seven-day period of such delay. The contract also provided that if the owner failed to close, he would forfeit his entire $175,000 deposit. The owner moved into the unfinished unit on January 20, 1997. An independently retained contractor testified that the work was incomplete and had been done in a slipshod, unworkmanlike manner. The developer walked off the job and the owner hired another contractor to finish the job. The owner sued. The trial court determined that the owner was entitled to summary judgment on the issue of enforceability of the liquidated damage clause of the closing agreement. The developer appealed.

The developer argued that the liquidated damages at issue were a penalty disproportionate to the owners' actual damages. The parties in the case negotiated for the purchase and sale of a luxury residence. Both sides were sophisticated and were represented by counsel. The appellate court agreed with the owner that the liquidated damage provision had been included because actual damages could not be determined at the time of closing and the owner would suffer significant inconveniences in the event the dwelling was not completed in a timely and workmanlike manner. For the eight to nine months that followed the closing, the owner was forced to live in a construction zone rather than a luxury apartment. Six months after closing, the developer abandoned the premises and the owner was forced to contract another construction company to finish the job. The prolonged inconvenience, discomfort, invasion of privacy, and renegotiation with another contractor to complete the job were exactly the things the parties considered when they agreed to the liquidated damage provision. The appeals court affirmed.

In *Carr-Gottstein Props. v. Benedict*, 72 P.3d 308, 2003 Alas. LEXIS 57 (2003), Carr-Gottstein Properties developed the Southport Subdivision. The developer executed a Declaration of Covenants, Conditions and Restrictions ("CC&R") that regulated the use of lots within the subdivision. One covenant required owners of lots within the subdivision to complete any construction of a dwelling within one year. A purpose of this one-year covenant was to protect the aesthetics of the subdivision. The CC&R contained a liquidated damages clause that provided a $25 daily fine for any violation of the CC&R. On September 20, 1999, Ruth

Benedict began construction on her lot. On October 31, 2000, the developer gave the owner written notice that the owner was in violation of the one-year construction limitation covenant. The developer filed suit to require completion of the construction and for assessment of liquidated damages. The owner challenged the validity of the liquidated damages provision. The trial court found that the owner was not in compliance with the covenant, but also that the CC&R's liquidated damages clause was impermissible. The developer appealed.

The court found that the liquidated damages were meant to address the aesthetic harm caused by delays in construction. It was generally accepted that injuries caused by construction delays were nearly always difficult to determine. The developer's clause also met the second part of the test, as a $25 daily fine was "a reasonable forecast of the damages" likely to occur as a result of this covenant breach, because injury caused by construction delay was nearly always difficult to determine. Furthermore, the per diem liquidated damages provision took into account the magnitude of the breach because the temporal length of the breach affected the magnitude of the breach. Each additional day of unfinished construction prolonged harm, and its magnitude, therefore, was correlated to its temporal length. In setting a per diem fee, the developer's liquidated damages provision attempted to reasonably forecast damages by assigning different damages to breaches of different magnitudes. Because the liquidated damages provision in question met both prongs of the test and was not a penalty, it was enforceable. The superior court was ordered to grant the developer liquidated damages. The court's grant of summary judgment to the owner was reversed.

§ 2.16 NO DAMAGES FOR DELAY CLAUSE

Page 2-86, add at end of first paragraph:

A no damages for delay clause may also apply to delays before the contract time commences. In *Daniel E. Terreri & Sons, Inc. v. Mahoning County Board of Commissioners*, 786 N.E.2d 921 (Ohio Ct. App. 7 Dist. 2003), the contract scope was for the renovation of a building that was to be completed within 120 days after the owner issued the notice to proceed. The owner awarded the contract to Terreri, but did not issue notice to proceed. The contractor demanded that the owner issue a notice to proceed, demanded a 15-percent increase in the contract price, and later stated that it considered the contract terminated. Finally, the contractor sued the owner, alleging delay because of the owner's failure to issue notice to proceed. In its defense, the owner relied upon a broadly worded no damages for delay clause that stated the contractor "shall have no claim against the Owner for an increase in price" based on "any damage, loss or additional expense . . . that it may suffer as the result of any delays." The clause further provided that "the contractor assumes all risks of delays in prosecuting or completing the work under the contract." The court held that the no damages for delay clause did apply to the owner's delay in issuing the notice to proceed and denied the contractor's delay costs.

[A] Enforceability

Page 2-90, add after first full paragraph:

For example, in *Landis & Gyr Powers, Inc. v. Berley Industries, Inc.*, 298 A.D.2d 435, 750 N.Y.S.2d 82 (2002), Berley Industries, Inc. contracted with the Dormitory Authority of the State of New York for heating, ventilation, and air conditioning work at a renovation project at Queens College. Berley subcontracted the automatic temperature control work to the plaintiff, Landis & Gyr. Due to delays allegedly caused by Berley's mismanagement of the project and failure to compel another subcontractor, General Sheet Metal, Inc., to timely and properly perform work upon which the plaintiff's work depended, the plaintiff's work was not completed until approximately 418 days after the original contract completion date. Alleging that it was damaged by the delay, the plaintiff commenced its action against Berley seeking to recover the increased cost of the work occasioned by the delay and other damages under theories of breach of contract, tort, and quantum meruit. The subcontract between Berley and the plaintiff contained a clause providing that the plaintiff was not entitled to compensation for delay caused by Berley. On its motion for summary judgment, Berley argued that the plaintiff's cause of action, which sought to recover for extra expenses as a result of delays allegedly caused by Berley, was barred by the no damages for delay provision of the subcontract. The plaintiff failed to raise any factual issue that the delays were not contemplated by the parties, or that Berley's conduct was intentional, willful, or grossly negligent, thus, summary judgment dismissing the cause of action was properly granted.

The delay of a third-party subcontractor may not be a reason not to enforce a no damages for delay clause against another subcontractor. For example, in *J.A. Jones Construction Co. v. Lehrer McGovern Bovis, Inc.*, 89 P.3d 1009 (Nev. 2004), the concrete subcontractor on and exposition center expansion project was delayed by the failure of the excavation subcontractor to complete its work in a timely manner. The concrete subcontractor alleged that it was forced to use less efficient methods and was delayed by over eight months. The concrete subcontractor submitted a delay claim against the general contractor, and then filed suit against the general contractor when the claim was denied. The general contractor asserted the no damages for delays clause in the subcontract as a defense. Under Nevada law, only three exceptions are recognized to the enforcement of a no damages for delay clause: (1) unreasonably long delays amounting to abandonment; (2) delays caused by fraud, misrepresentation, concealment, or bad faith; and (3) delays caused by active interference. The delay caused by a third-party subcontractor did not fall within one of these exceptions, and the concrete subcontractor's claim against the general contractor was barred by the no damages for delay clause in the subcontract.

A subcontract's no damages for delay clause may bar a subcontractor's claim for damages against a construction manager, even if the subcontractor falls behind schedule through no fault of its own. For example, in *Kiewit Constr.*

Co. v. Capital Elec. Constr. Co., 2005 U.S. Dist. LEXIS 23621 (D. Neb. Oct. 12, 2005), Kiewit Construction (construction manager) was hired for a convention center and arena project. The construction manager agreed that it would substantially complete the work by a specified date for a guaranteed maximum price (GMP). The construction manager negotiated GMP contracts with its subcontractors, including Capital Electric (subcontractor). The construction manager directed the subcontractor to accelerate its work so that the project could be completed on time. The subcontractor complied, but the work was not completed on time. The construction manager brought suit against the subcontractor, who counterclaimed against the construction manager. The parties filed cross-motions for summary judgment.

The subcontractor sought payment for sequencing changes, active interference, and work disruption that allegedly hampered its progress before the acceleration directive was issued. The construction manager argued that because the subcontractor failed to give written notice as required under the subcontract, the subcontractor's claim was barred. The subcontractor countered that the written notice requirement was irrelevant because the requirement was triggered only if the subcontractor requested more time to complete the work. The subcontractor pointed out that it requested more money, not more time. The construction manager also asserted that the subcontract contained a valid no damages for delay clause that barred the subcontractor's claim. The subcontractor argued that the clause did not bar its claim for damages because, at the time the acceleration directive was given, the project was already behind schedule. The subcontract unambiguously stated that damages and additional compensation were not available to the contractor based on delays or interference with its work caused by the owner, construction manager, or other subcontractors. The subcontractor's only relief from delays and interference caused by others was to seek an extension of time from the construction manager. The subcontract's no damages for delay clause was valid and thus barred the subcontractor's claim for additional money for delays and interference caused by others. Even if the subcontractor was behind schedule before the acceleration directive through no fault of its own, its only recourse was to seek a time extension. There was no evidence of malfeasance by the construction manager that would have rendered the clause unenforceable. The construction manager's motion for summary judgment was granted with regard to the subcontractor's counterclaim.

Page 2-90, add new footnote after second sentence in last paragraph:

[532.1] L&B Constr. Co. v. Ragan Enterprises, Inc., 482 S.E.2d 279 (Ga. 1997); Burt Welding & Automotive Repair, Inc. v. U.W. Marx, Inc., 272 A.D.2d 737, 707 N.Y.S.2d 548 (2000).

Page 2-91, replace note 533 with:

221 Ga. App. 543, 472 S.E.2d 88 (1996), *rev'd*, 482 S.E. 279 (1997).

Page 2-91, add to note 536:

Blake Constr. Co. / Poole & Kent v. Upper Occoquan Sewage Auth., 266 Va. 564, 587 S.E.2d 711 (2003) (Contract provisions barring a contractor's claim for unreasonable delay damages, except upon an authority's bad faith, malice, gross negligence, or abandonment, are void and unenforceable as against Virginia's public policy.).

[B] Exception for Delay Not Within the Terms

Page 2-92, add after carryover sentence:

In *Triple R Paving, Inc. v. Broward County*, 774 So. 2d 50 (Fla. Dist. Ct. App. 2000), the standard form contract used by the owner contained a no damages for delay clause that permitted recovery only for hindrances or delays due to fraud, bad faith, or active interference on the part of the county or its agents. Triple R Paving successfully bid on a road construction contract, which included widening a bridge that spanned a canal. During construction, delays resulted from a horizontal sight distance design flaw, a Florida Power & Light (FP&L) utility relocation, and detention pond elevation problems. Triple R filed suit against the owner for delay damages, including lost home office overhead and lost efficiency. The owner, in turn, filed a third-party complaint against the designer for indemnification.

Triple R discovered that the plans did not meet horizontal sight distance standards. The owner then refused to allow the bridge to open. Triple R advised the designer that it would move its manpower and equipment off the job until a solution to the horizontal sight distance problem was discovered; however, it was directed not to do so. While a solution was being worked out, Triple R worked on a portion of the project north of the bridge, but the inability to open the bridge and switch traffic to the other lane impeded its ability to proceed efficiently with its work.

A second delay occurred when FP&L did not show up to remove the power lines because of maintenance problems with its equipment. The third delay resulted from problems involving the detention pond that was to drain into a canal. The canal elevation was higher than depicted in the design and higher than the pond; however, no proof was elicited to show that either the owner or the designer was aware of this design flaw prior to actual construction. Had the construction proceeded as designed, a backward flow would have resulted.

The entire project was finished within the extension period granted for the delays, but not within the original contract period. The work was never completely suspended on the project. According to Triple R, it was never able to become more efficient in performing the remainder of the work. Triple R claimed that during the delays and extended performance period, the company obtained no substitute work to compensate it for lost overhead. The designer moved for directed verdict at the close of Triple R's case, claiming that Triple R had failed to prove that the construction delays were the result of fraud, bad faith, or active interference. The jury returned a verdict that awarded damages to Triple R for loss of efficiency in the amount of $112,929.31, but nothing for home office overhead. The designer claimed that the trial court erred in denying its motion for directed verdict. The court agreed with the designer with respect to the delays occasioned by the FP&L utility relocation and the detention pond elevation. However, the court found that the delay resulting from the horizontal sight distance design flaw was sufficient to allow a jury to decide the question of fraud, bad faith, or active interference. According to the reviewing court, evidence of the designer's knowledge of the design flaw and the subsequent failure to apprise Triple R of the problem were sufficient to constitute "willful concealment of foreseeable circumstances which impact timely performance," such that the no damages for delay clause may be overcome. However, in view of the owner's and designer's lack of control over FP&L and the lack of proof of knowledge as to the detention pond flaw, the appellate court found the evidence insufficient to show fraud, bad faith, or active interference, and so reversed in favor of the owner and the designer on those damage claims.

There are other arguments that may be raised to avoid the application of no damages for delay clauses. For example, in *PYCA Industries, Inc. v. Harrison County Waste Water Management District*, 177 F.3d 351 (5th Cir. 1999), the claimant argued that the no damages for delay clause was superseded by federal contract provisions that authorized increases in contract time and price for changes in the work. The project was a waste water treatment facility that was partially funded by the U.S. Environmental Protection Agency (EPA). The EPA special conditions were incorporated by reference into the contract documents. They provided for time extensions and the recovery of time-related costs, and further provided that the special conditions would supersede other conflicting contract provisions. A no damages for delay clause was included in the contract documents. In dealing with the issue, the court reasoned that the EPA special conditions were minimum requirements that EPA grantees must include in the contract documents, but the grantees were expressly authorized to impose more stringent terms. The no damages for delay clause was an additional requirement, not a conflicting provision, and the court enforced the no damages for delay clause to preclude the claimant from recovering its delay costs.

The contract clause preventing recovery for delays caused by a utility does not bar contractor claims based on an owner's own performance problems. In *Scoccolo Construction Co. v. City of Renton*, 9 P.3d 886 (Wash. Ct. App. 2000), Scoccolo Construction, Inc. was awarded a contract with the City of Renton to widen Park Avenue. The contract provided that the relocation of utilities would be effected by the relevant utility companies, not the contractor, although the contractor was required to coordinate relocation efforts with the utility companies. The contract further provided that no additional compensation would be made for delay caused by actions of any utility company. A franchise agreement between the city and the utility gave the city the authority to compel the utility to relocate utilities at the utility company's cost and expense. The contractor claimed that the utility did not complete its obligations under the contract with the city in a timely manner, which delayed the contractor's work on Park Avenue. The contractor filed suit against the city to recover damages due to delays in the contract, and the city moved for partial summary judgment in light of the contract language providing that the contractor would not receive additional compensation for delays caused by the actions of any utility company. The trial court granted the city's motion for summary judgement. The parties appealed.

The contractor argued, among other things, that summary judgment was inappropriate because: (1) the city knew that the utility was delaying the contractor's work, but did not exercise its right to direct the utility to complete the work or to have others complete the work if the utility failed to perform; (2) the city breached its implied duty to cooperate with the contractor and facilitate completion of the project; and (3) the contractual language precluding additional compensation for delays caused by the actions of utility companies was invalid under a statute. The city argued that the statutory prohibition applied only to the actions of the city or its agents and that the utility was not the city's agent. The appellate court noted that the statute was a legislative reaction to state cases upholding the protections afforded by a no damages for delay clause. The court then stated that the contractor's claims were based on the actions of the city and not the utility. The court that indicated claims against the city were not barred by the no damages for delay provision. The fact that the utility companies contributed to delays did not shield the city from potential liability based on its own contractual breaches. The statute by its plain language voided a particular type of construction contract clause that purported to prevent delay damage claims. The appellate court noted that even if the statute invalidated the language in the contract, the contractor could not necessarily recover for delays related to the conduct of the utilities. However, summary judgment was inappropriate because it was based solely upon the existence of the no damages for delay clause in the agreement. The trial court's summary judgment order was reversed.

[C] Exception for Delay Not Contemplated by the Parties

Page 2-93, add after note 548 in first carryover paragraph:

In *Clifford R. Gray Inc. v. City School District*, 277 A.D.2d 843, 716 N.Y.S.2d 795 (2000), an electrical contractor had a contract to renovate and expand a school. The parties' contract included a no damages for delay clause. The contractor filed suit for money damages for delays. The trial court concluded that the clause did not preclude the contractor's suit and awarded money damages. Both parties appealed.

The construction schedule contemplated three phases. Phase II (the construction of a new addition to the school) was to continue for 10 months and be finished in February 1995. However, it was not substantially completed until September 1995. Similarly, although phase III (renovation of an existing building) was anticipated to take six months, with completion in August 1995, the work was not concluded until July 1996. Among the various causes of the delays were the owner's failure to timely obtain easements for electrical and drain sewer installations; the failure of the owner's construction manager to adequately supervise and coordinate the work of the various contractors, including the failure to prepare coordinated construction schedules and drawings; the termination of both the construction manager (in December 1994) and the general contractor (in June 1995); and the owner's decision to hire 30 subcontractors in lieu of replacing the general contractor. The appellate court found that the project impediments were wholly unanticipated and of a character and magnitude not ordinarily encountered or anticipated by parties to a contract of this nature. Moreover, the evidence established that the owner's construction management team failed to substantially fulfill the express contractual obligation to schedule and coordinate the work, resulting in extensive work interruptions or delays and inefficient labor deployment. The court saw no basis to disturb the state supreme court's determination that there was a pervasive and ongoing breach of the contract.

Page 2-93, add to note 552:

Other types of delays that may not have been contemplated by the parties include the failure of the general contractor to expedite the revisions of drawings needed for the subcontractor to perform the foundation work due to the encountering unanticipated subsurface rock. MacQuesten General Contracting, Inc. v. HCE, Inc., 296 F. Supp. 2d 437 (S.D.N.Y. 2003).

Page 2-93, add after note 552 in first carryover paragraph:

and delays resulting from the owner's mismanagement of the multiple trade contractors. *Clifford R. Gray, Inc. v. City Sch. Dist. of Albany*, 277 A.D.2d 843, 716 N.Y.S.2d 795 (2000).

Page 2-93, add after note 553 in first carryover paragraph:

The exception to the no damages for delay clause for unanticipated delays is also found in some statutes (Cal. Pub. Cont. Code § 7102 (1985 & Supp. 1990)), and may be found in the contract provisions as well. *Howard Contracting, Inc. v. G.A. MacDonald Construction Co.*, 83 Cal. Rptr. 2d 590, 71 Cal. App. 4th 38 (1999).

Page 2-94, add after note 558 in first carryover paragraph:

In one contract, the owner expressly advised the contractor of the possibility of unmarked or inaccurately located utility lines. The contract also contained a specific clause disallowing damages for delay due to unanticipated utility interferences. The contractor encountered 26 different unmarked or mislocated utility lines that interfered with its contract work to install three miles of water main, and claimed that they resulted in 18 weeks of delay. The contractor filed suit to recover its delay costs. Although the trial court directed the verdict in favor of the contractor, the appellate court concluded that the contract clearly contemplated this type of delay and the no damages for delay clause was enforceable for this type of delay. *DiGioia Brothers Excavating v. City of Cleveland Dep't of Public Utilities*, 734 N.E.2d 438, 135 Ohio App. 3d 436, *appeal denied*, 727 N.E.2d 134 (2000).

[D] Exception for Unreasonable Delay

Page 2-95, add after note 564 in first carryover paragraph:

The jury is to decide whether the delay was so unreasonable as to constitute abandonment. *Mississippi Transportation Comm'n v. SCI, Inc.*, 717 So. 2d 332 (Miss. 1998).

Page 2-95, add after note 565 in first carryover paragraph:

For example, where the owner failed to obtain the necessary permits to allow the work to proceed, a four-month delay was held to be unreasonable. *Howard Contracting, Inc. v. G.A. MacDonald Constr. Co.*, Inc., 71 Cal. App. 4th 38, 83 Cal. Rptr. 2d 590 (1999).

[E] Exception for Active Owner Interference

Page 2-95, add after note 569 in first full paragraph:

or where the owner issued the notice to proceed to the contractor knowing that the required preliminary work had not been completed. *U.S. Steel Corp. v. Missouri Pacific Railroad Co.*, 668 F.2d 435 (8th Cir. 1982).

Page 2-96, add after note 574 in first carryover paragraph:

Ordering the mechanical subcontractor to perform its work in areas that were not ready for the plumbing and mechanical work may constitute active interference by a general contractor to the subcontractor's work such that the no damages for delay clause may not preclude the subcontractor from recovering its costs associated with the general contractor's actions. *Dennis Stubbs Plumbing, Inc. v. Travelers Casualty and Surety Co. of America*, 67 Fed. Appx. 789, 2003 U.S. App. LEXIS 11231 (2003).

Page 2-96, add after carryover paragraph:

The subcontractor also proved other conduct that was indicative of bad faith and malicious intentions, including backcharging the subcontractor for delays when the owner already granted time extensions for the matter; grossly inflating the amount of subcontractor backcharges; wrongfully removing excavated material that the subcontractor planned to use for backfill; not making subcontractor payments allegedly because the general contractor had not been paid by the owner when in fact the general contractor had been paid; and terminating the subcontract just one day before completion on the grounds that the subcontractor did not pay its insurance, when in fact the insurance premiums had been paid.

Page 2-97, add after note 578 in first carryover paragraph:

Some jurisdictions examining the issue of active owner interference require more than negligent conduct by the owner. For example, in *P.T.& L. Construction Co., Inc. v. State of New Jersey Department of Transportation*, 531 A.2d 1330 (N.J. 1987), the court refused to find active owner interference in the Department's failure to coordinate the work of the utilities subcontractor, stating that active interference involves more than negligence and "contemplates reprehensible behavior beyond a 'simple mistake, error in judgment, lack of total effort or lack of complete diligence.'" *Id.* at 1343. One commentator noted that other jurisdictions appear to allow "simple negligence" by the owner to be adequate to overcome a no damages for delay clause, even though they do not explicitly apply the negligence standard. Lesser & Wallach, *Risky Business in the Active Interference Exception to the No-Damages-For-Delay Clause*, 23 Constr. Law. 26, 28 (Winter 2003). Such "simple negligence" actions include improper sequencing of work, *Id.*, citing *Blake Constr. Co., Inc. v. C.J. Coakley, Inc.*, 431 A.2d 569 (D.C. Ct. App. 1981); tendering defective plans and specifications, *Id.*, citing *Felhaber Corp. v. State of New York*, 410 N.Y.S. 290 (3d Dep't 1978); failure to coordinate the work of other contractors, *Id.*, citing *Housing Auth. of Dallas v. Hubbell*, 325 S.W.2d 880 (Tex. App. 1959) (failure to provide access to the work site, *Id.*, citing *Commonwealth of Pennsylvania, State Highway & Bridge Authority v. General Asphalt Paving Co.*, 405 A.2d 1138 (Pa. 1979)); and failure to grant time extensions in a timely manner, *Id.*, citing *Mississippi Transp. Comm'n v. S.C.I., Inc.*, 717 So. 2d 332 (Miss. 1998). The commentator suggests revising the no damages for delay clause to deal with the concerns that simple negligence may be deemed an exception to the application of the clause by including language such as "the provisions of this paragraph shall apply if the loss or damage results from the negligence of the owner, its agents or employees." *Id.*

Page 2-97, add after note 579 in first full paragraph:

or "inexcusable incompetence" (*John E. Gregory & Sons, Inc. v. A. Guenther & Sons, Inc.*, 432 N.W.2d 584 (Wis. 1998)), or "intentional or gross fault" (*Pellerin Constr., Inc. v. Witco Corp.*, 169 F. Supp. 2d 568 (E.D. La. 2001)). The actions of the owner may need to be more than negligent; they may be required to reach the level of gross negligence or intentional misconduct. *Landis & Gyr Powers, Inc. v. Berley Indus., Inc.*, 750 N.Y.S.2d 82 (App. Ct. 2 Dep't 2002).

Page 2-98, add at end of carryover paragraph:

Also, the court in *Mississippi Transportation Commission v. SCI*, 717 So. 2d 332 (Miss. 1998), held that the owner's refusal to grant time extensions on a timely basis could be interpreted as active interference or bad faith preventing the application of the no damages for delay clause from barring the contractor's recovery of delay damages.

[F] Exception for Waiver

Page 2-99, add after note 589 in first carryover paragraph:

Waiver of the right to enforce the no damage for delay clause has also been found where the owner repeatedly promised the contractor that the owner would personally be liable to the contractor for the delays and where the owner stated that the no damages for delay clause did not apply to the extraordinarily long delays incurred on the project. *Chicago College of Osteopathic Medicine v. George A. Fuller*, 776 F.2d 198 (7th Cir. 1985).

CHAPTER 3

RESPONSIBILITY FOR DELAY

§ 3.02 OWNER-CAUSED DELAY

Page 3-5, add after first sentence in first paragraph:

The owner may also have implied duties, obligations, and warranties that relate to delay claims despite the absence of language in the contract. For example, it is often stated that there is an overall duty of the owner not to delay, hinder, or interfere with the contractor's progress or timely completion of the work. *Metropolitan Paving Co. v. United States*, 325 F.2d 241 (Ct. Cl. 1973); *Howard P. Foley Co. v. J.L. Williams & Co.*, 622 F.2d 402 (8th Cir. 1980). Courts have found other implied owner duties and obligations in the following areas:

- The implied duty to provide access to the project site. *Gasparini Excavating Co. v. Pennsylvania Turnpike Comm.*, 187 A.2d 157 (Pa. 1963).
- The implied duty to perform reviews in a timely manner and make decisions on the contractor's submittals. *Fehlhaber Corp. v. State of New York*, 410 N.Y.S.2d 920 (N.Y. App. Div., 3d Dep't 1978).
- The implied warranty of the adequacy of the contract plans and specifications. *United States v. Spearin*, 248 U.S. 132 (1918).
- The duty to coordinate separate and multiple prime contractors. *Shea-S&M Ball v. Massman-Kiewit-Early*, 606 F.2d 1245 (D.C. Cir. 1979).

Page 3-5, add after first full paragraph:

New York has developed a line of cases that contractors are not liable to subcontractors in the event delays arise from circumstances outside the contractor's control, including the acts of the owner. In *Triangle Sheet Metal Works, Inc. v. James H. Merritt and Co.*, 79 N.Y.2d 801, 588 N.E.2d 69, 580 N.Y.S.2d 171 (1991), the New York Court of Appeals denied a subcontractor recovery for delay damages, after finding that the delay was not the fault of the general contractor. Triangle, a subcontractor on a construction project for New York City, sought recovery against Merritt, the prime contractor, for performance delays, which were contributed to by the City, other contractors, and weather delays. The trial court dismissed the claim and the appeals court affirmed, holding that:

> This case falls squarely within the general rule that, absent a contractual commitment to the contrary, a prime contractor is not responsible for delays that its subcontractor may incur unless those delays are caused by some agency or circumstance under the prime contractor's direction or control. Contrary to Triangle's contention, there is no basis for concluding that a prime contractor—which oftentimes lacks control over much of the work to be performed at a particular project—has implicitly agreed to assume responsibility for all delays that a subcontractor might experience—no matter what their cause. *Id.* at 802 (citations and quotations omitted) (emphasis added). See also Norcross v. Wills, 198 N.Y. 336, 91 N.E. 803, 805 (1910) (holding that a party to a construction contract will be "answerable for all losses caused by delays, which his control of the work should make him responsible for . . .").

The *Triangle* court concluded that a subcontractor unhappy with this rule "should bargain for the inclusion in its subcontract of a provision" to the contrary. *Id.* at 803.

Similarly, in *Regional Building Systems, Inc. v. Planning Commission of Regional Building System*, 320 F.3d 4821 (4th Cir. 2003), Regional Building Systems (RBS), a manufacturer of modular housing units, contracted with Aspen Knolls Construction Corporation to manufacture, deliver, and install 1,000 housing units on Aspen Knolls' property in Staten Island. RBS subcontracted with Bedford Construction Corporation under which Bedford was responsible for transporting the modular units manufactured by RBS to the Aspen Knolls building site, and then erecting and completing the structures. Aspen Knolls experienced financial difficulties and defaulted on a number of payments to RBS. Without these payments, RBS experienced a severe cash flow problem and was unable to meet its contractual obligation to deliver the requisite number of housing units to Bedford. Consequently, RBS suspended its work under the subcontract. Aspen Knolls later ceased paying RBS altogether, forcing RBS to terminate the Aspen Knolls contract as well as the subcontract with Bedford. RBS filed a voluntary petition for relief under Chapter 11 of the Bankruptcy Code. Bedford filed a claim in the amount of $1,448,226.49. The bankruptcy court determined that Bedford was only entitled to recover an additional $22,067.82. The bankruptcy court denied Bedford's claim for delay damages, holding that New York law prevented such a recovery when the contractor was not responsible for the delays. The district court affirmed the bankruptcy court's decision. Bedford appealed the holding that New York law did not permit a subcontractor to recover delay damages from a general contractor.

Bedford argued that financial difficulty did not excuse performance under a contract and that delay damages were proper. In its opinion, the bankruptcy court concluded that Aspen Knolls' breach of its payment obligations to RBS was the cause of RBS's suspension of work. The court found that RBS acted reasonably, suspending deliveries only after Aspen Knolls had defaulted on several payments and resuming them once payments were again made. RBS terminated the contract only after Aspen Knolls had defaulted on two invoices totaling over $3,000,000.00, which the court determined so severely hampered RBS's cash flow that RBS was justified in suspending work. During this time, RBS kept Bedford fully informed about Aspen Knolls' financial troubles, knowing that Bedford also had direct contracts with Aspen Knolls. Having determined that the delay was caused by Aspen Knolls' failure to pay RBS, not by any wrongdoing of RBS, the bankruptcy court concluded that RBS was not liable to Bedford for delay damages. Bedford could not claim delay damages for circumstances beyond RBS's control. Accordingly, the denial of delay damages to Bedford was affirmed.

Whether the contractor is liable to the subcontractor for acts out of its control or for acts of the owner should depend on what the subcontract states about owner-caused and third-party-caused delay. Without reference to the risks of owner delay in the subcontract, the court's determination that the contractor is, as a matter of law, not at risk for the owner's actions may be appropriate. However, where a contract contains language indicating otherwise, such a contract should be enforced.

[A] The Project Site

Page 3-9, add after note 19 in first carryover paragraph:

A contractor can be compensated by the state for utility delays due to unknown utility appurtenances. *Quality Asphalt Paving v. State*, 71 P.3d 865 (Alaska 2003). Quality Asphalt Paving, Inc. and the state contracted for a highway construction project. However, after performance began, both parties believed a buried utility cable conflicted with the project plans. The state terminated the contract under the termination-for-convenience clause. The contractor submitted to the state a claim for damages and costs the contractor claimed the termination caused the contractor to incur. The contractor filed a late claim for the utility delay that held up the project. The state appointed a hearing officer. The hearing officer's decision recommended awarding the contractor damages plus prejudgment interest. The hearing officer included the "utility delay component" in awarding the contractor recovery for mobilization and demobilization, standby, and overhead. The contractor appealed the commissioner's decision to the superior court, which vacated the contractor's prejudgment interest award but affirmed the commissioner's decision in all other respects. The contractor and the state appealed.

The state argued that the hearing officer erred in considering the claim because the contract precluded monetary recovery for utility delays. The state emphasized the portion of the contract provision that stated "no additional compensation will be allowed for any delays," but the cited provision clearly referred to delays caused by known utility appurtenances. Because the delay arose due to an unforeseen (although ultimately nonexistent) utility conflict, the court rejected the state's argument that the contract precluded recovery for the utility delay. The hearing officer based his decision to allow the contractor's delay claim to go forward on the state's recognition of the perceived utility conflict and the state's efforts to resolve it. The hearing officer observed that, even though the utility delay claim was not separately presented, the state was aware generally of the nature of the claim. Substantial evidence therefore supported the decision that the contractor was delayed as a result of the utility conflict and that delay was a compensable damage item. Affirmed.

Page 3-9, add after note 20 in first full paragraph:

Even if there is a no damages for delay clause, the failure to relocate utilities in a timely manner may result in the contractor recovering for its delay costs. For example, in *Mississippi Transportation Commission v. Ronald Adams Contractor, Inc.*, 753 So. 2d 1077 (Miss. 2000), the contractor was awarded a contract to widen a highway, and the contract contained an affirmative statement that all the utilities in the right-of-way would be relocated prior to the contractor receiving notice to proceed (NTP), and NTP was stated as a specific "not later than date." The contractor mobilized equipment at the project site in anticipation of the stated NTP date, but because there was a water main running throughout much of the contract area that had not yet been relocated, the owner provided a restricted NTP for areas not affected by the water main relocation work. The contractor declined to proceed, stating that inefficiencies would result if it proceeded on such a restricted basis. The owner's no damages for delay clause addressed the situation in which third parties failed to timely relocate utilities. However, the contractor maintained that the failure to relocate utilities was "active interference," a recognized exception to the enforcement of a no damages for delay clause. The issue ultimately was taken to the Mississippi Supreme Court, which determined that the failure to relocate utilities could be active owner interference that would allow the contractor to recover its delay costs despite the owners no damages for delay clause.

Page 3-9, add at end of first full paragraph:

In *Thomas & Associates, Inc. v. The Metropolitan Government of Nashville*, 2003 Tenn. App. LEXIS 425 (June 6, 2003), Thomas & Associates (contractor) entered into two contracts worth $3.7 million with the Tennessee Department of Transportation Department to serve as general contractor for the construction of the road improvements on Project No. 4105, sometimes called the "Little Bell Road Project" and Project No. 4622, sometimes called the "Murfreesboro Road Project." Project 4105 was scheduled to take 265 calendar work days for completion by December 15, 1996. Project 4622, which had a later start date, was scheduled to take 427 calendar work days and be completed by June 15, 1998. Neither project was completed on schedule. Ultimately Project 4105 was extended by 536 calendar work days (twice its original length), and Project 4622 was extended by 690 calendar work days, more than doubling the length of that project. The principal difficulties encountered were delays in relocating utilities, and failure to obtain all necessary right-of-way for the construction. Both these items were the

Department's responsibility. Thomas filed a claim against the Department for damages with the Tennessee Division of Claims Administration. The Department asserted that all of Thomas's claims were barred by the express terms of the parties' no damage for delay clause and that, furthermore, the Department enjoyed sovereign immunity against all the contractor's claims. The circuit court granted summary judgment in both cases. Thomas has appealed.

Thomas contended that the Department failed to use due care in coordinating the relocation of utilities on the projects. Thomas alleged specifically that the Department's actions were a violation of the relocation of utilities provisions in Tenn. Code Ann. §§ 54-5-801 through 856 (1993). Contracts 4105 and 4622 both included a note on the cover page of each stating that "the Plans and Specifications for Road and Bridge Construction and applicable sections of Title 54 and Chapter 4 of Title 12 of Tennessee Code Annotated are made a part hereof fully and completely as if attached hereto." This language incorporated the utility relocation statutes into the contracts.

Thomas maintained that its negligence and negligence per se claims fell under Tenn. Code Ann. § 9-8-307(a)(1)(I) (Supp. 2002), which in effect waived the State's sovereign immunity for "negligence in planning and programming for, inspection of, design of, preparation of plans for, approval of plans for, and construction of, public roads, streets, highways. . . ." The Department's duty to Thomas on both counts was, if anything, a contractual duty related specifically to the two construction projects. To the extent that the Department merely breached a contractual duty toward Thomas, Thomas could not maintain an action against the Department for negligence. Unless Thomas could identify some other source of duty, it has no claim against the State for negligence within the exception to sovereign immunity carved out by Tenn. Code Ann. § 9-8-307(a)(1)(I). Thomas identified one other potential source for duty in this case: the utility relocation statutes found at Tenn. Code Ann. §§ 54-5-801, -856 (1998 & Supp. 2002).

The statutes set forth specific procedures for both the Department and utility owners to follow to facilitate timely, economical relocation of utilities when necessitated by road construction. Thomas alleged that the Department failed to require and approve a meaningful schedule of working days for utility relocation on these two projects.

Tenn. Code Ann. § 54-5-854 provided that the Department would send copies of its proposed construction plans to the affected utility, who would then map where its lines or pipes are located within the project along with "any proposed new location of such facilities." The utility owner must submit a schedule of working days to accomplish any relocation. Once submitted, the Department could approve that relocation plan and accompanying work schedule "if reasonable." If the utility owner refused to reasonably cooperate, the Department and its contractors might go ahead with the project without liability for any damage to utilities located in the construction area. Nothing about Tenn. Code Ann. § 54-5-854(c) imposed a public policy duty on the Department to act in any certain way toward a road contractor regarding scheduling utility relocation work. Whatever duty the Department owed to Thomas came not from the statute but from the

Department's contract with Thomas. Because Thomas failed to establish that the Department owed it any duty other than a contractual one, the court affirmed summary judgment against Thomas on all of Thomas's claims for negligence.

Thomas maintained also that the parties' contracts expressly provided that the Department had acquired all the needed right-of-way for the road construction with only certain identified exceptions. As to those exceptions, the contracts provided month-and-day-specific anticipated acquisition dates, after which those right-of-way parcels would be available. Thomas contended that the Department breached the contracts by not having certain of those parcels available for construction by the specified anticipated availability dates, thus delaying Thomas in performing its part of the contracts. Based on the utility relocation statutes, Thomas argued that the Department was required to determine, prior to commencing construction to prepare and submit to the Department "a plan and schedule of working days" to accomplish the relocation of conflicting utilities. After the utility owner submitted its plan and schedule of working days, the Department had to approve them, if reasonable, and notify the owner. The Department also had to notify utility owners reasonably in advance of the beginning date for utility relocation. Finally, the Department's resident engineer was supposed to act as liaison between the utility owner and the contractor on the construction project.

Thomas complained that the Department never required the utilities to submit a meaningful schedule of working days with specific start and end dates for the two projects and that the relocation plans submitted by the utilities and approved by the Department did not contain vertical locations for existing underground lines. Also Thomas maintained that the Department's resident engineer failed to properly act as a liaison between Thomas and the utility owners. While the relocation of utilities statutes anticipated that the Department would do these things, the parties' contracts expressly shifted these obligations partially back to the contractor. The contract made clear that the contractor was responsible to deal directly with the involved utilities to secure timely relocation of affected lines.

Thomas also alleged that part of its delay damages flowed from the Department's failure to timely acquire certain right of way parcels needed for the road construction. These breach of contract claims fell outside the contract's exculpatory clause related to utility delays. The contracted-for construction work on both the Little Bell Road and the Murfreesboro Road projects required the State to acquire additional road right-of-way. The contracts on both projects stated that on each, all the needed right-of-way would be obtained before the effective date of each contract. Construction was delayed, at least in part, by the Department's failure to timely acquire all needed right-of-way. The Department had a contractual duty to timely acquire all right-of-way as part of its duty to furnish its contractor with a work site so the contractor could perform its side of the bargain. Thomas created a triable issue that should not have been concluded by summary judgment. The trial court erred in holding as a matter of law that this contractual breach by the Department was excused under the two contracts' "no damages for delay" clauses. While those causes covered delays attributable to utility relocation problems, they did not cover general delays due to the Department's failure to provide the contractor the promised work site.

The owner may not be able to limit the contractor's ability to recover for delayed utility relocations by disclaiming liability in the contract. For example, in *Scoccolo Construction, Inc. v. City of Renton*, 9 P.3d 886 (Wash. Ct. App. 2000), the city awarded a contract to Scoccolo to widen a city street. The contract provided that the city would have the utility company relocate their utility lines, and that the contractor would not be compensated for any delays caused by the utility company. A separate agreement between the city and the utility company provided for the relocation of utility lines at the utility company's cost and pursuant to a schedule agreed upon by the city and the utility company. Scoccolo was delayed in its work by the failure to timely relocate the utility lines. Scoccolo sued the city for its delay costs, based upon the city's failure to act to direct the utility company to perform rather than the utility company's delays. The court found the city liable to the contractor for the delay costs incurred for the city's breach of an implied duty to cooperate with Scoccolo and to facilitate the relocation of the utility lines. In ruling in this manner, the court avoided the issue of the enforceability of the no damages for delay clause, as such clauses are not enforceable under Washington state statutes.

Page 3-10, add after note 29 in first carryover paragraph:

In the contract, the owner may put other restrictions upon the contractor's access to the project, but this should be weighed against the owner's duty not to hinder the contractor's performance. For example, in *Ryco Construction, Inc. v. United States*, 55 Fed. Cl. 184 (2002), the contractor was awarded a contract by the U.S. Forest Service to repair a road in a national forest, including stream crossings and reinforcing embankments. The contract was to be performed in 168 calendar days, and the contract drawings contained a note limiting the stream bed work to when the stream was dry, construction season limitations, or when approved by the contracting officer. Start on this work was delayed and later suspended because of the presence of an endangered frog species. When the contractor filed an action for delay damages, the government moved for summary judgment, arguing that the contract language noted above gave the contracting officer discretion to restrict access to the site, so that the contractor could not recover its delay costs for such matters. The contractor argued that in mandating a 168-day performance period, the government warranted that the contractor would have access to the site in order to perform in the required contract time. The Court of Federal Claims, in reviewing the matter, held that the contractor was not entitled to unimpeded access during the contract performance period. Although the government was entitled to limit site access, it had to be done in a reasonable manner consistent with the government's implied duties to cooperate and not hinder the contractor's performance. The claims court denied summary judgment, leaving this as an issue to be determined by the trial court.

Page 3-11, add at end of subsection:

The general contractor may also be liable to a subcontractor for failing to provide proper access to the project site for the subcontractor's work. For example, in *Advance Iron Works, Inc. v. John Edwards Construction Co.*, 2004 U.S. Dist. LEXIS 18066 (N.D. Ill., East. Div. Sept. 9, 2004), Advance Iron Works, Inc. (Advance) and John Edwards Construction Company (JECC) entered into an agreement whereby Advance was to furnish and install certain structural steel for the VA Hines Reactivation Center Renovation. The schedule called for the Project to be completed in November 2000. The dates of commencement and substantial completion were not set forth in the Advance Subcontract; rather, the parties agreed that the dates would be set according to the schedule from JECC. The parties agreed that time was of the essence in the completion of the Advance Subcontract. One mobilization was included for the Advance work. Additional mobilizations required additional compensation from JECC. Advance agreed to "cooperate with [JECC] in scheduling and performing their Work to avoid conflict or interference with the work of others." As incorporated by the Advance Subcontract, JECC was to provide suitable, level compacted steel lay-down area and access for cranes, equipment, and trucks to point of pick, and erection work was to be based upon OSHA standards for steel erectors. Advance was allowed to stop work if JECC failed to make payment. Advance fabricated the steel, began delivering it in April of 2000, and installed the steel for Courtyard A between April 19 and April 25, 2000. Courtyard B was unavailable for steel erection until Saturday, April 29, 2000. JECC notified Advance that it was to continue the erection work at that time. However, Advance performed no work on the Project from April 26 through June 5, 2000. On June 26 and 27, 2000, Advance erected the steel for Courtyard B, only three days after the area experienced .11 inches of rain on June 23rd and .94 inches of rain on June 24. Between April 29 and May 7, 2000, the only rain in the area was .32 inches on May 1. While the eight remaining days were clear, Advance did not take advantage of those days to erect the steel. However, four of those days fell on weekends. Advance was unable to complete the work until the end of June because it experienced scheduling conflicts with either the weather or the crane rental. JECC sent a Notice of Termination to Advance pursuant to the subcontract and made arrangements for another steel erector to complete the work. Advance claimed that it was unable to erect and deliver all of the steel due to delays in the project. Advance also claimed that JECC improperly terminated Advance's subcontract even though Advance was ready, willing, and able to complete the work on the project. JECC claimed that Advance breached the subcontract by failing to properly complete the subcontract work on a timely basis.

The evidence showed that had there been a proper road at the project, the work could have been done sooner because the crane could be stationed on a stable base even shortly after a rainy day. JECC failed to provide a proper road as required under the terms of the Advance Subcontract and OSHA regulations. JECC materially breached the Advance Subcontract by its failure to provide adequate access to the project site. This breach delayed the progress of the Project. JECC also materially breached the Advance Subcontract by its failure to make timely payments to Advance. Because JECC had breached the contract, Advance was entitled to stop work under the subcontract and recover the value of the work it had already performed. Judgment was entered in favor of Advance Iron Works in the amount of $13,322.75.

[B] Approvals

Page 3-11, add after note 36 in first full paragraph:

On some projects, there may be numerous permits, approvals and other submissions to government agencies that are required. For example, in *Designer Direct, Inc. v. Deforest Redevelopment Authority*, 313 F.3d 1036 (7th Cir. 2002), the agreement between the redevelopment authority and the developer provided that the authority would engage a full-time liaison officer to handle and expedite zoning applications, required permits, and other relevant governmental submissions. The authority did not do so, but reimbursed the developer for its costs to engage a consultant to handle these matters. The developer sued the authority under breach of contract, and was allowed to recover its economic losses because reimbursement of its consultant fees was not adequate because of the "disorganization and inefficiency that resulted from the failure of the authority to have a full time liaison officer to deal with permitting."

Page 3-12, add at end of subsection:

A contractor has no basis to decline to proceed with work if the contractor has no reasonable basis for believing that the work would cause a material violation of law. *Empire Energy Management Systems, Inc. v. Roche*, 362 F.3d 1343, 2004 U.S. App. LEXIS 5512 (Fed. Cir. 2004). Empire and the Air Force entered into a contract to provide cogeneration of electricity to MacDill Air Force Base. Empire was required to "furnish, install, own, operate and maintain a cogeneration plant" at its own expense. The Air Force contracted only to pay for the utility services at a discount from the market price. The contract specified a commercial operation date (COD) by which Empire was required to have substantially completed construction of the cogen plant. The contract also required Empire "to comply with all Federal, State and local environmental and archeological laws and regulations." The contract contained the standard FAR default clause. Shortly after groundbreaking, the parties provided a new completion schedule for the cogen site and a facility charge to be paid monthly by the Air Force after COD. The COD was set at 300 days. The Air Force also agreed not to terminate the contract for default within 480 days of the starting date. Shortly after it had started working, Empire discovered oil-based contaminants. Empire stopped working and requested a stop work order from the Air Force. The contracting officer denied Empire's request for a stop work order. The Air Force subsequently investigated the soil conditions at the site. The EPA representative stated in a letter that it did not object to the construction activities planned and that it did not believe that any such activities would interfere with the contamination investigation. The Air Force provided this letter to Empire and directed Empire to return to work. During the work stoppage, Empire repeatedly refused the Air Force's demands that it return to work. The Air Force issued a cure notice. Empire responded to the cure notice, asserting that it had continued performance of the contract by continuing its investigation into the cogen site's "environmental status" and that a fast-track construction schedule would enable Empire to achieve the COD. Empire resumed work, but it remained substantially behind schedule. Empire requested a schedule extension, which the Air Force refused. The contracting officer terminated the contract for default. Empire appealed the contracting officer's decision to the Board of Contract. The Board also rejected Empire's claims that it had been justified in refusing to work on the site because of environmental concerns. The Board found that Empire could not have had, and did not in fact have, any reasonable environmental concerns that its action would create a material violation of environmental regulatory requirements, and that Empire's claimed environmental concerns were pretextual.

Empire argued that it was not required to resume work until the date of EPA approval of construction activity on the cogen site, giving it a total of 362 days of excusable delay. In the alternative, Empire contended that it was entitled to 217 days of excusable delay, or until the date when Empire was notified of the EPA's initial statement that the cogen site did not need the contamination investigation. The Air Force was only required to notify the EPA of planned physical alterations, not that the EPA specifically approved such alterations. Empire was not required to delay work until EPA approval. Empire said that it was precluded from engaging in unlawful conduct in connection with the contamination investigation. Under these circumstances, says Empire, it could reasonably insist on EPA approval before resuming work to avoid the risk of environmental liability. However, the mere assertion of a colorable claim by the contractor (later found to be meritless) that its actions would violate some regulatory requirement did not excuse performance. Nor would performance be excused by the mere existence of a regulatory investigation. The Federal Circuit affirmed the Board's decision.

A township's 65-day delay in issuing construction permits due to a property owner's failure to conform to barrier free requirements although it delayed work on the project, did not violate the owner's due process rights. In *Cherry Hill Towers, L.L.C. v. Twp. of Cherry Hill*, 2006 U.S. Dist. LEXIS 230 (D.N.J. Jan. 6, 2006), Cherry Hill Towers (owner) applied for construction permits for a proposed reconstruction project involving two dilapidated apartment buildings. Cherry Hill Township initially denied the application, allegedly due to Barrier Free Code (code) compliance issues. The township eventually issued the permits, three months after the application was submitted. The owner alleged that the director of the township's code enforcement department intentionally delayed the issuance of the permits to pressure the owner into using union contractors. The owner alleged that the township's delay in issuing the permits caused it to lose over $930,000. The owner filed a 42 U.S.C.S. Section 1983 complaint against the township, the code enforcement department, and the department's director (collectively, township). The township filed a motion for summary judgment.

The township argued that the permits were delayed due to problems with the owner's application. The owner asserted that the township's delay in issuing construction permits was arbitrary, irrational, and tainted by improper motive. The owner asserted that the delay violated its substantive due process rights under the Fourteenth Amendment. The owner contended that the township delayed the issuance of the permits to force the owner to use union labor. Although the owner clearly had a property interest protected by due process, the owner failed to demonstrate that the township's actions shocked the conscience. Even though the owner incurred extensive costs during the period between the submission of its application and the issuance of the permits, the actual delay was not significant. A 65-calendar-day delay in issuing the permits did not shock the conscience. Even if the owner's allegations that the township delayed the issuance of the permits to pressure the owner to use union labor were accurate, substantive due process concerns were not implicated. The director of code enforcement's alleged warnings that the owner would encounter problems with the project if it did not use union labor did not reflect an egregious abuse of power. The owner ultimately received the permits and did not use union labor. There was no allegation that the township sought to profit from the delay in issuing the permits to the owner. The issuance of permits was delayed, at least in part, because the owner's original plans failed to conform to the code. Neither party suggested that the revised plans that were eventually approved were the same as the plans originally submitted by the owner. The township could not be faulted for strictly interpreting the code's requirements. The owner's substantive due process rights were not violated.

[C] Owner's Financial Obligations

Page 3-13, add after first full paragraph:

Absent a specific clause in the contract, a general contractor may not be liable to a subcontractor for delays caused by the owner's breach of its financial obligations to the general contractor. *In re Regional Building Systems, Inc.*, 320 F.3d 482 (4th Cir. 2003), Regional Building Systems (RBS) was awarded a contract to manufacture, deliver, and erect 100 modular housing unit on property owned by Aspen Knolls. RBS subcontracted with Bedford to deliver and erect the units on the Aspen Knolls property in Staten Island, New York. Aspen Knolls developed financial problems and stopped paying RBS on the prime contract. RBS was unable to pay the manufacturer of the housing unit without cash flow from Aspen Knolls, and later terminated its subcontract with Bedford. Bedford submitted a claim to RBS for delay damages, and RBS responded that it was not responsible to Bedford for the owner's delays and financial problems. The trial court determined that Aspen Knolls's default to RBS was the cause of RBS's inability to timely produce the modular units to Bedford for transport and erection, and that there was no clause in the subcontract between RBS and Bedford that made RBS responsible to Bedford for the owner's delays. In reviewing the matter, the court of appeals held that absent a provision in the subcontract, a prime contractor like RBS was not responsible to a subcontractor for the delays of the owner, as this was beyond the prime contractor's direction and control.

[D] Owner's Responsibility for Design

Page 3-14, add to note 48:

In Alaska Dep't of Natural Res. v. Transamerica Premier Ins. Co., 856 P.2d 766 (Alaska 1993), the court articulated the owner's responsibility as follows:

> Unlike design professionals, project owners owe purely contractual duties to the accuracy of the designs. When providing plans and specifications to a contractor, an owner makes an implied warranty that they will be sufficient for their particular purpose. If the defective specifications cause the contractor to incur extra costs in performing the contract, then the contractor may recover these costs that result from breach of the implied warranty. The implied warranty is part of the bargain between the owner and independent contractor and does not exist outside the contract.

Owners often take a proactive approach in the time period between contract award and the start of construction to hold preconstruction meetings to discuss scheduling, design issues, mobilization, safety plans, special construction requirements, submittals, and coordination of the trades. Delays may often be averted by discussions and problem resolution at preconstruction meetings. Although preconstruction meetings may be good practice for owners and are often the custom in the construction industry, absent a contractual requirement, the owner may not be liable to the contractor for delays for the sole reason of not timely scheduling and holding such meetings. For example, in Program & Constr. Mgmt. Group, Inc., GSBCA No. 14178, 00-1 B.C.A. (CCH) ¶ 30,641 (1999), the contractor requested that the government hold a preconstruction meeting to discuss schedules, safety, submittals, and other issues. There were no contract provisions or other requirements for preconstruction meetings. There was a meeting, but it was not held until nearly seven weeks after contract award. The contractor alleged that it incurred more than a month in front-end delays because of the government's failure to schedule a preconstruction meeting earlier. The board, in ruling in favor of the government, reasoned that without a specific requirement for a preconstruction meeting, it had to reject the contractor's claim and find that there is no implied duty to hold a preconstruction meeting.

Page 3-14, add after first full paragraph:

Some jurisdictions in the United States do not impose a warranty for the sufficiency of plans and specifications on the owner unless indicated in a contract. *Alamo Cmty. College Dist. v. Browning Constr. Co.*, 131 S.W.3d 146 (Tex. App.

2004). Alamo College Community District and Browning Construction Company entered into a contract for the construction of a new campus. The contract between the owner and contractor provided that the contractor would carefully study and compare the contract documents with each other and with information furnished by the owner, and would at once report to the architect errors, inconsistencies, or omissions discovered. The contractor would not be liable to the owner or architect for damage resulting from errors, inconsistencies, or omissions discovered. After disagreements over delay, the contractor sued the owner for breach of contract and won damages. The key issue in the suit was whether the contractor could collect damages for delay when the contract had a no damages for delay clause. The jury answered in the affirmative because the owner had waived its right to rely on that clause; the owner was estopped from relying on that clause; the owner and contractor had agreed to modify the clause; the delays encountered by the contractor were caused by the active interference of the owner; the owner committed unreasonable delay such that the contractor would have been justified in abandoning the contract; and the owner committed fraud, misrepresentation, or other bad faith. The jury assessed damages against the owner because it found the owner breached the contract in several ways, including failing or refusing to correct design errors. The owner appealed.

The owner argued that the damages were unenforceable because the contract did not impose a duty on the owner for such design errors. Texas law provided that unless so expressed in the contract, an owner that furnished a prime contractor with plans and specifications was not a guarantor of the sufficiency of the plans and specifications. If there was any obligation resting upon the owner, it should have been expressed in the language of the contract, or there should have been contract language that would have justified the court in concluding that the parties intended that the owner should have guaranteed the sufficiency of the specifications to the contractor. The contract indicated that the owner was to be responsible for design errors. Additionally, it was nearly identical to the language that another Texas appellate court found to intend the owner guarantee the sufficiency of the specifications. Because the contract contained language indicating the owner had a duty to the contractor for design errors, the owner's second issue was overruled. Affirmed.

Page 3-15, add at end of first full paragraph:

If multiple proprietary products are specified, the owner warrants that each and every specified item is suitable for its intended purpose, and it makes no difference whether the contractor selected the particular product from various alternatively specified items. *Hawaiian Bitumuls & Paving v. United States*, 26 Ct. Cl. 1234 (1992).

Page 3-16, add after first fuller sentence of first carryover paragraph:

Generally the use of performance specification shifts more of the responsibility to the contractor for the design. However, the owner may have responsibility for whether it is possible to meet the design criteria. In *Edsall Construction, Inc.*, ASBCA No. 51,787, 2001-01 B.C.A. (CCH) ¶ 31,425 (2001), the government awarded a contract to Edsall for the construction of two aircraft hangars. The specifications for the hangar doors provided that the door loads were distributed evenly among three specified lift points. However, the shop drawings submitted by Edsall and approved by the government depicted a fourth lift point. Edsall requested to be paid for the additional costs involved in a fourth lift point because the design requirements could not be met using only three lift points. The government refused to pay the extra costs arguing that the contract provided performance specifications for the door, and this shifted the cost risk to the contractor. In considering the matter, the board found that the government warranted that the design load could be distributed to three lift points, and because this was not possible the government's performance specifications were defective.

Another situation where the owner may have responsibility for performance specification is where there is a "mix" of detailed design and performance specifications. In *Trataros Construction, Inc.*, GSBCA No. 14875, 01-1 B.C.A. (CCH) ¶ 31,306 (2001), portions of the specifications for the renovation of a government building were detailed and prescriptive, but other specifications were performance in nature. Specifically, the contractor was required to complete the design and to submit shop drawings sealed by a professional engineer for the fiberglass panels for roof cornices. The cornice work also contained detailed specifications for support structures and attachment points. Some of the government's detailed specifications were defective in that the contractor's structural engineer was required to make modifications to carry the required loads of the fiberglass. The contractor requested additional compensation for the cost of these structural revisions, but the government maintained that the requirement for a professional engineer's sealed shop drawings for the fiberglass panels was clearly proof that the contract was essentially in the nature of performance specifications. In reviewing the contractor's claim for the structural redesign costs, the board looked at both the performance requirements and the detailed specifications, and held that the contractor was not contractually required to correct design problems contained in the government's drawings.

[E] Owner's Contract Administration Responsibilities

Page 3-18, add after note 73 in first full paragraph:

The owner's contract administration duties include providing the contractor with a complete set of plans and specifications. *Overstreet Elec. Co.*, ASBCA No. 51654, 00-1 B.C.A. (CCH) ¶ 30,588 (1999).

What is reasonable depends upon the circumstances. In one situation, 10–14 days was held to be reasonable for reviewing submittals (*Azerind, Inc.*, ASBCA No. 34294, 88-3 B.C.A. (CCH) ¶ 21,047 (1988)); in another, 21 days was held to be reasonable (*C. Walker Constr. Co.*, VABCA No. 1527, 82-2 B.C.A. (CCH)

¶ 15,799 (1982)); and on a highway project, 10–12 weeks was determined to be the norm (*Department of Transp. v. Brayman Constr. Corp.-Bracken Constr. Co.*, 513 A.2d 562 (Pa. Cmwlth. 1986)). The contract may provide a specific time period for the architect's review, and an average turnaround time that exceeds this period may still be held to be reasonable. *C. Walker Constr. Co.*, VABCA No. 1527, 82-2 B.C.A. (CCH) ¶ 15,799 (1982) (21-day turnaround time was reasonable even though the design contract provided for a 15-day submittal review period). If the shop drawing review time is delayed, the reasons for any delays in the turnaround should be examined. The problem may be in the design or owner's specifications (*Malone Constr. Co.*, ASBCA No. 41483, 91-2 B.C.A. (CCH) ¶ 23,694 (1991); *J.S. Alberici Construction Co., Inc.*, GSBCA No. 10788, 1991 GSBCA LEXIS 45 (1991)), the owner's changes to the contract drawings that require shop drawing revisions by the contractor (*Fehlhaber Corp. v. State of New York*, 410 N.Y.S.2d 920 (N.Y. App. Div., 3d Dep't 1987)), the contractor's failure to provide required information with its submittals (*Santa Fe Eng'rs*, ASBCA No. 45523, 93-2 B.C.A. (CCH) 25,744 (1993); *Twigg Corp. v. General Services Admin.*, GSBCA No. 14386, 2000-1 B.C.A. (CCH) ¶ 30,772 (2001); *Jackson Electric Co., Inc.*, ENG BCA No. 6238, 97-1 B.C.A. (CCH) ¶ 28,848 (1997)), or the contractor's failure to coordinate shop drawings from its subcontractors (*William F. Wilke, Inc.*, ASBCA No. 28742, 87-1 B.C.A. (CCH) ¶ 19,390 (1987)).

Page 3-18, add after note 74 in first full paragraph:

Some industry commentators observe that the formal manner in which many written Requests For Information (RFIs) are processed results in an "adversarial atmosphere." Rather than providing an expeditious procedure to resolve design-related questions that affect the ongoing construction activities, a formal written process imposes a more "circuitous route" than face-to-face meetings or other verbal discussions. Design professionals may even believe that the RFI process provides contractors with an avenue to increase the contract amount by change orders or provide the basis for claims based upon design deficiencies. For example, the mere submission of a large number of RFIs may bog down the contract administration, provide the justification for field delays and reflect numerous instances where the design was deficient. *Bumpier Road to Finish Line,* Engineering News-Rec. 56 (May 14, 2001).

[F] Changes in the Work

Page 3-22, add after note 103 in second full paragraph:

The owner may also be responsible to changes to the work resulting from code or environmental regulations changes, provided the owner is given proper notice. For example, in *Maryland State Highway Administration v. Engineering Management Services, Inc.,* 2002 Md. App. LEXIS 166 (Ct. Sp. App. 2002), Maryland State

Highway Administration (SHA) requested bids for the removal of lead-based paint and the repainting of five bridges over I-95. Engineering Management Services, Inc. (EMS) was the successful bidder. The contract required that written notice of a claim for extra compensation to the procurement officer within 30 days after the basis for the claim is known or should have been known. General Provision 4.06 of the contract contained provisions relating to contract changes. It provided, in part:

> B. Any other written order or oral order including a direction, instruction, interpretation or determination from the procurement officer that causes any such change, shall be treated as a change order under this clause, provided that the Contractor gives the procurement officer written notice stating the date, circumstances, and source of the order and that the Contractor regards the order as a change order. . . .
>
> E. If the Contractor intends to assert a claim for an equitable adjustment under this clause, he shall, within 30 days after receipt of a written change order under A. above or the furnishing of written notice under B. above, submit to the procurement officer a written statement setting forth the general nature and monetary extent of such claim, unless this period is extended by the State. The statement of claim hereunder may be included in the notice under B. above.

The contract required EMS to comply with the EPA National Ambient Air Quality Standards which established a general permissible exposure limit for particulate matter of 150 micrograms per cubic meter. The contract further required EMS to comply with "all Federal, State, and local laws, regulations and ordinances applicable to its activities and obligations under this contract." Before the promulgation of the new regulations, OSHA regulations for lead exposure did not apply to construction workers. The new regulations, which were applicable to construction workers, imposed a maximum permissible exposure limit for lead of 50 micrograms per cubic meter, and also required special precautions such as protective clothing and equipment and special hygiene facilities and practices.

The new OSHA regulations were published in the Federal Register on May 4, 1993 and had an effective date of June 3, 1993. The contract obligated EMS to comply with OSHA regulations. EMS did not begin performance under the contract until September 30, 1993. By letter dated April 27, 1994, SHA informed EMS that the more stringent OSHA standards governed the contract. Finally, by letter dated May 2, 1994, EMS informed SHA that it would complete its evaluation of compliance with the new OSHA regulations "in a few days." Pursuant to General Provision 5.14 of the contract EMS was required to file a written notice of claim relating to the contract within 30 days after the basis for the claim is known or should have been known, whichever is earlier.

On March 6, 1995, EMS wrote to SHA requesting an 180-day extension on the contract due to "the increased loss of productivity resulting from new Health and Safety Regulations that have been enacted after the start of this contract." EMS explained that "the current production rate under these new conditions is approximately 550 square feet per day. Previously our rate was approximately

1,500 square feet per day. This 64% decrease has greatly impacted our schedule and is the basis for our extension request." On March 23, 1995, SHA replied to EMS, taking issue with the complaints raised in EMS's letter of March 6, 1995. SHA denied the EMS request for a 180-day extension on the project, but agreed to a shorter extension. On June 13, 1995, EMS again wrote to SHA requesting $1,244,564 in additional compensation due to changes in the Lead Exposure Regulations codified at 29 C.F.R 1926.62. The parties met to discuss a possible resolution to their conflict, but no final agreement was reached. EMS later requested additional compensation in the amount of $2,377,341, but subsequently reduced its request to $764,036. By letter dated June 28, 1999, SHA denied EMS's claim for additional compensation and a time extension. EMS filed a timely appeal to the Maryland State Board of Contract Appeals (the Board) seeking the time extension and an equitable adjustment to the contract in the amount of $764,036. SHA moved for summary disposition of the appeal on the grounds that EMS failed to file a timely notice of claim. The board agreed with SHA and dismissed the appeal. EMS filed a petition for judicial review in the Circuit Court for Baltimore City. The circuit court issued a memorandum opinion reversing the board's decision. This appeal followed.

EMS contended that the board improperly granted summary disposition because the issue of whether timely notice of the claim was given to SHA involved a factual dispute as to when EMS knew or should have known of the basis for its claim. The January 12, 1994 letter from EMS did not serve as notice of a claim resulting from the new OSHA regulations and the potential impact of those regulations on EMS's work. In that letter, EMS requested a time extension because of difficulties experienced in securing appropriate equipment. The letter did not mention the new OSHA regulations or their potential impact on the work required under the contract. The Board acted properly when it granted summary disposition and found that EMS was barred from any equitable adjustment because the notice of claim should have been filed within 30 days after the OSHA regulations became effective on June 3, 1993 and, certainly, within 30 days of EMS's completion of a cost evaluation of the effect of compliance, in May 1994. The failure of EMS to file a notice of claim within the required time limit was sufficient basis for the summary disposition.

Page 3-22, add after note 105 in third paragraph:

In some situations, the owner may grant a time extension to deal with the time associated with the change order work. If a time extension is given by the owner relating to changes in the work, there is a rebuttable presumption that the time extension is compensable. For example in *Gottfried Corp.*, ASBCA No. 51041, 98-2 B.C.A. (CCH) ¶ 30,063 (1998), the contractor was awarded a contract for the construction of a classroom building at an Air Force base. During the course of construction, the government issued a contract modification to the contractor extending the contract time by 16 days to deal with a structural steel design defect. Later, when the contractor sought to recover its delay costs for extended overhead, the government argued that contractor-caused delays were concurrent with the design-related delay such that the contractor was not entitled to compensation for the time extension. The board held that the granting of a time extension created a rebuttable presumption that the delay was solely due to government-caused delay, was not concurrent with contractor-caused delay, and was critical (delayed the overall completion of the contract). A similar result was reached in *Sherman R. Smoot Corp.*, ASBCA No. 53115, 2003 ASBCA LEXIS 25 (2003), where the government made several changes to the contract work, and even though the changes were not on the critical path of the schedule, the contractor alleged that administering the modifications delayed by project. The government issued a contract modification extending the contract time, but did not provide for any time-related compensation. In the hearing before the board on the claim for its delay costs, the contractor did not provide any critical path analysis to prove that the change orders caused any delay, but relied upon the time extension granted that related to the changes. The board agreed that in granting a time extension associated with changed work, the government creates a rebuttable presumption that the delays resulting from the changes were critical and that the government was responsible for the delay-related costs. The government may be able to rebut this presumption with a schedule and entitlement analysis, but did not do so in the instant case. Even if the change order is issued unilaterally by the owner and a time extension is given, there may be a rebuttable presumption that the owner is responsible for the delay costs associated with the time extension.

In *Gavosto Associates, Inc.*, PSBCA No. 4058, 01-01 B.C.A. (CCH) ¶ 31,389 (2001), Gavosto was awarded a fixed-price construction contract for a truck wash facility to be completed within 150 days. During the course of construction, the Postal Service issued numerous unilateral change orders, some of which extended the performance period, but did not include an overhead markup or extended field or home office overhead costs. Gavosto brought its claim for extended field overhead to the board, which in finding for Gavosto noted that when an owner directs additional work by unilateral modifications and grants time extensions for the work, it creates a rebuttable presumption that the owner is responsible for the delays and the contractor is entitled to compensation for time-related costs.

[H] Failure to Coordinate Separate Prime Contractors

Page 3-27, add after note 144 in first full paragraph:

A contractor is not responsible for an owner's delay damages if the delays were caused by the acts or failure to act of other contractors. In *Portland Utilities Construction Co. v. Chase Creek, L.L.C.*, 2004 Tenn. App. LEXIS 211 (Apr. 7, 2004), Portland Utilities Construction Company, L.L.C. contracted to remove two barns from the property, to remove a pond by filling it with compacted onsite materials, and to install storm drains, water lines, and sewer lines for owner-developer Chase Creek, L.L.C., Grading of the roads was also included in the contract, but paving was to be done by another contractor. Grading and preparation of the lots themselves were also to be done by other contractors. The developer acted as the general contractor for the project. The contract did not contain a liquidated damages provision. The infrastructure work took far longer than the developer thought reasonable, and far more solid rock had to be removed than was anticipated. The contractor submitted a final bill to the developer for a remaining balance of over $300,000. The developer declined payment. The contractor filed a complaint. According to the developer, delays resulted from the contractor's failure to do the work correctly the first time, leading to delays and to additional expenses for repairing or redoing defective work. The contractor denied that it was at fault, and pointed instead to a number of other factors, including errors by surveyors, engineers, and other contractors. The trial court found that the delays the developer complained of were due to the heavy rains, the errors of surveyors and engineers, and the activities of other contractors. The developer appealed.

The developer must prove that it had been damaged by the contractor's delay. The developer argued that it was entitled to an offset of $65,661 for all loan interest payments it made after July 1998, because the contractor's delay reduced the proceeds from the sale of the residences. The developer argued that because of site conditions, no sales were made between November 20, 1998, and March 12, 1999, other than one sale to a member of the developer's staff. However, damage to the graded roadway by construction traffic and another prime contractor's own lack of promptness were significant factors in the delay of road completion. Thus, if the developer suffered any damages as a result of this delay, they could not be attributed to the contractor's actions alone. Progress on the subdivision was impeded by lack of coordination between the activities of various contractors who worked on different parts of the same project at the same time, difficulties which were exacerbated by unusually wet weather. The developer had chosen to attribute all its damages to only one of those contractors, despite proof showing that many others (including itself) were also at fault. The judgment of the trial court was affirmed.

Page 3-27, add to note 147:

See Bolton Corp. v. T.A. Loving Co., 380 S.E.2d 796 (N.C. App. 1989); Biemann & Rowell Co. v. The Donohoe Companies, Inc., 556 S.E.2d 1 (N.C. App. 2001).

Page 3-28, add at end of subsection:

The owner's damage apportionment burden shifts to the contractor where all of the damage suffered by the innocent owner is caused by the concurrent negligence of multiple contractors. In *O'Brien (Newark) Cogeneration, Inc. v. Automatic Sprinklers Corp. of Am.*, 2003 N.J. Super. LEXIS 218 (June 24, 2003), O'Brien (Newark) Cogeneration, Inc. hired engineers and construction companies including Automatic Sprinkler Corporation of America (contractor 1) and GEC Alsthom International, Inc. (contractor 2). Contractor 2 manufactured and supplied steam turbine equipment and related piping, expansion joints, and other equipment for the plant. Contractor 1 was to provide a fire protection and detection system for the plant, including an automatic firewater deluge system for the lube oil unit, hydraulic unit, and gear box for contractor 2's steam turbine/generator. An oil fire began in the plant. A rubber expansion joint in the steam turbine's lube oil system, which had been manufactured by contractor 2, separated. No sprinklers had been installed in the area housing the lube oil pump and numerous supply and return lines. The owner contended that because of the absence of sprinklers or a deluge water spray system that could discharge a high volume of water necessary to put out an oil fire, the fire was not contained and progressed through the plant. The owner sustained lost profits during the time it took to repair the plant. As part of the total repair costs, the owner incurred additional labor costs for expediting the repairs, which according to the owner were necessary to avoid breaching the owner's long-term power purchase agreement with its sole electric customer. The owner began suit against contractors 1 and 2 to recover its repair and labor costs and lost profits. The owner settled with contractor 2. The trial judge dismissed the owner's entire complaint against contractor 1 because the judge concluded that the owner was required to apportion its damages caused by contractor 1's failure to provide sprinklers under the turbine pedestal. The owner appealed.

Contractor 1 argued that the owner was more knowledgeable regarding apportionment proofs than defendants. However, no eyewitness to this fire survived. The defendants had better understanding and access to the materials, equipment, and procedures that were employed in operating the plant, including any limitations on the equipment, products, and procedures. The defendants were better situated to interpret the fire investigations and attempt to limit the role their various materials, equipment, and systems may have played in this catastrophe. The owner's proof indicated that contractors 1 and 2's negligence and other defendants caused fire loss throughout the plant. According to contractor 1, however, the owner did not suffer a unitary harm because the absence of a sprinkler system in no way contributed to the initial oil leak and explosion, which caused some damage besides fire damage. The trial court concluded that the harm suffered by the owner was not unitary but instead was divisible. However, a properly designed sprinkler system would have extinguished the fire. The fire developed and spread to other parts of the plant. Thus, the innocent owner sustained unitary harm by way of fire damage to the entire plant and because contractor 1 was jointly responsible with the other defendants for the damage and because it was more knowledgeable than the owner about its fire suppression system, it bore the burden of limiting by apportionment its otherwise joint responsibility. The apportionment burden was shifted where all of the fire damage suffered by the innocent owner was caused by the concurrent negligence of multiple defendants, who were responsible for plant maintenance and installation of the various equipment, machines, and systems that were involved in the fire. The trial court should not have dismissed the owner's case, so the matter was reversed.

§3.03 DESIGNER-CAUSED DELAY

Page 3-30, add to note 159:

Cases and jurisdictions allowing the contractor to bring an action directly against the designer include: Gulf Contracting v. Bibb County, 795 F.2d 980 (11th Cir. 1986); Malta Constr. v. Henningson Durham & Richardson, Inc., 694 F. Supp. 902 (D. Ga. 1988); Berkel & Co. Contractors, Inc. v. Providence Hosp., 454 So. 2d 496 (Ala. 1984); State v. Tyonek Timber, Inc., 680 P.2d 1148 (Alaska 1984); Donnelly Constr. Co. v. Oberg/Hunt/Gilliland, 677 P.2d 1292 (Ariz. 1984); Danforth v. Acorn Structures, Inc., 608 A.2d 1194 (Del. 1992); Comtech Int'l, Inc. v. Milan Commerce Park, 711 So. 2d 1255 (Fla. Dist. Ct. App. 1998); NBD Bank v. Krueger Ringier, Inc., 686 N.E.2d 704 (Ill. App. Ct. 1997); Normoyle-Berg & Assocs., Inc. v. Village of Deer Creek, 39 Ill. App. 3d 744, 350 N.E.2d 559 (1976); Koss Constr. v. Caterpillar, Inc., 960 P.2d 255 (Kan. Ct. App. 1998); Gurtler, Hebert & Co. v. Weyland Mach. Shop, Inc., 405 So. 2d 660 (La. Ct. App. 1981); Milton Co. v. Council of Unit Owners of Bentley Place Condos., 708 A.2d 1047 (Md. Ct. Spec. App. 1998); Prichard Bros., Inc. v. Grady Co., 428 N.W.2d 391 (Minn. 1988); Jim's Excavating Serv., Inc. v. HKM Assocs., 878 P.2d 248 (Mont. 1994); Conforti & Eisle, Inc. v. John C. Morris Assocs., 418 A.2d 1290 (N.J. Super. Ct. Law Div. 1980), *aff'd*, 489 A.2d 1233 (N.J. Super. Ct. App. Div. 1985); Davidson & Jones, Inc. v. County of New Hanover, 255 S.E.2d 580 (N.C. Ct. App. 1979); Tommy L. Griffin Plumbing & Heating v. Jordan, Jones & Goulding, Inc., 463 S.E.2d 85 (S.C. 1995); Midwestern Elec., Inc. v. DeWild Grant Reckert & Assocs., Inc., 500 N.W.2d 250 (S.D. 1993); John Martin Co. v. Morse/Diesel, Inc., 819 S.W.2d 428 (Tenn. 1991); Thomson v. Espey Huston & Assocs., 899 S.W.2d 415 (Tex. App. 1995); American Towers Owner's Ass'n v. CCI Mech., 930 P.2d 1182 (Utah 1996); Daanen & Janssen, Inc. v. Cedarapids, Inc., 573 N.W.2d 842 (Wis. 1998).

Cases in which the contractor has not been able to recover directly against the designer for economic losses without being in privity of contract include: Bowling Green Mun. Utils. v. Thomasson Lumber Co., 902 F. Supp. 134 (W.D. Ky. 1995); City Express, Inc. v. Express Partners, 959 P.2d 836 (Haw. 1998); Duffin v. Idaho Crop Improvement Ass'n, 895 P.2d 1195 (Idaho 1995); Fleischer v. Hellmuth, Obata & Kassabaum, 870 S.W.2d 832 (Mo. Ct. App. 1993); Floor Craft Floor Covering, Inc. v. Parma Cmty. Hosp. Ass'n, 54 Ohio St. 3d 1, 560 N.E.2d 206 (1990); Gerald M. Moor & Sons, Inc. v. Drewry, 467 S.E.2d 811 (Va. 1996); Rissler & McMurray Co. v. Sheridan Area Water Supply Joint Powers Bd., 929 P.2d 1228 (Wyo. 1998).

Page 3-30, add after note 159 in first full paragraph:

The immediate effect of design errors upon the contractor may be the reason that some courts hold the designer directly responsible to the contractor for economic losses. One court characterized the issue in terms of a "special relationship" that exists between the contractor and the designer—the contractor relies upon the designer's plans and specifications in bidding and performing its work; the damage to the contractor from design errors is foreseeable; the designer is in the best position to prevent errors in the plans and specifications; and, thus, the designer owes a duty to the contractor to guard against negligence concerning certain design errors. *Eastern Steel Constructors, Inc. v. City of Salem*, 549 S.E.2d 266 (W. Va. 2001).

Page 3-30, add to note 162:

Miami Heart Inst. v. Heery Architects & Eng'rs, Inc., 765 F. Supp. 1083 (S.D. Fla. 1991) (owner recovered delay damage paid to contractors and subcontractors as a result of architect's delay to project; in addition, court allowed owner to recover additional architectural/design expenses and out-of-sequence construction costs incurred to modify building due to designer's errors and omissions). *See also* R. Crewdson, *Recovering Delay Damages Against an Architect: An Owner's Approach*, 9 Constr. Law. 24 (Jan. 1989).

Page 3-30, add after note 165 in second full paragraph:

Designers may attempt to limit their liability for delays and other damages resulting from their performance. The most common types of clauses limit the designer's exposure to the available insurance coverage, to a specified amount, or to a duty to reperform the services that caused the problem. The effectiveness of such limitation of liability clauses depends upon a variety of considerations embodied in statutory and judicial law, such as whether the parties negotiated and expressly agreed on liability limitations, whether the clause was an indemnification or a limitation of liability, the sophistication and bargaining powers of the parties, whether the damages were tort or contractual in nature, the reasonableness of the liability cap, the clarity of the language, and the level of negligence or intentional actions by the designer. Zetlin & Chillemi, *Building a Safe Haven: Clauses Imposing Monetary Limits on Designer Liability*, 20 Constr. Law. 5 (Jan. 2000).

[A] Design Defects

Page 3-32, add to end of subsection:

Design problems like the ones mentioned are often avoidable, especially through methods such as peer reviews, in which an independent multidisciplinary panel is engaged to review the proposed design to obtain the insights and experiences of other architects, engineers, and similar professionals. Not only are errors often discovered and resolved, but additional value can also be obtained for owners, cost savings explored, and a longer-term perspective developed to minimize the obsolescence of the project. *See Owners and Designers Are Airing Experiences Through Peer Review*, 254 Engineering News-Rec. 66 (Nov. 13, 2000), and *Face-to-Face Peer Reviews May Have No Peer in Quality*, 254 Engineering News-Rec. 104 (Nov. 13, 2000).

A contractor whose work is directly delayed by defective specifications is entitled to a remission of liquidated damages. In *R.P. Wallace, Inc. v. United States*, 2004 U.S. Claims LEXIS 329 (Fed. Cl. Dec. 15, 2004), the Navy and R.P. Wallace contracted for repairs to a historic building. Among other duties, the contractor was responsible for replacing the exterior windows. The Navy's architect prepared the plans and was responsible for reviewing the contractor's submittals. The contract provided that if the contractor failed to complete the work within the time specified in the contract, liquidated damages would be assessed against the contractor for each day of delay. The specifications required that the replacement windows were historically accurate duplicates of the original windows, with full muntins between each pane of glass in a multi-pane window, and that the windows met wind load requirements to remain intact through a hurricane. These two requirements were incompatible because the penetration of the window by true muntins would have weakened the glass to such an extent that it could not meet the wind load requirement. The Navy's architect incorrectly assured the contractor that the windows could be found. There were ultimately 305 days of project delays. The Navy assessed liquidated damages against the contractor. The contractor brought suit against the Navy, seeking remission of the liquidated damages.

The contractor contended that the Navy's defective window specifications delayed its performance. While the Navy did not contest that the defective specifications were harmful, it asserted that the contractor was not entitled to more than the 36 days of excusable delay already granted in a contract modification. During the period leading up to the approval of the window submittal, the contractor was delayed in finding a window manufacturer for the unavailable windows. The Navy's expert agreed that it could have taken up to three weeks for a contractor to conclude that a product could not be obtained. Therefore, for the submittal period, the contractor was entitled to an additional period of excusable delay totaling 21 days, beyond the 36 days already allowed. The contractor also contended that once its efforts to meet the original window specifications led it to Alenco (fabricator), it was committed to the window fabricator's delivery schedule. The contractor produced no evidence, however, that it had to subcontract with the fabricator. The contractor also asserted that it had anticipated installing the windows while it re-roofed the building, but as the roofing was completed before the windows were received, there were fewer individuals on site to assist with the installation. However, there was no critical path relationship between the roofing and window installation tasks, as the contractor used different subcontractors for each task. The defective specifications did not directly delay the fabrication or installation of the windows, so the contractor's excusable delay claim for the fabrication or the installation period was not supported. The contractor was entitled to remission of liquidated damages for 21 days during the submittal period.

The advent of computer-aided design and drafting (CADD) helps in many ways to identify and minimize many of the types of design deficiencies, errors and omissions that may lead to construction delay claims. For example, design coordination of the various design disciplines (architectural, structural, mechanical, etc.) may be enhanced with the "electronic" overlay capabilities of many of the CADD programs. Errors in calculations may be more thoroughly checked and coordinated by CADD programs. Information concerning the project site may be collected from a variety of sources and integrated electronically into the design to minimize problems with the interface of the design with the project site that often is the source of problems resulting in delays. However, it has been noted that the current computer technologies will not eliminate design defects, and may result in different types of design deficiencies, and these problems in turn may result in construction delays. For example, the computer tools themselves may have "latent bugs" that may produce errors in the design. The use of computers may lull designers into a false confidence that all matters have been verified electronically and discourage other types of "checking" and reviews of the design. Senior designers may delegate design-related tasks to less experienced designers who have more expertise in computer skills but who may lack engineering judgment. The lack of complete training in the use of computerized design programs may lead to inadvertent misuse of the programs resulting in deficiencies in the design. Therefore, new quality control and assurance procedures may be necessary to guard against computerized design defects. Lurie, *Responsibility for Computer-Aided Defects: CAD in the Construction Industry*, Con. Law. 23 (Fall 2001).

[B] Slow Correction of Design Error

Page 3-32, add after note 176 in first (carryover) paragraph:

When a deficient contract drawing and a delay in government response to a request for information cause a contractor to incur increased costs, the contractor is entitled to equitable adjustment under the solicitation's changes clause. In *Maroon Constr. Co.*, 2005 ASBCA LEXIS 22 (Mar. 10, 2005), Maroon Construction was awarded a contract by the government to procure and install an electrical substation. The proposed substation was to be located near an existing substation. The substations were to be connected via conduit and wire. The solicitation issued by the government contained a changes clause, which provided that the contracting officer could at any time makes changes within the general scope of the contract. The contractor discovered that one of the government's contract drawings showing that the existing space and conduit configuration was compatible with the specified transformer housing was in error. The contractor submitted a request for information, which described the problem, proposed a solution, and requested a government reply. The government took one month to make its reply, which did not resolve all issues. As a result of the government's delay, the contractor was forced to proceed with the manufacture of the equipment, which was resized to fit the allotted space, in order to meet the schedule requirements. The contractor also had to resolve the remaining conduit and configuration issues in the field. The contractor submitted a claim for an equitable adjustment. The contracting officer denied the claim. The contractor appealed.

The contractor argued that the government should not have denied its claim for an equitable adjustment for the deficient drawing and failure to timely act on the request for information. Given the relative simplicity of the problem, the contractor's proposal for a solution, and the absence of any government explanation for its month-long delay in reply, a reasonable time for a complete government reply was no more than 14 days. Because the contractor and its subcontractor were forced by the construction schedule to procure resized equipment and resolve conduit and configuration problems in the field, the government's deficient drawing and delay in responding to the request for information caused an increase in the cost of the contractor's performance. Under the changes clause, if any of the government's changes caused an increase in the cost of performance, the contracting officer was to make an adjustment in the contract price and modify the contract. The contractor was entitled to a price adjustment under the changes clause for any increased cost of performance caused by the deficient drawing.

[C] Tardy Shop Drawing Review

Page 3-34, add to note 191:

The general contractor may also have the duty to coordinate the shop drawings and related information from its own subcontractors, and may be responsible for delays in failing to provide proper coordination. Some of the contract clauses may appear onerous, as in W.M. Schlosser Co. v. United States, 767 F.2d 870 (Fed. Cir. 1985), in which the shop drawing coordination clause was enforced to the detriment of the contractor. The shop drawings were required to include great detail concerning location, layout, and alignment, a higher standard than the contractor believed was the custom and requirement.

[D] Delay Due to Tests and Inspections

Page 3-35, add at end of section:

Architects can delay the release of retention to a contractor by unjustifiably withholding certificates of substantial completion. Architects can take the position that, because the contractor has not achieved substantial completion, the contractor is liable for late completion and liquidated damages. R. Lowe & E. Walthall, *When Architects Withhold Certificates of Substantial Completion, and Other Problems*, 19 Constr. Law. 5 (No. 4, Oct. 1999).

In case law, the terms and doctrines of substantial completion and substantial performance are used interchangeably. Substantial completion is a term of art used in the construction industry to measure the time commitment of the contract. The definition of substantial completion used by the parties, however, is similar to the definition of substantial performance used by the courts. Like the doctrine of substantial completion, the doctrine of substantial performance provides for performance of a contract that, although possibly not full performance, is so close to contract terms that it would be unreasonable to deny the promisee the full contract price—subject however, to the promisor's right to recover damages that result from the failure to render full performance. Lowe & Walthall, at 5.

Although a contractor may be substantially complete in terms of work on the project, defects in the work may prevent the owner from using the project for its intended purpose and thus, prevent a finding of substantial completion. Lowe & Walthall, at 8.

The court in *Utley James of Louisiana, Inc. v. Louisiana Division of Administration*, 671 So. 2d 473 (La. Ct. App. 1995), held that a prison construction project was not substantially complete because of a 48-page punch list that included aesthetic issues, an inoperable fire alarm system, an inoperable sewage lift system, and coordination problems with subcontractors. The contract defined substantial completion as the date the architect issued certification of substantial completion. The court affirmed the trial court's holding that substantial completion did not occur until the architect issued the certificate. Lowe & Walthall, at 8.

Once substantial completion has been attained, the architect is under a legal duty to issue a certificate. This principle was reflected in *Haugen v. Raupach*, 260 P.2d 340 (Wash. 1953), in which the court held that the contractor was entitled to the balance under the contract, without having to obtain a certificate of substantial completion. Although the architect felt that there was substantial performance, he yielded to the owner's insistence that work was not done properly. The court held that:

> [I]f the architect is satisfied that there has been a substantial performance of the contract, it then becomes his duty to issue the certificate of completion, and if he does not do so, his conduct is regarded as arbitrary and capricious. If the architect is in collusion with his principal, or yields to his opposition to the issuance of the certification when such opposition is not justified, then in such cases the contract has a legal excuse for not obtaining the certificate as a condition precedent to recovering on his contract.

The *Haugen* decision, instead of resolving the problem, created two more complicated and unanswered questions: (1) whether the "arbitrary and capricious" standard applies even when the architect has not yet determined whether substantial completion has been obtained; and (2) whether a contractor has a cause of action against an architect who acts arbitrarily and capriciously by withholding a certificate of substantial completion. Lowe & Walthall, at 8–9.

Although it appears that liquidated damages cannot be assessed against the contractor, several courts have interpreted contractual provisions to allow them, even if the contractor obtains substantial completion. In *Ledbetter Bros., Inc. v. North Carolina Department of Transportation*, 314 So. 2d 761 (N.C. Ct. App. 1984), the North Carolina Court of Appeals held that, under terms of the contract, liquidated damages could be assessed for the time period between substantial performance and final completion. In *Ledbetter*, the liquidated damages provision expressly recognized that failure to complete the work would result in increased inspection and administrative costs to DOT including the inspection costs arising between substantial performance and final acceptance. The contract provisions stated that no part of the project would be accepted until final acceptance. Lowe & Walthall, at 9.

The Pennsylvania Supreme Court, in *Sutter Corp. v. Tri-Boro Municipal Authority*, 487 A.2d 933 (Pa. 1985), issued a similar holding. The court held that the authority was entitled to liquidated damages for each day past the agreed completion date, despite the fact that it was able to begin using the sewage treatment plant, because the contract stated that liquidated damages accrued for any work that should remain uncompleted beyond that time. In *Vrgora v. Los Angeles Unified School District*, 200 Cal. Rptr. 130 (Ct. App. 1984), the California Court of Appeals relied on prior case law in holding that use and occupancy of a building by the owner did not prevent the assessment of liquidated damages until the contract was fully performed. Lowe & Walthall, at 9.

Although substantial completion entitles a contractor to the balance of the contract price (less any damages caused by the contractor's breach), AIA contracts modify this approach. Under A201-1997, the contractor is not entitled to the balance until the architect issues the final payment certificate. In theory, the owner could withhold the entire retainage, but in practice courts interpreting the AIA document would apply the substantial completion doctrine to determine whether the amount was reasonable. Lowe & Walthall, at 10.

§ 3.04 CONTRACTOR-CAUSED DELAY

[B] Contractor Management Problems

Page 3-38, add to note 218:

Elter S.A., ASBCA No. 52451, 2001-1 B.C.A. (CCH) ¶ 31,373 (2001) (a construction contract's Quality Control (QC) contract provisions stated that "the only work that is authorized prior to the approval of the QC plan is mobilization of storage and office trailers and surveying," although the notice to proceed and commencement of work clauses did not mention the QC plan approval requirement. The government rejected the contractor's initial submission and several revisions of its QC plan, and notice to proceed was delayed until the QC plan was approved. The contractor's claim for improper delay in issuing the notice to proceed was found to be without merit).

Page 3-38, add to note 219:

Cape Romain Contractors, Inc., ASBCA No. 50557, 00-1 B.C.A. (CCH) ¶ 30,697 (1999).

Page 3-38, add after fourth full paragraph:

The contractor is responsible for the time needed to develop plans that comply with a performance contract's terms. *P.R. Burke Corp. v. United States*, 277 F.3d 1346 (Cal. Fed. Cir. 2002). Burke's contract to repair and improve a sewage treatment plant required that it keep the plant in operation during the entire construction period. Portions of the trickling filter pump station and chlorine contact tank that Burke was required to construct appeared in the same location as the existing trickling filter. Burke's draft demolition plan entailed demolishing the trickling filter at the beginning of construction to make way for the new structures that were to be built in its place. The government, however, indicated that this would cause "operational" problems and provided a suggested alternate sequence of tasks. Burke submitted a new demolition plan, but asserted that adapting its plans to conform to the government's suggested sequence would cost Burke approximately $400,000 and would extend the estimated completion date approximately nine months. The government approved Burke's demolition and construction plan but rejected Burke's adjusted performance schedule and sued the government.

Burke contended it incurred delay damages for the period the government kept requesting contract-compliant demolition plans. Burke, in addition, argued that the government's suggestions about the demolition and construction sequence were binding changes to the contract's design specifications and that Burke therefore had no choice but to follow them. The contract showed that the contractor, not the government, had to design the trickling filter. The time needed to develop plans that complied with contract terms was chargeable to the contractor, especially when the contract expressly required submission of design plans and work schedules for approval. Therefore, because it was Burke's responsibility to determine how to perform the demolition and construction and because the contract could be performed as written, Burke did not require a change order to be able to perform its obligations under the contract. The government's subsequent actions did not convert the agreement from a performance contract to a design contract. None of the asserted actions indicated that Burke had to abide by the government's suggested sequence. The government never actually did anything to prevent Burke from immediately submitting its own contract-compliant demolition and construction plan. In addition, nothing suggested Burke had no discretion to deviate from the suggested demolition and construction sequence.

Burke was contractually required to prepare a demolition plan, submit it to the government, and obtain the government's approval. The government had a right to review this demolition plan and ensure that it conformed to the contract requirements. On October 31, the government received Burke's first demolition plan. The government had no contract obligation to issue change orders, modifications, or suggestions on sequencing or do anything else before October 31. When it submitted this plan, Burke was aware that the government would almost certainly reject the request. Ten days after receipt, as expected, the government returned the demolition plan to Burke for correction, because the plan included taking the trickling filter off-line for the duration of the contract, a sequence that conflicted with the "remain in operation" term. As a matter of law, the ten-day period could not be attributable solely to the government because it was caused by Burke's failure to comply with the contract. The government was merely exercising its right under the contract to receive, review, and approve a plan that conformed to the contract. Burke could not recover damages for the ten-day period after October 31.

Page 3-40, add at end of first full paragraph:

Also in *Jenkins, Inc. v. Walsh Brothers, Inc.*, 776 A.2d 1229 (Me. 2001), the contractor was found to have breached its duties to the subcontractor, allowing the subcontractor to recover for its delay and disruption costs. Jenkins was the drywall subcontractor to Walsh for the construction of a health sciences center for the University of New England. During the course of construction, problems with the design and site conditions delayed the project. To overcome these delays, the owner agreed to pay for certain acceleration costs of the general contractor and certain subcontractors. Walsh directed Jenkins to work in areas where the roofing and enclosure was not complete or where the preceding work was not completely ready for the drywall installation. The cost of Jenkins' performance was increased because it had to "standby" or work around the other trades. The trial court held that by directing Jenkins to work out-of-sequence, Walsh had breached the subcontract, and this finding was upheld in appeal.

Page 3-41, add after note 42 in first full paragraph:

In one case, the general contractor was found not to be responsible to a subcontractor for its alleged failure to coordinate the work of the other subcontractors. In *United States ex rel. Virginia Beach Mechanical Services v. SAMCO Construction Co.*, 39 F. Supp. 3d 661 (E.D. Va. 1999), the mechanical subcontractor sued the general contractor building a warehouse for the Navy in Virginia Beach, Virginia. The project initially was delayed by the late delivery of a prefabricated building by another subcontractor. The mechanical subcontractor demobilized from the project site without completing its work and filed suit claiming that the general contractor failed to make timely payments and to coordinate the work of the subcontractors. The general contractor filed a counterclaim that included a request for a portion of the liquidated damages assessed against the general contractor by the Navy for late completion. The court found that the failure of the mechanical subcontractor to complete its work was a material breach of the subcontract, excusing the payment obligation of the general contractor. The court also addressed the issue of the responsibility of the general contractor to coordinate the work of the subcontractors:

> In a project involving multiple contractors, however, the duty to coordinate work and thereby avoid hindrances does not necessarily fall on the general contractor. Instead, the terms of the various contracts involved will determine

> who assumed that obligation. . . . While [the general contractor] may have, in fact, inherited this duty by virtue of its status as general contractor, the Court will not make that assumption absent evidence proving such. *Id.* at 674–75.

The finding of the court may have been the result of its consideration of the proof on the matter rather than a general statement of the duties of a general contractor.

Page 3-43, add after note 257 in second full paragraph:

A contractor is responsible to the owner for the nonperformance of a supplier even though the supplier is one of two manufacturers identified by the owner in the contract specifications. For example, in *A.G. Cullen Constr., Inc. v. State Sys. of Higher Educ.*, 2006 Pa. Commw. LEXIS 128 (Mar. 15, 2006), the State System of Higher Education for the Commonwealth of Pennsylvania (State) contracted with A.G. Cullen Construction, Inc. (contractor) for renovation of an historic school administration building. The contract contained a scheduled project completion date and provided for liquidated damages in the event the contractor failed to meet that date. As a result of delays, the contractor completed the project 172 days after the scheduled completion. The State reduced its payments to the contractor based on the liquidated damages provision. One of the delays was due to a window manufacturer's failure to meet its production schedule or to provide windows that met contract specifications. The window manufacturer was one of two providers specifically identified by the State in the contract specifications. The specifications allowed the use of other manufacturers as long as the windows met specifications. The contractor filed complaints against the State with the Board of Claims (Board). One of the claims was for breach of contract for failure of a manufacturer identified by the State in the contract to provide windows that met the design specifications. The Board determined the State was not responsible for the delays and costs caused by the manufacturer's failure to perform. The contractor appealed.

The contractor argued that, by identifying the window manufacturer as an approved manufacturer in the contract specifications, the State impliedly warranted the manufacturer could produce the required windows. The contractor maintained that the identification of a manufacturer that proved unable to adequately perform constituted a defect in the contract specifications for which the State was responsible. The Board stated that the contractor retained discretion to select a manufacturer from the list of two identified by the State or, alternatively, to select another manufacturer that could produce substantially similar windows. The contractor possessed the right to enforce its agreement with the named manufacturer by insisting on timely performance or terminating the agreement. Unlike typical "design" specifications, the window replacement specifications did not describe in precise detail the materials to be employed and the manner in which the work was to be performed. Rather, the specifications merely specified the result to be obtained, which was replacement of the existing windows, and gave the contractor discretion to determine how to obtain the results. While the specifications required that the windows installed satisfy various "performance requirements," the contractor retained sole responsibility for the means and methods of the replacement of the windows. Consequently, the Board did not make an error in its determination, and the appellate court affirmed the Board's decision.

Page 3-43, add after second full paragraph:

Absent a specific clause in the contract, a general contractor may not be liable to a subcontractor for delays caused by the owner's breach of its financial obligations to the general contractor. *In re Regional Building Systems, Inc.*, 320 F.3d 482 (4th Cir. 2003), Regional Building Systems (RBS) was awarded a contract to manufacture, deliver, and erect 100 modular housing unit on property owned by Aspen Knolls. RBS subcontracted with Bedford to deliver and erect the units on Aspen Knolls' property in Staten Island, New York. Aspen Knolls developed financial problems and stopped paying RBS on the prime contract. RBS was unable to pay the manufacturer of the housing unit without cash flow from Aspen Knolls, and later terminated its subcontract with Bedford. Bedford submitted a claim to RBS for delay damages, and RBS responded that it was not responsible to Bedford for the owner's delays and financial problems. The trial court determined that Aspen Knolls' default to RBS was the cause of RBS's inability to timely produce the modular units to Bedford for transport and erection, and that there was no clause in the subcontract between RBS and Bedford that made RBS responsible to Bedford for the owner's delays. In reviewing the matter, the court of appeals held that absent a provision in the subcontract, a prime contractor like RBS was not responsible to a subcontractor for the delays of the owner, as this was beyond the prime contractor's direction and control.

[D] Construction Defects

Page 3-46, add after note 285 in carryover paragraph:

Further, a contractor's substandard work that causes excessive delays may constitute a lack of good workmanship and a breach of contract. For example, in *Austin Homes, Inc. v. Thibodeaux*, 2002 La. App. LEXIS 1264 (May 8, 2002), the Thibodeauxes (buyers) contracted with Austin Homes, Inc. (contractor) for the construction of a new home. The buyers obtained plans and specifications from a designer. The contractor had not weatherproofed the house seven months after construction began. The wooden framing rotted from excessive water exposure. The contractor submitted a draw request, but the buyers refused to pay. After the contractor remedied the problems, the house passed inspection. The buyers paid the contractor $20,000 with an agreement to make weekly payments thereafter. The contractor stopped working because the buyers refused to pay per the agreement. The buyers hired a consultant to inspect the home, and the consultant found excessive water damage. The jacuzzi did not have a drain or water source and the contractor substituted waferboard for plywood sheeting on the walls without permission. The extensive water damage caused it to reek of mildew, and the buyers could not get rid of the odor. Because of the smell and other problems, the buyers could not live in the home. The contractor submitted another draw request for the amount due, but the buyers made no payment. The buyers terminated the contract and hired the consultant to take over construction. The contractor filed suit to recover costs that it had expended to build the home. The buyers counter-sued, seeking damages for breach of contract. The trial court awarded the contractor $58,388.55 for the work that it had done properly, and $18,570 in lost profits because of the buyers' breach. The court did not award the buyers damages. The buyers appealed.

The buyers asserted that the contractor's failure to supervise delayed construction. The contractor argued that certain occasions required waiting until the ground dried, but the buyers contended that if the home had been expediently weatherproofed, the contractor could have continued working inside on other projects. The contractor had promised to make the buyers' house a priority, and it did not do so. The contractor was not present to direct construction, and only came to the site to ask for money. The contractor failed to prove that the leaks resulted from design errors. The contractor continuously substituted materials without the buyers' or the architect's authority. The contractor's substandard work caused excessive delays in weatherproofing the home, and its failure to expediently weatherproof the home constituted a lack of good workmanship and a breach of contract. Because of all of the problems and disillusionment, the buyer's blood pressure escalated, which created a possibility for a stroke or heart attack. The buyers suffered a non-pecuniary loss and the contractor should have known that its failure to perform would cause it. The matter was reversed and remanded to the trial court.

§ 3.05 SUBCONTRACTOR-CAUSED DELAY

Page 3-47, add after note 295 in first full paragraph:

In some instances, the owner may designate a sole-source subcontractor or supplier for a part of the work. The contractor may claim that the delays caused by the subcontractor are the responsibility of the owner because of the owner's sole-source requirement. For example, in *Appeal of Protech-Atlanta, Inc.*, 02-1 B.C.A. (CCH) ¶ 31,736 (2002), the government awarded a contract for the replacement of a portion of a fire alarm system at a VA medical center and required that certain components be furnished and supplied by Johnson Controls. During the course of performing the contract work, progress became delayed and the government terminated the contract for default. The contractor challenged the termination on the grounds that it experienced excusable delay as its work was delayed by Johnson Controls' late submission of shop drawings, and abandonment of its subcontract. The contractor claimed that because the "sole-source" subcontractor was the cause of a portion of the contract delay, this portion of the delay was excusable, and the default termination should be set aside. In examining the facts of the case, the board found that the contractor as a "business decision" withheld payments from Johnson Controls, and that was the reason Johnson Controls stopped performing. The failure of a subcontractor to perform due to the prime contractor's business decision not to pay the subcontractor is not an excusable delay, and thus the board did not have to address the legal responsibility of the owner for delays of a sole-source subcontractor.

Page 3-47, add after note 296 in third (carryover) paragraph:

A subcontractor or supplier may attempt to disclaim responsibility to the general contractor for delays in its bid or proposal to the contractor. On occasion, the subcontractor's or supplier's disclaimers may be found to govern the relationship. For example, in *Eldor Contracting Corp. v. County of Nassau*, 708 N.Y.S.2d 447 (App. Div. 2d Dep't 2000), Centrum, the general contractor for improvements to a wastewater treatment facility, sent a purchase order to General Electric (GE) to rebuild the electric motors that were part of Centrum's responsibility under its prime contract with the owner. GE responded to Centrum's purchase order with a letter agreeing to rebuild the motors conditioned upon Centrum's acceptance of GE's standard terms and conditions. One paragraph of GE's standard terms provided that GE would not be liable for consequential damages. Centrum did not respond to GE's letter in writing. Centrum shipped the electric motors to GE and paid GE when the motors were rebuilt. Later, Centrum and the project owner became involved in a dispute over delays to the overall WWTP improvement project, and Centrum, in turn, alleged that some of the delays to the overall project were due to GE's delays in rebuilding the electric motors. In court, GE argued that its terms and conditions applied to the allegations of delay brought by Centrum, and Centrum countered that it never agreed to GE's standard terms and conditions. In reviewing the trial court's determination, the Appellate Division held that by its "acquiescent conduct" to GE's letter (that is, shipping the motors to be rebuilt and paying for the work upon completion), Centrum accepted the counteroffer tendered by GE that included the limitation of liability provisions.

Page 3-48, add at end of carryover paragraph:

Further, in *Jeffrey M. Brown Associates, Inc. v. CRK Contracting of Suffolk, Inc.*, 124 F. Supp. 2d 895 (E.D. Pa. 2000), the general contractor was unable to recover its delay costs from the subcontractor because the delays were found to be beyond the control of the subcontractor. Brown subcontracted the electrical work to CRK. During construction, the project was delayed by several causes: crane accidents, safety shutdowns, owner revisions to the electrical work, long procurement time for lighting fixtures required by change orders, and the late performance of the concrete, drywall, and mechanical work. In any event, Brown determined that CRK was behind schedule, terminated the CRK subcontract, and retained another electrical subcontractor to complete the electrical work. Brown brought suit against CRK to recover damages for delay, the cost of completing the electrical work, and damage to the work of the other trades, allegedly caused by CRK. However, the U.S. District Court found that Brown did not prove that CRK caused any of the problems, rather the "whole job was a disaster" for reasons that were not attributable to CRK.

Page 3-49, add at end of section:

In one subcontracting arrangement, the terms of the agreement allowed the subcontractor to avoid the costs of delay resulting from its allegedly defective and negligent work. In *Sunbond Engineering Ltd. v. Konwall Construction & Engineering Co., Ltd.* 29 BLISS 10 (High Court of Hong Kong 2004), Konwall held the contract for substructure construction for a residential development. Konwall subcontracted the piling work to Sunbond. Sunbond's tender to Konwall stated in Clause 3 that if a pile could not be driven to the required level or had to be abandoned for any reason, Sunbond was to be paid for the installation cost. Konwall accepted Sunbond's tender subject to liquidated and ascertained damages of $20,000 per day, but did not take issue with tender Clause 3 concerning payment for piles that could not be driven to the required level or were abandoned for any reason. After piling commenced, cracks appeared in adjacent houses, and work was suspended by the local building department, and later the piling work was allowed to resume. There were also numerous abandoned piles that had to be replaced. Konwall applied to the employer for an extension of time. Konwall did not pay Sunbond for the cost of many of the abandoned piles and sought to withhold liquidated damages as well. Konwall and Sunbond were unable to reach an agreement concerning the disputed costs and court proceedings ensued. At trial, an expert witness for Konwall testified that the high rate of abandoned piles was due to negligent piling work by Sunbond. Konwall also maintained that project delays also resulted from the abandonment of piles negligently driven by Sunbond and the time required to drive replacement piles. Konwall maintained that Sunbond was under an obligation to follow good building practices when driving the piles, and that Clause 3 should not be construed to entitle Sunbond to payment of piles that were abandoned through a failure to follow good building practices. As Konwall was facing liquidated and ascertained damages of $60,000 per day in its contract with the owner, Konwall argued that it would be absurd if Konwall had agreed with Sunbond to bear the risk of Sunbond's delays due to negligent pile driving by Sunbond. The court considered Konwall's arguments but determined that Clause 3 unambiguously placed the risk of piles being abandoned for "whatsoever any reason" on Konwall, and that Konwall must bear the cost and consequences of the same. Further the court noted that the copy of the Konwall/Employer Main Contract that attached to the Konwall/Sunbond subcontract blotted out the amount of the liquidated damages payable by Konwall to the Employer for delays. The court stated that where a fact is known only by one party at the time of contracting, why is should be taken into account when objectively assessing what both parties intended when entering into an agreement. The court concluded that even if Sunbond had been negligent in driving piles and caused many of them to be abandoned, the loss and delay arising must be borne by Konwall.

A contractor's delay claim settlement with an owner does not have to allocate delay among subcontractors that contributed to the delay. In *Suitt Construction Co. v. Ripley's Aquarium*, 2004 U.S. App. LEXIS 17806 (6th Cir. Aug. 17, 2004), Ripley's Aquarium, LLC entered into a contract with Suitt Construction Co. (Suitt) for the construction of Ripley's Aquarium. Suitt subcontracted with Blalock Lumber Co. (Blalock) to supply the concrete. The contract originally established the date of substantial completion as July 10, 2000. Due to delays for which Ripley's was responsible, the parties agreed to push the substantial completion date back to July 31, 2000. The contract provided for liquidated damages of $15,000 per day for each calendar day after the date of substantial completion until the work actually was substantially completed. The contract also provided for a grace period and a liquidated damages cap. The grace period allowed that no liquidated damages would accrue during the first 14 days after the substantial completion date specified in the contract. The cap provided that liquidated damages would in no event exceed $705,000, or 47 days of liquidated damages at $15,000 per day. Because of the defective concrete, Suitt had either to coat the concrete walls with a liner with the brand name "Polibrid" or remove and replace the concrete. Before the Polibrid liner could be installed, the company supplying an acrylic tunnel attached to the walls, had to conduct tests to determine whether the acrylic and the liner were compatible and would create a watertight seal. The delay involving the acrylic installation delayed painting the tanks and installing simulated underwater plant life. Suitt asserted that Blalock's defective concrete delayed substantial completion by 92 calendar days, but that by rescheduling projects not on the "critical path," Suitt reduced the delay to 79 days. Suitt issued a revised construction schedule which showed that the substantial completion date had been pushed back to October 18, 2000, as a result of delays associated with the concrete. Blalock admitted that its defective concrete caused the entire 79-day delay. Ripley's sent Suitt a letter informing it that the full amount of liquidated damages was being assessed against Suitt for the period from August 15, 2000, through September 30, 2000. Ripley's sued, and then settled with Suitt for $400,000 in liquidated damages. Suitt filed a complaint against Blalock for damages which included $400,000 in liquidated damages and $261,535.03 for extended general conditions. The jury returned a verdict in favor of Suitt in the amount of $806,000.

Blalock challenged the jury's verdict to the extent that it held Blalock responsible for the $400,000 in liquidated damages Suitt paid and to the extent the jury determined that Suitt was entitled to recover from Blalock its full general conditions costs. Blalock argues that Ripley's should have been required to present proof allocating the liquidated damages settlement between Blalock and Suitt. After the grace period ended, liquidated damages accrued to their contractual maximum of $705,000. Significantly, Blalock was solely responsible for the delay during this period. By the time the delays caused by the other parties began, the entire amount of the liquidated damages had already been incurred. Based upon this fact, the jury could reasonably have concluded that Blalock was responsible for the entire amount of liquidated damages which Suitt negotiated down to $400,000.

Blalock also argued that Suitt was barred from collecting liquidated damages from Blalock because Suitt contributed to the delay for which the liquidated damages were assessed. Because Blalock admitted that it caused the entire 79-day delay, Suitt was not barred from recovering liquidated damages. According to Blalock, Suitt performed tasks not on the critical path and worked on a repairs caused by its own improper installation of a window. Blalock alleged that "Suitt used the float time created by Blalock's delay to perform changes and corrective work that otherwise would have delayed the critical path, and Suitt worked and received a benefit from the general condition expenses incurred during the 79-day delay attributable to Blalock." Blalock argued that "Suitt provided no proof as to how many days the project would have been delayed absent Blalock's delay, and without such proof, there was no way for the jury to determine how many days of extended general conditions expenses were incurred as a result of Blalock's delay." Suitt met its burden of proving the damages caused by Blalock's admitted delay. Blalock could have presented evidence of any benefit Suitt derived from the work it performed during the delay, and how this benefit should offset Suitt's damages. Blalock, however, put forth no evidence in support of its contention that the 79-day delay in some way benefited Suitt. Suitt could have completed whatever work it did during the delay at some other time without impacting the project's substantial completion date. Because Blalock has presented no such evidence, the district court did not err in denying Blalock's challenge.

§ 3.06 DELAY NOT CAUSED BY PARTIES TO THE DESIGN AND CONSTRUCTION PROCESS

Page 3-50, add after note 322 in carryover paragraph:

The owner is not liable for gate access delays experienced by a contractor at an airport construction site when the unforeseen delays are caused by increased security measures following the terrorist attacks of September 11, 2001. In *Broward County v. Brooks Builders, Inc.*, 2005 Fla. App. LEXIS 9099 (June 8, 2005), Broward County (county) hired Brooks Builders (contractor) to build a fire station adjacent to an airport runway. Locating the fire station near an active runway presented challenges and resulted in numerous delays. Due to the sensitive location, the contract expressly required strict compliance with all airport security measures, including entry and exit procedures. Those security measures increased significantly after the terrorist attacks carried out on September 11, 2001, causing even greater delays. The contractor's workers spent significant time gaining access to the airfield through the security gates. The project was not substantially completed until approximately nine months after the scheduled completion date. The contractor filed suit against the county alleging breach of contract and seeking damages for post-9/11 delays, among other compensation. The county counterclaimed for defective work and delay damages. The trial court entered a final judgment for the contractor and a damages award against the county, which was not awarded any damages on its counterclaim. The county appealed.

The county argued that the contract did not provide for damages for post-9/11 gate access delays. The contractor argued that it should receive compensation for the daily delays that occurred when its workers attempted to gain access to the airfield through the security gates. The contractor argued that the contract authorized an award of damages for post-9/11 gate access delays, pointing to a provision that stated if the contractor was ordered by the engineer to suspend work for some unforeseen cause, the contractor was to be reimbursed. However, the post-9/11 gate access delays were not orchestrated by any orders of the county's engineers. The contract did not anticipate any extraordinary delays resulting from a terrorist attack, but there were numerous provisions suggesting that any risk of loss for unexpected changes in conditions was to be fully assumed by the contractor. The portion of the final judgment that awarded damages for post-9/11 gate access delays was reversed and remanded.

Page 3-50, add to first full paragraph:

For example, in *Brown v. Olin Chemical Corp.*, 231 F.3d 197 (5th Cir. 2000), the owner bore the cost of excusable delay that resulted from a third party's air pollution, because the owner failed to prove that the delay was due to the unreasonable acts of others. Intervenor Horseshoe Entertainment (owner) procured an unfinished vessel to replace an old casino. A number of contractors and subcontractors were hired to finish the vessel. During the course of the finishing work on the riverboat, the construction workers fell ill. This forced the owner to shut down the finishing project until the workers could recover and monitoring and safety systems could be put in place. The workers' illnesses were apparently caused by exposure to sulphur dioxide and sulfuric acid. The owner alleged that emissions from the manufacturer's plant were the cause of the chemical exposure, and sued the manufacturer for the increase in the cost of the construction of the riverboat caused by the delays and lost profits for delay in placing the riverboat into service as a casino. The manufacturer filed a motion for summary judgement, which the district court granted. The owner appealed.

The owner argued that summary judgment should not have been granted because the manufacturer was responsible for the emissions and thus should be liable for the damages incurred because of the construction delay. The appellate court stated that the evidence did not show that the manufacturer exceeded acceptable emissions limits at any of the times at issue, nor that the manufacturer had exercised anything less than reasonable care in the operation of its plant. The owner argued that proof of negligence was not necessary because strict liability pursuant to a statute applied. The owner also argued that the doctrine of *res ipsa loquitur* applied, relieving the need to prove the manufacturer's negligence. The appellate court pointed out that amendments to the statute required a showing of negligence. The appellate court also stated that the workers' illness was not the type of injury that did not ordinarily occur in the absence of negligence. The owner had not shown that the workers fell ill as a result of chemical exposure or that they could not have fallen ill from chemical exposure when the nearby chemical plants were operated with the exercise of reasonable care. The owner had also failed to negate the possibility that the chemical exposure could have been caused by another source. Several other chemical plants are located in the vicinity, and any one of these plants could conceivably have been the source of the exposure. The appellate court affirmed the district court's grant of summary judgment.

Page 3-51, add to end of first carryover paragraph:

When a contractor fails to protect against a risk that is reasonably foreseeable, the contractor is not entitled to additional costs or time. *Fraser Construction Co. v. United States*, 2003 U.S. Claims LEXIS 166 (June 25, 2003), the Army Corps of Engineers awarded a contract to Fraser Construction Company for flood control work. The contractor began work on the project on May 17, 1993, and the contract was scheduled to be completed by September 1, 1993. The contract's default clause absolved the contractor of responsibility for any delays in contract completion that arose "from unforeseeable causes beyond the control and without the fault or negligence of the Contractor." During the months of June, July, and August, the contractor experienced adverse weather conditions and high water flows that impeded its progress. In response to those conditions, the contractor repeatedly sought both time extensions and additional contract compensation. The Corps granted 33 of the days sought for unusually severe weather. With regard to the time extensions based on high water flows, however, the Corps denied all claims. The Corps additionally refused all requests for increased compensation. The contractor brought suit, seeking the additional days and funding it had been denied. The court concluded that the high water flows did not result in an excusable delay since they constituted neither unusually severe weather nor a flood as those terms were used in the contract's default clause. On appeal, the Federal Circuit rejected the trial court's conclusion. The court explained that unusually high water flows could have been found to constitute excusable delay even if they did not fall into the category of either a flood or unusually severe weather. The action was therefore remanded for a determination of "whether [the contractor] established a right to excusable delay based on an unforeseeable rate of water in the river leading to inundation of the lake bed for an extended time during the performance period."

The contractor's statistics revealed a pattern of water flows that were unusually high. While the contractor made a clear case for the proposition that the daily river flows were, on average, greater in volume for a longer period than at virtually any other point in the prior years of record, that data revealed virtually nothing about the instantaneous peak flows experienced at the site. Yet, the flooding of the work site occurred when the contractor's diversion system was overtopped by instantaneous peak flows, and that high daily mean flows were at no point responsible for the breaching of the dam. Although the prolonged high average flows may indeed have caused difficulties in repairing the diversion system, those residual problems would have been avoided had the dam not been overtopped in the first instance. In attempting to identify the flows for which the Corps should have granted relief the Court was bound by the contractual limitation that the delay was occasioned by causes that were "unforeseeable." The contractor did not provide the court with a way of determining what level of instantaneous flows should be deemed unforeseeable. To the contrary, the highest peak flow experienced during the contract period fell well within the two- to ten-year-event frequency range that experts for both sides testified represented the industry standard for anticipating risk. A flow frequency chart revealed that a high flow could be expected to occur with a 20 percent frequency, or every five years, over the period of June, July, and August. The instantaneous peak flows encountered at the site were reasonably foreseeable, and the court concluded that the contractor assumed the risk for the overtopping it experienced. Accordingly, the court could not identify the effects of those overtoppings—the expenditure of additional time, effort, or money—with any action of the government.

[A] Weather

Page 3-52, add after note 343 in third full paragraph:

Further, a suspension clause may allow a public agency to suspend a contract for weather unsuitable for performance without cost obligation to the contractor. In *CEMS, Inc. v. United States*, 2003 U.S. Claims LEXIS 324 (Fed. Cl. Nov. 5, 2003), the Department of Transportation, Federal Highway Administration (FHA) awarded a contract to CEMS for the construction of the bicycle path on a portion of the Historic Columbia River Highway. The contract included the Suspension of Work clause. The agency directed that all work was to be suspended effective November 25, 1998 through April 19, 1999 due to unsuitable weather related conditions in the Columbia River Gorge. Liquidated damages were not assessed during this period of suspension. The plaintiff claimed delay damages arising out of agency's decision to suspend the work on the bicycle path project from November 25, 1998 to April 19, 1999. The plaintiff stated that due to the FHA's suspension of work, it was unable to complete the project until May 6, 1999. The plaintiff further claimed that prior to work beginning on the project, the FHA recognized that weather conditions were adversely impacting CEMS' schedule. Following the denial of many of plaintiff's claims by the contracting officer, CEMS filed its complaint in court for $1,241,203.70 in compensation for numerous claims.

The plaintiff alleged that the suspension unreasonably delayed work on the bicycle path during this period because the project was substantially complete as of October 30, 1998 and "given the minimal amount of paving left to be done and time required, it was patently unreasonable for FHA to suspend the work for almost five months." The agency testified that the suspension of the project until April 19, 1999 was warranted due to the climactic conditions prevalent during the winter months in the Columbia River Gorge. The agency explained that it did not consider the cost of the suspension period and focused on the "base rock work to be done and paving, and end of November and December, it just wasn't good weather to pave in." The agency considered the work remaining on the project and determined that the paving work and the amount of work remaining on the project could not be completed during the months of the suspension period, although the number of days necessary for completion were not known. The agency's judgment that the weather conditions prevalent during the winter months in the Columbia River Gorge would have prevented the final paving to be completed in an acceptable manner was reasonable. In addition, the definition of the Substantial Completion required that the public use the bicycle path "safely and effectively . . . without further delays, disruption, or other impediments." Although the remaining pavement work on the bicycle path ultimately required only 11 full days of additional work, prior to completion, the public could not have used the bicycle path "safely and effectively" without "delays, disruption, or other impediments." Moreover, the FHA contracted with CEMS for a paved bicycle path, and the court found that substantially completion required the completion of the pavement work. The plaintiff's claim that the FHA's suspension of work on the project from November 25, 1998 through April 19, 1999, was unreasonable was denied.

Page 3-53, add after note 358 in second full paragraph:

In *Pete Vicari, General Contractor, Inc. v. United States*, 47 Fed. Cl. 353 (2000), Pete Vicari, General Contractor, Inc., entered into a firm fixed-price contract with the United States Coast Guard (USCG). During the course of performance, adverse weather conditions at the site delayed project completion. Consequently, Vicari submitted a series of requests for equitable adjustment to the contract, all of which were granted. The contractor submitted additional requests for three-day, nine-day, one-day, and four-day extensions of time due to weather delays. The contracting officer denied all requested modifications except for the four-day extension.

Count I of the complaint requested compensation for weather delays through August 31, 1998, constituting 58 days at $378.56 per day, for a total of $86,564.42 in damages. Count III sought reimbursement in the amount of $72,961.71 for the cost of obtaining builder's risk/wind insurance, which the plaintiff claimed was necessary for adverse weather conditions at the job site. Vicari sought the cost of maintaining the additional insurance.

Rather than proceed under the clause dealing with adverse weather, Vicari argued that, in order to comply with the safety and health clause of the contract, it had been required to keep workers from the job site when dangerous conditions were present. Vicari maintained that excessive rain presented a danger to the safety of workers on the job site, thereby necessitating exclusion of employees from the project on those days of adverse weather, and that Vicari was therefore entitled to an equitable adjustment to the contract price. Although artful, Vacari's argument ignored general principles of contract interpretation. The plain meaning of the contract established that delays arising from severe weather would be governed by the delay clause, not the safety and health clause. To interpret the contract any other way would render the FAR provision governing severe weather meaningless and superfluous. The sole remedy for severe weather, provided in FAR § 52.249-10, is a time extension for completion of the project. Moreover, severe weather and rain are acts of God and presumptively not within the control of the USCG. The USCG cannot be held responsible absent clear contractual terms allocating liability for acts of God.

Vacari alternatively contended that the contracting officer acted arbitrarily and capriciously in denying certain requests for extensions of time due to weather delays. The defendant responded that the contractor failed to provide the requisite documentation to support these requests and that published meteorological records indicated normal weather conditions during the days in question. Vicari rejoined that the contract did not require requests for additional time extensions to be supported by published records of the National Weather Service. Rather, the contract required the officer to establish a reliable method for determining site conditions and to monitor conditions on a daily basis, which plaintiff charged he failed to do. Vicari kept its own record of weather conditions in a project diary and used this information to document its claims to the contracting officer. The contracting officer ultimately rejected this diary. Even assuming the correctness of Vicari's documentation and the failure of the contracting officer to consider the proffered weather records, monetary recovery was precluded both by FAR § 52.249-10 and the decisional law ascribing delay due to acts of God to a cause other than the USCG.

Page 3-54, add after note 364 in first carryover paragraph:

If a contractor enters into a contract to be performed during a time period when harsh weather conditions are normally incurred, it may have difficulty demonstrating that the conditions it actually encountered could not be anticipated. For example, in *Blue Ribbon Remodeling Co. v. Meistrich*, 709 N.E.2d 1261 (Ohio Mun. Ct. 1999), the contractor was to install concrete piers under the foundation of an existing apartment building in Cincinnati, Ohio. The contract was signed at the end of December and provided that the work was to be completed 30 days later, on January 29. The contract stated that the completion date was subject to "weather conditions or other circumstances beyond the control" of the contractor. The general contractor experienced difficulties in performing the work and its subcontractors refused to work because of the conditions at the project site. The contractor did not complete the work in the time allowed by the contract, and in March left the project without completing the work. The general contractor filed suit to recover payment for the work it performed, alleging that it was unable to meet the completion date because of snow and rain. The court ruled that the weather conditions it encountered could have been anticipated. The work was to be performed during the winter time frame, and the actual weather conditions were typical for the area in the winter months.

Page 3-56, add after note 381 in first full paragraph:

In some contracts, the owner may stipulate a specific number of weather days allowed for time extensions and disclaim any liability for other weather-related delays. Such a provision was used in *Randolph & Co.*, ENG B.C.A. No. 6373 99-2 B.C.A. (CCH) ¶ 30,472 (1999), where the Corps of Engineers awarded a contract for the construction of a dike that provided for up to 36 days of time extensions for inclement weather and disclaimed responsibility for additional weather impact, including time extension. During contract negotiations, the successful bidder attempted to increase the number of excusable weather days, but the Corps refused to agree to an increase. The contractor experienced more than 36 days of weather-related delays and was assessed liquidated damages. The contractor appealed the matter to the Board and the government filed a summary judgement motion on the issue. In denying the motion, the Board relied upon the Federal Acquisition Regulations provision requiring identification and estimates of performance uncertainties. The Board allowed the contractor the right to pursue the appeal further if it could prove that the Corps refused to negotiate in good faith.

CHAPTER 4

EFFECTS OF DELAY

§ 4.02 PREVENTING EARLY COMPLETION

Page 4-5, add after first paragraph:

For example, in *Keeney Construction v. James Talcott Construction Co.,* 309 Mont. 226, 45 P.3d 19 (2002), the University of Montana (UM) awarded a general construction contract to James Talcott Construction Co. The general contract contained an initial completion date of May 31, 1997. Representatives from Keeney Construction and Talcott attended a preconstruction meeting to present a preliminary bar chart schedule that showed a December 1996 finish date. Two days later, Keeney and Talcott entered into a subcontract for the site work. The subcontract stated that Keeney agreed "to complete the work of this Subcontract as required by job progress or within the following time limits: As directed by James Talcott Construction, Inc. Time is of the essence on this project." The project architect stated that the UM did not want an early completion date and requested that Talcott prepare a schedule to reflect the full contract period. The UM and Talcott later entered into a change order which added three additional housing units and extended the completion date to June 30, 1997. Taylor ultimately issued a Certificate of Substantial Completion on October 1, 1997. Keeney filed a complaint against Talcott and the UM for delay claiming that it had an early completion date of October 31, 1996. Talcott filed a motion for partial summary judgment. The district court granted the motion and Keeney appealed.

The subcontract stated that Keeney agreed to work at the direction of Talcott. The subcontract clearly authorized Talcott to set Keeney's completion schedule. In order for an early completion date to apply to Keeney, the parties would have had to modify the subcontract. Keeney argued that the minutes of the preconstruction meeting did just that. During that meeting, the parties discussed an early completion date. Keeney explicitly agreed in the subcontract, however, to complete its work at the direction of Talcott. Because the language of the subcontract was clear, Keeney could not rely on extrinsic evidence to support its contention that the parties intended an early completion date. Alternatively, Keeney argued that the preconstruction meeting minutes, along with various conversations, represented an executed oral agreement, which modified the written contract. The preconstruction meeting, however, took place two days before Talcott and Keeney executed the subcontract. Modifying a contract that the parties have not yet entered into was logically impossible. The district court concluded that Keeney could not recover damages for delay from Talcott because Keeney worked "as directed" by Talcott.

Similarly, *McDaniel v. Ashton-Mardian Co.*, 357 F.2d 511, 516 (9th Cir. 1966) (interpreting a contract, which stated that a subcontractor agreed to complete a project "as directed by the Contractor," to mean that the contractor had the right to direct and control the time of doing the work) involved a subcontractor that worked "as directed" by the general contractor. In concluding that the subcontractor could not receive an award of delay damages, the *McDaniel* court held that the "as directed" language accorded the general contractor the right to direct the time and manner of the subcontractor's work.

§ 4.07 THIRD-PARTY CLAIMS

Page 4-13, add after carryover paragraph:

On the other hand, without a duty, a third-party cannot bring a delay claim against a party. For example, in *Jenne v. Church & Tower, Inc.,* 814 So. 2d 522 (Fla. Dist. Ct. 4th Dist. 2002), the court concluded that a course of dealing could not create third-party beneficiary status. Church & Tower, Inc. contracted with Broward County to design and construct a detention center. The contract provided that the project was to be substantially completed within 548 days of the "project initiation" date. The anticipated completion date was October 4, 1997. Although the county began to occupy the facility in February 1998, a final Certificate of Occupancy was not issued until October 28, 1998. Sheriff Ken Jenne sued the contractor for breach of contract, alleging that the delay in completing the facility caused him to incur labor costs and expenses associated with transporting inmates to other counties, resulting in over $13 million in damages. The sheriff contended that he was the intended third-party beneficiary of the contract between the county and the contractor. The contractor moved to dismiss the complaint, arguing that the sheriff was not an intended third-party beneficiary and that damages were barred by the liquidated damages clause of the contract. The trial court dismissed the complaint with prejudice. The sheriff appealed.

The sheriff alleged that the parties intended that he primarily and directly benefitted from the contract. There was a federal court decree that ordered the sheriff to develop reorganization and jail population management plans for the jail system. The county requested construction proposals, which stated, "It is the intent of the County and the Broward Sheriff's Office that the program and schematic plans comply with the foregoing applicable laws." There was a proposal submitted by the contractor that the sheriff claimed "acknowledges that the contractor needed the sheriff's approval for performance under the contract and acknowledged that it would be working together with the sheriff throughout the project." The sheriff alleged that he purchased furniture for the jail, appointed his office personnel to the selection committee, and made decisions on site location and management. The factors did not confer third-party beneficiary status on the sheriff. Negotiations and dealings between the parties cannot modify a written contract to create the parties' "intent," when the lack of such intent was evident from the contract. The parties to the contract were the county and the contractor, and the contract involved the construction of a public facility funded by the county. The contract was intended to directly and primarily benefit the citizens of the county, not the sheriff. The sheriff received only incidental or consequential benefit from the contract's enforcement, so he was precluded from suing for a breach of it. An article concerning "Defective Work," gave the county, not the sheriff, the right to notify the contractor to make corrections. The liquidated damages clause allowing the county to recover delay damages indicated that the parties to the contract did not intend that the sheriff enforce a separate remedy for the same delay. Affirmed.

§4.08 ABANDONMENT AND TERMINATION

Page 4-16, add to first paragraph:

However, in *Fuschino v. Smith,* 2001 Ohio App. LEXIS 21 (2001), an Ohio appellate court concluded that an owner did not waive the completion date when the owner permitted the contractor to continue working on the project, when there was no evidence that the owner acquiesced in the delay. The mere fact that the owner attempted to get the project completed by the contractor did not indicate acquiescence. Instead, the evidence showed that the owner recognized that the delay constituted a breach (in fact, the owner sought the architect's approval for termination), and that the owner's purpose in permitting the contractor to continue working was not to excuse the breach, but instead to pursue the most expedient course to complete the project. The completion date provisions were not waived. The owner was entitled to terminate the contract and to seek damages for the cost of completion of the project.

Page 4-16, add after note 52 in fourth paragraph:

If a contractor does not achieve final completion of a project by a predetermined date, and default is tied to final completion, the owners may terminate the contract. For example, in *Jeffrey M. Brown Assocs., Inc. v. Rockville Center, Inc.,* 2001 U.S. App. LEXIS 5575; 7 Fed. Appx. 197 (4th Cir. April 3, 2001), Jeffrey M. Brown Assocs., Inc., entered into a fast-track contract with Rockville Center, Inc., to design and build a retail pavilion and theater on the owner's land. The contract provided that time was of the essence, because the owners had to meet an occupancy schedule for their incoming tenants. The contractor completed enough work on the theater portion of the project to receive a temporary occupancy permit from the city, while work on the retail portion of the project continued. Disputes arose concerning the quality of the contractor's work and delays in construction. The owner hired consultants to determine the scope of the deficiencies. The consultants prepared a deficiency list, and the contractor agreed to complete its work by a fixed date. A 15-day cure period allowed the contractor to cure any breaches after completion. The owner notified the contractor that it had not achieved substantial completion. This triggered the 15-day cure period. On the completion date, the city's chief inspector wrote a letter to the owners that a final certificate of occupancy would not be issued until a defect in an expansion joint was repaired. The owners deemed the contractor to be in default and terminated the contract. The contractor sued, alleging that the owners breached the agreement by terminating the contractor and by refusing to pay the contractor the escrowed funds. The district court granted the owners' motion to dismiss. The contractor appealed.

To achieve final completion, the contractor was contractually required to obtain a final certificate of occupancy. The contractor admitted that it did not obtain a final certificate of occupancy. The exhibits the contractor attached to its complaint revealed that it was not entitled to an extension of time. The consequence of the contractor's failure to achieve completion was the contractor's default, and the risk that it could be terminated with no notice or opportunity to cure. The owners were free to terminate the contract and take back the escrowed funds when the contractor defaulted. The contractor failed to state a claim for breach of contract.

§ 4.09 OTHER EFFECTS OF DELAY

Page 4-18, add at end of section:

As discussed in the supplement to **§ 1.01[D]**, one type of concurrent delay is a pacing delay. For example, a contractor may adjust the pace of its work in light of delays in owner-furnished equipment, delays by other multiple prime contractors, delays in permits, limited site access, or differing site conditions. The party slowing its performance takes the risk that the other party will not quickly resolve its delay and that the slow down will then become a critical delay or become a concurrently critical delay. In such a case, the decelerating party may become liable for the impact of its decision to slow down the pace of its performance.

There are several practical effects of such pacing delays. First, the party slowing down its performance has in one sense caused noncritical delay to certain activities. Second, the party decelerating its work may be unaware that the decelerated activities are actually on the critical path or on a parallel critical path. Third, even if the deceleration only consumes float on noncritical activities, these activities may later become critical. With the float consumed, it now has little flexibility, and may face exposure to delay damages from other parties.

CHAPTER 5

DELAYS, DISRUPTION, AND LOST LABOR PRODUCTIVITY

§ 5.04 JUDICIAL RECOGNITION OF LOST PRODUCTIVITY CLAIMS

Page 5-6, add at beginning of note 10:

Delay claims are based upon the premise that the project duration has been longer than anticipated; delay damages focus on the cost of extended performance, such as extended site costs, extended or unabsorbed home office expenses, and escalation of labor and material costs. See **Chapter 12**. Lost efficiency relates to additional expenditure of labor, materials, equipment, or other resources made without achieving an increase in productivity. It is possible to experience a loss of productivity without experiencing delay.

Page 5-8, add before last sentence in section:

See John E. Green Heating & Plumbing v. Turner Construction Co., 742 F.2d 965 (6th Cir. 1984); *S.L. Harmonay, Inc. v. Binks Manufacturing Co.*, 597 F. Supp. 1014 (1984); *Clark Concrete Contractors, Inc. v. General Services Administration*, GSBCA No. 14340, 99-1 B.C.A. (CCH) ¶ 30,280 (1999).

§ 5.05 CAUSES OF LOST EFFICIENCY

[D] Adverse Weather

Page 5-15, add after second paragraph:

A recent study in the *Journal of Construction Engineering and Management* detailed the impact of material delivery practices and adverse winter weather conditions on structural steel erection activities on three similar projects. The original schedule had called for a four-week erection period in November. Because of delays in the shop drawings, and other factors related to the completion of other buildings at the site, the erection did not begin until the following January. Erection took place in harsh winter conditions. H. Thomas, D. Riley, & V. Sanvido, *Loss of Labor Productivity Due to Delivery Methods and Weather*, 125 J. Constr. Eng'g & Mgmt. 39, at 40 (No. 1, Jan./Feb. 1999).

The most significant factor affecting the construction plan was the complexity of the steel. Considerable time was spent shaking out steel as it arrived. Instead of setting the steel directly from the delivery trucks, all the steel was off-loaded and sorted prior to resuming the erection process. This practice resulted in double-handling of the steel. Thomas, Riley, & Sanvido, at 40. It also appeared that an orderly sequence of the work was not maintained. For instance, while all 579 pieces of steel were erected, only a few pieces were aligned, bolted, and tightened. Also, all 304 joists were in place, but no welding was done. No decking was installed. The absence of joist welding and decking was caused by fabrication errors to some joint connection plates. Access to these joints had to be maintained until the errors were corrected. The completion of several other connections was delayed because pieces of steel were lost in the snow. The remaining work on the connections resulted in a delay to the welding of metal joists and the erection of metal decking. Once these connections were completed, the joists and decking were completed. Thomas, Riley, & Sanvido, at 41.

The daily and baseline productivity showed the best that the contractor could do based on actual performances. Once the effect of repetition dissipated on workday 7, the baseline productivity was about 1.25 wh/pc. In the absence of the disruptions, the contractor should have performed at this baseline rate. The contractor worked at or near this baseline for 11 of the 21 workdays. The baseline was an estimate of the best productivity that a contractor could achieve on a particular project. It was based on actual performance and was the productivity when there were the fewest or no disruptions to the work. Therefore, the baseline was mainly a function of the design complexity. Thomas, Riley, & Sanvido, at 42.

Various numerical schemes have been tried to calculate the baseline productivity. The best approach is to visually fit a curve to the lowest and most consistent daily productivity values. These values need not be consecutive. Thomas, Riley, & Sanvido, at 42.

There were a number of fabrication errors with the steel members. This may have been because the fabricator subcontracted the detailing of the shop drawings, and the quality of these drawings seemed sub par. As a result, many connections required corrective measures in the field. On some days, these problems simply slowed progress. On other days, considerable time was spent making corrections. The fabrication problems were observed primarily in the latter half of the erection process, after workday 21. Because the steel was delivered while the work was in progress and shakeout was done upon arrival, the steel deliveries interrupted the erection process. Assessing the influence of the factors described earlier is not a straightforward process because of the early effects of improvements from repetition (learning curve) and multiple adverse factors occurring on the same day. The analysis showed that the effects of wind and crane relocations were insignificant, and these factors were deleted from the model. Thomas, Riley, & Sanvido, at 43.

The work was affected by weather events, such as high wind, snow, and cold temperatures. On three weather days, the crew worked less than a full day. Interestingly, on each of these days, productivity was near or better than the baseline. Overall, weather affected 12 of the 21 workdays. Of these days, the productivity was near or better that the baseline on three days. These are the only days when the crew worked less than a full day. Thomas, Riley, & Sanvido, at 42.

The most significant weather event was snow, which reduced the crew efficiency by 35 percent. This value is less severe than losses reported in an earlier publication of weather events (losses of 65 percent) related to rain. Cold temperatures also had an effect, resulting in a loss of efficiency of about 30 percent. This value was greater than losses estimated by the National Electrical Contractors Association, which estimated that for a relative humidity of 3–5 percent and temperatures of 21°C, the loss of efficiency was in the range of 12–14 percent. The loss of efficiency (about 30 percent) due to temperatures less than 27°C was greater than that stated in other literature sources. The work days on which material deliveries were made showed the largest reduction in efficiency, by almost 40 percent. Thomas, Riley, & Sanvido, at 43–44.

Erecting structural steel directly from the delivery truck was determined to be the most efficient erection method. The double-handling of steel on the Research Building project led to a productivity increase of 0.27 wh/pc, which equates to an additional 108.9 work-hour overrun. Thomas, Riley, & Sanvido, at 46.

To achieve good productivity, projects must experience relatively few disruptions. Projects that experience many disruptions are subject to the ripple effect, where even the best undisrupted days are worse than the baseline, which is based on observations from other comparable projects. To assess the ripple effect, the baseline productivity of the three projects was compared to the baseline of 1.25wh/pc. When compared to the baseline productivity, the Research Laboratory project experienced little or no losses of productivity beyond those attributed to the specific factors discussed before. Thomas, Riley, & Sanvido, at 44.

Multiple regression modeling is essential in productivity analyses because workdays often have multiple events occurring that are of interest. In many instances, the consistent term in the regression model must be defaulted to zero for the model to be valid. Binary variables are frequently used to represent the presence of potentially adverse factors. Thomas, Riley, & Sanvido, at 45–46.

[E] Out-of-Sequence Work

Page 5-17, add to end of first full paragraph:

In *The Sherman R. Smoot Company of Ohio v. Ohio Department of Administrative Services*, 736 N.E.2d 69 (Ohio App. 10th Dist. 2000), Smoot was the general trade contractor for Phase III of the construction of a correctional facility that was being constructed on a multiple contract basis. Smoot's masonry crews were unable to perform their work tasks on an uninterrupted sequence (in which the completion of one task would be followed by the next logical activity) because the separate prime contractor for the mechanical work did not complete its mechanical work activities in a timely manner as provided in the overall project schedule. Smoot allegedly moved crews in and out of areas, and back and forth between jobs, and this reduced the efficiency of Smoot's masonry crews. The disruptive effect of the "back and forth" crew movements was deemed by the court to be the basis for Smoot's recovery of certain disruption costs, such as the additional costs for mason-helpers and increased costs for preparatory work.

In *Clark Construction Group, Inc.*, No. VABCA-5674, 00-1 B.C.A. (CCH) ¶ 30,870 (2000), Clark originally planned to construct VAMC West Palm using a "horizontal" construction sequence. The planned sequence involved first constructing the piles, pile caps, and foundation of the hospital. Clark planned to erect the structure by moving horizontally from west to east. Instead of completing the construction of a hospital wing from the basement to the top floor before moving on to complete another wing, Clark intended to complete the structure of each floor of all three wings before constructing the next floor. For example, the third floor of the west wing would be constructed; the construction forces would then proceed to build the third floor of the center wing and immediately move on to the third floor of the East Wing. Clark planned to support this construction sequencing with three tower cranes deployed from west to east. The contract did not specify any particular construction sequence. A critical path method progress schedule was required for the project. A fully developed CPM employing horizontal construction scheduling logic was never approved by the VA. However, Clark developed an "interim" schedule that reflected scheduling logic using a horizontal construction sequence.

As a consequence of a stop work order, Clark re-sequenced the job from the planned horizontal to vertical construction. Consequently, Clark commenced vertical construction of the West Tower where the foundation was complete, while concurrently on a limited basis, continuing East and Center Wing foundation work. The Interim Critical Path Method schedule, still under development at the time a stop work order was issued, was abandoned. The actual CPM schedule approved by the VA incorporated the revised construction sequencing. In accordance with the contract requirements for CPM preparation, this initial CPM reflecting vertical construction did not reflect actual construction times related to the first 15 months of construction.

The change to vertical construction precluded two subcontractors from anticipated efficiencies in crew movement and material distribution. The vertical construction, as simple common sense indicated, and as the evidence showed, presented more difficult material handling, crew supervision, and work area access problems. These problems translated to lower labor productivity. The board found that the subcontractors' explanation of the effect of the changed sequence met the "fundamental triad of proof" to entitle it to recovery for the change in construction sequence.

Out-of-sequence work can also disrupt only one trade's work, causing that work to be performed in different time periods rather than in one continuous period. For example, in *Paliotta v. Department of Transportation*, 750 A.2d 388 (Pa. Commw. Ct. 1999), the contractor was awarded a contract to widen a highway. The scope of the work included constructing concrete curbs. The contractor's planned method of pouring the curbs was to use a slip-form machine in one continuous operation. However, the relocation of utility poles by the power company, under a separate agreement with the department, was performed over a period of time; thus, the contractor was not able to perform the curb work as planned, but rather had to do it in three phases as the utility poles were removed. This made use of the slip-form machine impractical, and the contractor hand-formed the curbs in a more labor-intensive manner with a significantly lower daily production rate. The contractor claimed the additional costs involved, but the department denied the claim. The dispute was taken to court and a ruling was issued in the contractor's favor based upon a finding that the change in the planned sequence as the result of interferences was compensable for the contractor. Expert testimony corroborated the contractor's planned production rates and estimated costs, and the court allowed the contractor to recover its increased performance costs for the curb work over its estimated costs.

[H] Cumulative Impact of Multiple Changes

Page 5-24, add after note 80 in third full paragraph:

However, the mere fact that there are numerous change orders, without further evidence of their affect on the unchanged work, the overall project completion or additional cost may not establish a claim for cumulative impact. In *Coates Industrial Piping, Inc.*, VABCA No. 5412, 99-2 B.C.A. (CCH) ¶30,479 (July 26, 1999), the Department of Veterans Affairs (VA) awarded a contract to Coates to upgrade boilers and replace an incinerator at the VA Medical Center in Salt Lake City. Coates claimed both delay and lost labor productivity under a cumulative impact theory due to numerous change orders. At a hearing before the VA Board of Contract Appeals, Coates was able to establish that design deficiencies resulted in over 150 change order proposals and contract modifications. However, Coates offered no evidence of either impact of the changes, cumulative effect, or an explanation of how its costs were affected by the cumulative impact. The numerous changes alone were the basis of its proof of the cumulative impact claim. The Board did not find this to be persuasive.

[1] Defining the Cumulative Effect

Page 5-26, replace note 87 with:

American Institute of Architects, *The Architect's Handbook of Professional Practice* 694 (1994).

Page 5-28, add at end of first full paragraph:

In *Centex Bateson*, VABCA Nos. 4613, 5162, 5165, 99-1 B.C.A. (CCH) ¶30,153 (1999), the Board required the contractor to provide a reasonable explanation of how the number of changes or other incidents caused the cost overruns that the contractor sought to recover in the claim. The Board further suggested that to demonstrate causation, the claimant may be required to show that there were no other reasons for the loss of productivity or cost overrun that the contractor is attempting to recover. This may amount to a requirement that the contractor prove a negative: that there were no other causes of the cost overrun. Keating & Burke, *Cumulative Impact Claims: Can They Still Succeed?*, 30 Constr. Law. (Apr. 2000).

[I] Constructive Changes

Page 5-33, add to note 121:

Sentinel Industrial Contracting Corp. v. Kimmins Industrial Serv. Corp., 743 So. 2d 954 (Miss. 1999) (subcontractor-claimant was able to recover for its costs in performing extra work without a written change order under an estoppel theory, as the jurisdiction did not accept the theory of constructive changes). Not all jurisdictions recognize the concept of constructive changes. However, there are legal theories and approaches that may allow a claimant to recover its costs nevertheless. For example, in Sentinel Industrial Contracting Corp. v. Kimmins Industrial Service Corp., 743 So. 2d 954 (Miss. 1999), the subcontractor-claimant was able to recover for its costs in performing extra work without a written change order under an estoppel-type theory, as the jurisdiction did not accept the theory of constructive changes. The subcontractor was awarded a subcontract to disable a portion of an ammonia refinery by the general contractor. The subcontract provided that written change orders were required for any adjustments to the terms or price. The general contract was incorporated by reference into the subcontract, and contained a provision requiring the contractor to comply with directives and instructions even without a written change order. The general contractor directed changes to the sequence and schedule of the subcontract work as well as requiring removal of equipment not in the subcontract scope. The subcontractor performed as directed and submitted change order requests to the general contractor who refused to approve them.

The subcontractor sued the general contractor to recover for the extra work costs and at trial was awarded $966,375. On appeal, the general contractor argued the lack of written change orders prevented any recovery by the subcontractor. The court found the subcontractor's arguments persuasive—that it had to comply with the general contractor's directives or face a possible default termination or breach of contract. Although the legal concept of constructive changes is not recognized under Mississippi law, its highest court allowed recovery for extra work without a written change order under the theory that one may not impose extra-contractual work while denying change order requests.

§ 5.06 PROVING LOST PRODUCTIVITY CLAIMS

Page 5-37, add after first sentence in section:

Basing a lost productivity claim upon "remedy-granting" clauses is another way to establish the owner's liability for lost productivity damages. For example, in *Youngdale & Sons Construction Co. v. United States*, 27 Fed. Cl. 516 (1993), the contractor's claim for lost labor productivity costs was based upon the differing site conditions clause. Other remedy-granting provisions include the changes clause and the suspensions clause that is used on federal government contracting. Patton & Gatlin, *Claims for Lost Productivity*, Constr. Law. 21 (Apr. 2000).

Page 5-37, add after third sentence in section:

Quantifying the loss of productivity may also involve examining quantities, establishing units of production, documenting labor hour expenditures and labor costs, calculating productivity, developing a productivity baseline, determining and measuring the causes of productivity losses, and quantifying the economic loss.

[B] Causation and Injury

Page 5-39, add after third full paragraph:

Proof that an event caused delay to the overall project completion is an essential ingredient in order to recover delay-related costs. However, the claimant need not prove that the disruptive event delayed the overall completion of the project in order to recover its disruption costs. The disruptive event may even be concurrent with the claimant's own delays or not delay the overall completion of the project, and yet still be the basis for recovery of disruption costs. In *Sauer, Inc. v. Danzig*, 224 F.3d 1340 (Fed. Cir. 2000), Sauer was awarded a contract to complete the interior shell of an industrial building for the Navy. A separate prime contractor was installing heavy-duty cranes in the same area. The crane contractor did not follow its schedule, and the Navy acquiesced to this, even though it disrupted Sauer's work. In a hearing before the Board of Contract Appeals, it was determined that some of the disruptions impacting Sauer were concurrent with Sauer's own delays, such that they did not delay Sauer's overall completion. The Board ruled that Sauer could only recover for productivity losses if the disruptive events were excusable delays that also delayed the overall completion date. On appeal, the United States Court of Appeals for the Federal Circuit ruled that recovery for disruption could be obtained if the event was the responsibility of the government, and that it caused an increase in the cost of performance due to disruption whether or not it delayed the overall completion of the project.

[C] Prereserving the Right to Collect Lost Productivity Costs

Page 5-41, revise note 151 to:

1990 GSBCA LEXIS 496

Page 5-43, add after carryover paragraph:

Courts may also ignore releases by distinguishing between disruption and delay claims. For example, in *Atlantic Coast Mechanical v. R.W. Allen Beers Construction*, 264 Ga. App. 680, 592 S.E.2d 115 (2003), Atlantic Coast Mechanical (ACM) worked as a subcontractor for R.W. Allen Beers Construction (Beers) on the Children's Medical Center in Augusta. ACM performed the heating, ventilation, air conditioning, and plumbing work for the project. Subcontract article 3(b) required completion within the contract times and sequences, and represented that the subcontractor had taken into consideration and made allowances for all hindrances and delays incident to its work. Article 10(b) required compliance with any schedule requirements imposed upon the contractor in its agreement with the owner and gave the contractor the right to decide the time, order, and priority, which the various portions of the subcontractor's work would be performed. Article 10(c) required the subcontractor to prosecute the work in accordance with the contractor's schedule and any modifications thereto issued by the contractor in order to achieve the completion date. In April 1998, ACM submitted a request for an equitable adjustment (REA) based on increased labor costs incurred as a result of disruptions to its work caused by Beers. In its REA, ACM claimed that it was forced to perform its work in a radically different manner, method, and sequence than that reflected in the project schedules or contemplated by ACM at the time it bid on the project. ACM claimed that the concrete operations on the project were performed late and out of sequence, and that reshores were not removed as scheduled. As a result, ACM had only restricted access to work areas and could not rough in the overhead pipe and equipment using rolling scaffolds and manlifts as planned. In addition, when the restricted areas became more accessible, ACM's work was constructively accelerated. ACM claimed that Beers required it to work on most floors simultaneously, which required ACM to more than double the size of its crews and required the oversized crews to work overtime frequently. Beers also instructed other subcontractors to work on floors where ACM was working, further obstructing ACM's access to ceiling work areas. ACM claimed that Beers allowed other subcontractors to stockpile drywall, curtainwall, and other materials on the floors, further restricting ACM's access to work areas. ACM claimed that Beers' failure to provide ACM access to work areas, active interference with ACM's work, and failure to schedule and manage other subcontractors properly resulted in a significant increase in labor costs. In a letter dated April 29, 1998, Beers responded that ACM's request was precluded by the terms of the trade contract agreement and various change orders, and asked ACM to withdraw the request. ACM ultimately filed suit. Beers sought summary judgment on grounds that Articles 3(b), 10(b), and 10(c) of the trade contract agreement barred ACM's claim. The trial court granted Beers' motion.

The change orders stated that they constituted compensation in full for all costs, claims, markup, and expenses, direct or indirect, attributable to this or any prior change orders, for any delays encountered by the subcontractor in the performance of the work through the date of the change order. Beers sought summary judgment on grounds that ACM had released any claim it held by signing certain change orders.

The change order language appeared to bar any claim for delays encountered by ACM in the performance of its work through the date of the change order. But ACM's claim was not merely a claim for damages based on delay. ACM's claim was based on substantial disruption of its work, resulting in loss of efficiency and increased labor costs. The fundamental distinction between the disruption claim and the delay claim was that "unlike the delay claim, the disruption claim is intended not to redress [the subcontractor]'s loss from being unable to work, but to compensate [the subcontractor] for the damages it suffered from [the contractor]'s actions that made its work more difficult and expensive than [the subcontractor] anticipated and than it should have been." The language in the change orders did not address a disruption claim like that asserted by ACM and thus did not entitle Beers to summary judgment.

[D] Experts

Page 5-44, add at end of carryover paragraph:

The use of expert witnesses to address "non-scientific" construction issues may come under increasing scrutiny by trial courts. Under recent case law, trial courts have a "gatekeeping" responsibility in allowing expert witness testimony, and may apply various factors before allowing expert witnesses to testify. The factors may include: (1) whether the theory or techniques used by the expert can be tested; (2) whether the theory or technique has been subjected to peer review and publication; (3) whether there is a high known potential rate of error; and (4) whether the theory or technique has general acceptance within the community of experts in the field. *See* Davidson, *Supreme Court Extends Trial Court's Role as "Gatekeeper" of Expert Testimony*, 1 Constr. L. & Bus. 99 (Winter 2000).

Page 5-45, add to note 172:

J.R. Roberts Corp., DOTCAB No. 2499, 98-1 B.C.A. (CCH) ¶ 29,680 (1998). Also, in Hensel-Phelps Constr. Co., ASBCA No. 49,270, 99-2 B.C.A. (CCH) ¶ 30,531 (1999), the Armed Services Board of Contract Appeals noted the fact that the expert testifying on the loss of productivity had not reviewed change orders, daily reports, and other project records. The Board thus did not accept the expert's testimony, because it was not based upon a reliable foundation.

Page 5-45, add after note 177 in second full paragraph:

In *Advanced Engineering & Planning Corporation, Inc.*, ASBCA Nos. 53366, 54044, 05-1 B.C.A. (CCH) ¶ 32,935; 2005 ASBCA LEXIS 33 (April 14, 2005), in a motion for reconsideration the Board examined the issue of causation in loss of productivity claim. The contractor (AEPCO) was awarded a contract for the overhaul of ships on a Restricted Availability basis (RAV), which required it to put the ship back together for service within 96 hours in case of emergency. One of the ships that it was engaged to overhaul was the flagship for the second fleet. During the course of the overhaul, the government required AEPCO to perform radiographic testing (RT) of piping that AEPCO claimed was not a contract requirement. AEPCO also was required to perform numerous changes that it claimed had a disruptive effect upon its performance. AEPCO submitted claims to the government for disruption in performing the work due to changes, radiographic testing, and inappropriate weld rejections. After rejection by the contracting officer, the matter was appealed to the Armed Service Board of Contract Appeals which did not find merit in many of AEPCO's claims related to disruption. Further, on the issue of disruption, the Board's initial finding was that formal and constructive changes had minimal impact, and that AEPCO, not the Navy was responsible for the bulk of the loss of productivity experienced in performing the work. The Board held:

> Because AEPCO has failed to establish a causal connection between the disruption factors or variables used in its methodology and the disruptive events that are the subject of its claim, and because AEPCO has failed to allocate its claim disruption costs between itself and the Navy, we hold that AEPCO is not entitled to recover any such costs. (05-1 B.C.A. (CCH) ¶ 32,806 (2005) at 162,326)

AEPCO's motion for reconsideration requested the Board to reconsider nearly every aspect of the entitlement portion of its decision—responsibility for the allegedly disruptive events, such as weld rejection, as well as the denial of disruption damages. AEPCO argued that in requiring specific proof of cause and effect relationship between the claimed changes and the disruption they caused, the Board failed to follow its precedents, such as *Coastal Dry Dock & Repair Corporation,* ASBCA No. 36754, 91-1 B.C.A. ¶ 23,324 and *Triple "A" South*, ASBCA No. 46866, 94-3 B.C.A. ¶ 27,194.

The Board was not persuaded, however, that the high rate of weld rejection was attributable to overinspection on the part of the Navy. Thus, AEPCO was not entitled to an equitable adjustment for any disruption that might have arisen out of the requirement to correct any rejected welds. As far as any disruption damages that might have arisen out of any formal and constructive changes (to which entitlement was found), the Board distinguished *Coastal Dry Dock* case, and discussed the application of the *Triple "A" South* case, in which the claimant was awarded a job order to overhaul the USS BAGLEY. In *Coastal Dry Dock*, sometime during the overhaul, the contractor began adding "Estimated

Contingency Delay Costs," to its change order proposals. The contingency factors, in different percentages, did not reflect a specific cause and effect relationship between the change orders issued and any resultant impact claimed. At the hearing, the contractor's evidence concerning the disruptive effect of changes on unchanged work consisted of theoretical discussions of the general subject of disruption in the shipbuilding and ship repair programs. Several Navy studies, not unlike the studies AEPCO used in the instant case, were used. The contractor also attempted to rely on a separate and unrelated Memo of Understanding (MOU) it reached with SUPSHIP San Diego for evaluating impact of change orders that would begin with one of the three specific vessels to be overhauled. The MOA established a methodology for forward pricing change order impacts with ranges of specific factor percentages applying to each impact variable. The *Triple "A" South* case held unless there was an applicable agreement to use impact variable-weighted factor methodology, such methodology was inappropriate for after-the-fact pricing of disruption costs because no causal relation between particular changes and their actual impact on other overhaul work was shown. Although the government later agreed to use such a methodology for forward pricing impact costs during the overhaul of one vessel and upon the particular percentages assigned to specific impact variables for that purpose, the parties never agreed to use factors for retrospective pricing of disruption costs on the overhaul of the other vessels and the validity of those factor percentages for that purpose was not demonstrated. Instead, the Board found that the factor percentages used by the contractor showed no causal relation between particular changes and their actual impact on other overhaul work and amounted essentially to a variation on the "total cost" method for computing damages. The Board held that the contractor failed to establish the appropriateness of applying its proposed factor percentages. *Triple "A" South, supra*, 94-3 B.C.A. at 135,542.

The AEPCO Board noted that while the parties were free to use disruption factor percentages to forward price local and cumulative impact in resolving formal (directed) sequences, it became a burden of proof issue for AEPCO once litigation started. As a litigation issue, pricing local and cumulative impact by disruption factor percentages without regard to any causal connection between the percentages and the actual disruptive event did not satisfy the burden of proof that AEPCO, as the claimant, had to carry by a preponderance of the evidence.

Southern Comfort Builders, Inc. v. United States, 2005 U.S. Claims LEXIS 225 (Fed. Cl. 2005), illustrates both a lost productivity calculation that was not persuasive and the effect of another expert's challenge. The National Aeronautics and Space Administration awarded a contract to Southern Comfort Builders for replacement of HVAC equipment in Mobile Launch Platforms for space shuttles. The contractor alleged that scheduling conflicts occurred throughout its work, delaying its performance and requiring submittal of numerous requests for information (RFIs). The contractor also encountered lead paint on the first platform, which required abatement before the contractor could proceed with the contract. In addition, the contractor alleged that it encountered defective contract specifications. The contracting officer issued a final decision, denying the contractor's certified claim for equitable adjustment. The contractor filed a complaint, seeking damages for loss of productivity for its work on the first platform, among other claims.

The contractor claimed that the acceleration, delays, and disruptions were caused by the government's failure to allow adequate access to the project site, to furnish complete and accurate specifications and plans, to timely and sufficiently respond to RFI's, to timely process change orders, and to disclose differing site conditions. The contractor claimed that it was entitled to compensation for overall loss of productivity. The contractor offered examples of delays caused by the government, but did not provide a critical path analysis of the alleged government-generated hindrances and their effect upon the productivity of the project. The contractor's consultant offered three alternatives for assessing the contractor's damages for loss of productivity: a modified total labor-cost method, a modified measured-mile analysis, and a method that averaged the first two calculations. The consultant did not state which method was most supportable and admitted that inaccuracies existed regarding his damage calculations. The government's expert concluded that the contractor underbid the contract and that the bid was unreasonable. The expert also concluded that the contractor's consultant failed to properly calculate the subcontractor's payrolls. The government argued that the contractor could not demonstrate the degree to which it was responsible for the overrun in price. The government argued that the contractor should have attended the site visit and prepared the required coordination drawings. The contractor's failure to produce coordination drawings and to attend the pre-bid site visit contributed to the contractor's failure to understand the challenges it would face in completing the contract. The court chose not to adopt the contractor's consultant's flawed analysis to support the contractor's calculation of damages for loss of productivity. The contractor did not prove that it was entitled to damages for loss of productivity and its claim was denied.

[E] Trade Publications

Page 5-46, add after first paragraph:

In *Clark Construction Group, Inc.*, No. VABCA-5674, 2000 VA B.C.A. LEXIS 4 (Apr. 5, 2000), the board stated that quantification of loss of efficiency or impact claims was a particularly vexing and complex problem. The board recognized that maintaining cost records and identifying and separating inefficiency costs was both impractical and essentially impossible. The board admitted that percentage estimates for lost efficiency may be an appropriate method to quantify such losses.

In *Clark*, the board used the productivity factors from the MCAA Manual as the best method to arrive at the percentage estimates for a subcontractor's undeniable productivity losses. As a foundation to their use in *Clark*, the MCAA

Executive Vice President had testified that the productivity factors contained in the MCAA Manual were developed by MCAA's Management Methods Committee, but were not based on any empirical study determining the specific factors or the percentages of loss associated with the individual factors. These factors were intended to be used in conjunction with the experience of the particular contractor seeking to use them because percentage of increased costs could vary from contractor to contractor, crew to crew, and job to job.

The board stated that the MCAA factors were widely used in the industry for estimating and productivity valuation purposes. In assessing productivity loss, the MCAA factors were generally used as a guideline as interpreted by experienced project personnel familiar with the specific circumstances of a particular job and contractor. As contemplated by the MCAA Manual, the board used their reasonable judgment of how the factors applied.

For the change from a horizontal to vertical construction sequence that resulted in reduced productivity because of more difficult internal communication within the building and more difficult control of labor forces arrayed over different floors, the board concluded that the applicable MCAA Factors were "Dilution of Supervision" and "Site Access." Because the sequence change occurred early in the project permitting the subcontractor to adjust to the condition, the board determined the percentage of loss for both conditions as "minor." MCAA attributed a 10 percent loss of efficiency for a minor Dilution of Supervision Condition and a 5 percent loss of efficiency for a minor Site Access problem. Thus, the productivity of the subcontractor's installation labor was adversely impacted by a factor of 15 percent. As the subcontractor acknowledged, however, concurrent with the inefficiencies resulting from the change in sequence were other causes affecting labor productivity not attributable to the VA. These other causes (late stair installation, late building dry-in, lack of CPM coordination or adherence, and man-lift congestion problems) lead the board to adjust the 15 percent inefficiency factors as contemplated in the Manual. When taken together, the non-VA causes of labor inefficiency equaled the impact of the change in sequence. The board adjusted the 15 percent indicated inefficiency factor for the construction sequence change and found that the change to vertical construction impacted the subcontractor's efficiency by 7.5 percent.

The various NECA studies and manuals have also been used in support of lost productivity claims. For example, in *Floura Corporation*, ENG BCA No. 2516, WL 435 (1968), the claimant's expert witness was an engineer who examined the disputed changes and referenced the NECA manual for productivity factors associated with the various working conditions, and then combined the factors into an overall factor used to determine the productivity loss. The board stated that while the NECA factors are a guide, the factors are not entirely accurate, nor should they be applied without a "judgment factor." The board recommended that additional factors pertinent to the project should be taken into account, but nevertheless accepted the expert's opinion. In *Hoffman Construction Co.*, 94-3 B.C.A. (CCH) ¶ 27,246 (1994), it was the government that referenced the NECA Manual as an industry standard to refute the contractor's claim for additional labor costs associated with changes on a prison construction project. The government's expert testimony used the NECA factors to establish the "reasonable cost of the work" and opined that since the labor hours claimed by the contractor exceeded the NECA standards, they were unreasonable. The board found that the estimates of the government's expert did not establish the costs of performance with greater reliability than the contractor's submission of its "adjusted actual costs." This board also reasoned that the application of the NECA factors requires the "exercise of subjective judgment" in selecting the factors to be used and adjusting the factors to the actual working conditions. The government's expert never visited the project site, had not personally supervised the installation of the equipment involved for over 25 years, assumed that there was only one appropriate installation method, based his opinion that the work was "typical" only on a review of the drawings, and then merely referenced the NECA factors. The board did not follow the government's expert's opinion over the contractor's proof of its "adjusted actual costs." In *John R. Hundley*, VABCA No. 3493, 95-1 B.C.A. (CCH) ¶ 27,494 (1995), the contractor based its damages claim for change order costs on the differences between its estimated material and the actual material used, plus labor costs calculated using the NECA manual. The board rejected the claimant's methodology as an unreliable estimating method. In *Havens' Steel Co. v. Randolph Engineering Co.*, 613 F. Supp. (W.D. Mo. 1985), the subcontractor sued the prime contractor alleging loss of productivity resulting from extensive overtime. The subcontractor's expert based his opinions concerning the loss of productivity on statistics from the NECA manual and the MCAA publications. The court found that the subcontractor's expert was not qualified to give an opinion concerning the calculation of labor efficiencies because of his lack of training and experience in the matter, rather than the lack of reliability of the NECA and MCAA factors.

[F] Historical Data

Page 5-47, add after note 187 in first carryover paragraph:

There have been other studies of labor productivity that purport to be based upon historical data. For example, in a master's thesis by Charles A. Leonard, The Effects of Change Orders on Productivity (M. Eng. Thesis, Concordia University, Montreal, Quebec, 1988) and a subsequent article, Moselni & Leonard, *Impact of Changes on Construction Productivity*, 18 Canadian J. Civil Eng'g 484 (1991), the authors endeavor to quantify the lost productivity impact of change orders upon the unchanged or "base" contract work. In his thesis, Leonard examined 57 construction projects and attempted to quantify through statistical analysis the cumulative impact of change order work upon the base contract work. The gist of the thesis and the article is that lost labor productivity is significant where the hours expended on change order work exceed 10–15 percent of the base contract labor hours. In what has become known as the "Leonard Thesis," the theory is that when the volume of change order work exceeds this percentage range, a loss of productivity is experienced due to "stop and go" operations, out-of-sequence work, loss of productive rhythm, "de-motivation" of the workforce, learning curve loss, unbalanced crew size or composition fluctuations in the labor force, lack of engineering support, and acceleration if time extensions are not given with the change orders in a timely manner. Some claimants on lost productivity claims have cited and relied upon the Leonard Thesis to support their contentions that numerous change orders are the cause of recoverable disruption costs. However, courts have not readily accepted the Leonard approach.

In *Aetna Casualty and Surety Co. v. 39th Street Ltd. Partnership*, 1998 U.S. Dist. LEXIS 22627 (E.D. Pa. 1998), the court rejected a formula based upon the Leonard Thesis where the number of change order hours was compared with the estimated base contract hours. The court opined that the formula was "wholly unreliable" because it was based upon "unfounded assumptions, biased and incomplete data and speculative assumptions" (*Id.* at 233–234). Specifically, the court noted that the formula based upon the Leonard approach failed to account for the impact associated with the timing of a specific change, what actually happened on the project concerning workforce and material availability, the ability to efficiently manage and schedule the changes, and the actual rate of progress before and after the change orders were issued. *Id.* at 229, n.12. In *J.A. Jones Construction Co.*, ENG BCA 6348, 6386–6391, 2000-2 B.C.A. (CCH) ¶31,000 (2000), the board also rejected the application of the Leonard approach because it was based upon relatively small projects (94 contracts totaling $220 million), not civil engineering projects like the flood protection project involved in the dispute before the board.

[G] Job-Specific Data

Page 5-53, add after carryover sentence in carryover paragraph:

In *Clark Construction Group, Inc.*, No. VABCA-5674, 2000 VA B.C.A. LEXIS 4 (Apr. 5, 2000), the board stated that contemporary records were required to prove lost labor productivity. In *Clark*, a subcontractor contended that three circumstances, for which the VA was responsible, adversely impacted the productivity of its labor constructing a VA hospital and caused the subcontractor's labor to be overrun by 97,594 man-hours. First, the subcontractor asserted that the change in construction sequence caused by stop work orders resulted in its labor being less productive and required excessive expenditures for preparation of coordination drawings. Second, the subcontractor contended that the site was excessively wet, both outside and inside the building, because of the stop work orders and the VA's improper proprietary roof specification. Finally, the subcontractor claimed that the VA's endemic failure to timely respond to Requests for Information (RFIs) adversely effected both the subcontractor's coordination drawing effort and the labor productivity of its installations, causing an increase in labor coordination drawing costs. The subcontractor provided various detailed quantum analyses and methodologies in support of its request for a judgment of $1,935,092.

The board found that the subcontractor made little, if any, effort to cite contemporaneous project records in support of the testimony. Given the labor overrun that the subcontractor knew had begun very early in the project, the board found it difficult to believe that the contemporaneous documentation contained in the record would not provide relevant evidence supporting both the fact that an impact on productivity occurred and the extent of that impact. Therefore, the board inferred that the contemporaneous project records did not support the subcontractor's position. The board found that the after-the-fact, conclusory assessments of the project managers or the opinion of its experts were not sufficient substitutes for the subcontractor's underlying obligation to contemporaneously document the severe adverse impact on labor efficiency it claimed resulted from the changes and RFIs. *Citing Centex Bateson*, 99-1 B.C.A. P30,153; *Fru-Con Construction Corp.*, 43 Fed. Cl. 306 (1999); *Triple "A" S.*, 94-3 B.C.A. ¶ 27,194, 135,529–30; Michael R. Finke, *Claims for Construction Productivity Losses*, 26 Pub. Cont. L.J. 326–28 (1997).

For example, in *Clark Construction Group, Inc.*, VABCA No. 5674, 01-1 B.C.A. (CCH) ¶ 30,870 (2001), the prime contractor for the general construction work sponsored a lost productivity claim for its mechanical subcontractor on a large, multi-year hospital project for the Department of Veterans Affairs. In presenting its case before the board, the claimant relied primarily upon the testimony of its project managers and expert witnesses; however, the board noted that the presentation of "contemporaneous project records" is also extremely helpful, if not a requirement, for successful recovery. For example, one aspect of the subcontractor's claim alleged labor disruption due to RFIs that were not responded to and resolved by the government in a timely manner. The board noted that if the late RFIs changed the working conditions and disrupted the labor productivity, the problem would likely be reflected at the time in the daily logs, the schedule updates, CPM fragnets, and correspondence between the parties. Because the expert opinion was not supported by such records, the board decided that the claimant failed to meet its burden of proof.

Page 5-53, add at end of section:

The case of *Amec Process and Energy Ltd. v. Stork Engineers & Contractors BV*, 1997 ORB 659 (Mar. 15, 2002), illustrates how keeping detailed job-specific records may assist in the recovery of damages for lost productivity, particularly if the claimant alleged "abandonment" of the change order procedures set forth in the contract. Single Buoy Moorings (SBM) entered into agreements to construct a Floating Production Storage Off-Loading Unit (FPSO). SBM subcontracted the design, construction, and commissioning of the topsides of the FPSO to Stork, which, in turn, subcontracted some of the work to AMEC. The subcontract between Stork and AMEC in article 13.1 provided that for each variation (change order) AMEC was to provide quantification of the direct cost consequences, and a demonstration of cause and effect between the costs and the changed work directed by Stork. SBM, Stork, and AMEC also entered into a partnering arrangement to assist in the performance of the contracts and subcontracts relating to the FPSO and to resolve any problems or disputes that might arise. Disputes did arise between Stork and AMEC concerning the subcontract schedule, and AMEC alleged that the schedule was abandoned because of the large number of variations, the large number of revised drawings that were issued in an untimely manner, and late and irregular delivery of drawings, materials, and equipment by Stork. AMEC claimed that these problems resulted in the doubling of the required labor effort, an increase of 180,000 labor hours quantified at seven million. Stork maintained that AMEC failed to establish a "causal link" between the variations and the alleged disruption and claimed costs. In considering the matter, the court noted that "productivity and disruption are part of the same coinage, that is, the rate at which the work is performed and the output of the work force over time." The court also noted that AMEC's tender (bid) was based upon a definable number of labor hours, which could be increased by variations on the basis of an "ascertainable and definable" number of hours. AMEC kept actual labor records on a week-by-week basis throughout the course of performance in a "contemporaneous manner." Although AMEC did not perform a detailed time and motion study for each task and variation, it did provide the court with a "mass of primary evidence." The court found that Stork did not request or require AMEC to individually quantify each variation provided in article 13.1, and that Stork had not followed the contract variation procedure. Rather, Stork required AMEC to comply with "general variation instructions," that is, to meet the deadlines and requirements of the various program (schedules) as necessary to achieve timely completion. Therefore, AMEC was not required to prove cause and effect for each individual variation, only that AMEC was not the cause of the variation or disruption, that the hours claimed were the reasonable consequence of the general directions from Stork, and that the claimed costs were not paid in any other way. The court allowed AMEC to recover the additional actual direct labor costs plus a markup for profit.

§5.07 LOST PRODUCTIVITY DAMAGES

Page 5-53, add after note 206 in carryover paragraph:

Both approaches attempt to quantify the differences between what the work would have cost without the disruptive events and what the work actually cost as a result of experiencing the disruptive events.

Page 5-54, add after note 209 in first full paragraph:

Another method to quantify lost productivity damages is to use the "measured mile" method. The goal of the measured mile is to compare productivity during periods (or locations) that were not disrupted with productivity during disrupted periods (or locations). The measured mile approach avoids the need to prove the reasonableness of estimated productivity because it is not used as the basis for comparison. The measured mile method of calculating lost productivity costs is discussed in more detail in **§ 12.08**.

CHAPTER 6
ACCELERATION

§ 6.02 DIRECTED AND CONSTRUCTIVE ACCELERATION

Page 6-7, add after first full paragraph:

A contractor may also allege that the owner "constructively" gave the contractor a directive or notice to accelerate. This is different than constructive acceleration. However, an owner's repeated emphasis on adherence to a contract schedule does not automatically give rise to a constructive notice to accelerate. For example, in *Fraser Construction Co. v. United States*, 2003 U.S. Claims LEXIS 166 (June 25, 2003), the Army Corps of Engineers awarded a contract to Fraser Construction Company for flood control work. The contractor began work on the project on May 17, 1993, and the contract was scheduled to be completed by September 1, 1993. The contract's default clause absolved the contractor of responsibility for any delays in contract completion that arose "from unforeseeable causes beyond the control and without the fault or negligence of the Contractor." During the months of June, July, and August, however, the contractor experienced adverse weather conditions and high water flows that impeded its progress. In response to those conditions, the contractor repeatedly sought both time extensions and additional contract compensation. With regard to the time extensions sought for unusually severe weather, the Corps granted 33 of the days sought. With regard to the time extensions based on high water flows, however, the Corps denied all claims. The Corps additionally refused all requests for increased compensation. The contractor brought suit, seeking the additional days and acceleration costs.

The contractor contended that its acceleration claim was not limited to the days for which it was denied additional time, but included costs associated with 29 days for which it was ultimately granted weather-based time extensions but on which it nonetheless continued the contract work. The contractor contended that although it was permitted to cease operations at the disposal site on days for which it received time extensions, the Corps nonetheless specifically directed the contractor to continue operations at the excavation site, thereby forcing it to incur additional costs and essentially rendering the time extensions meaningless. The contractor pointed to various exhortations by the Corps that the contractor not fall behind its contract schedule. The court did not read the Corps' repeated emphasis on adherence to the contract schedule as a constructive notice to accelerate. The Corps specifically informed the contractor that there would be no acceleration unless it was formally ordered. The court did not construe legitimate concerns by the Corps that the contractor adhere to its original schedule (or else be subject to default) as an indication that the contractor's obligations had increased. The conditions inherent to the site were responsible for impeding the contractor's progress. With such a built-in hindrance to the timely accomplishment of the work, the Corps was correct to be concerned. The contractor had already received exactly what it sought: time extensions for the days in question. The contractor could only hope to extend the contract schedule to account for the time lost as a result of problems encountered at the site. The fact that the Corps granted time extensions on the basis of delays experienced at the disposal site rather than delays at the excavation site was irrelevant: Having received an extension for a particular day, the contractor would have thereby exhausted the full extent of its relief under the contract. The court concluded that the contractor was not entitled to any additional compensation.

Directed acceleration that is litigated usually involves a problem with the direction to accelerate. For example, consider *K.P. Meiring Construction, Inc. v. La Quinta Inns, Inc.,* 2003 Tex. App. LEXIS 1048 (2003). K.P. Meiring Construction, Inc. contended that the following language in an agreement accelerating the completion date in a construction contract constituted an enforceable agreement, "[i]n either case, La Quinta agrees to negotiate a future project with K.P. Meiring Construction, Inc." La Quinta Inns, Inc. moved for summary judgment on K.P.'s fraud claims on the basis of the economic loss rule and the absence of an enforceable contract. The parties entered into an acceleration agreement as evidenced by La Quinta's letter to K.P. An agreement that did not contain all material terms was not enforceable. The agreement to negotiate a future project was simply an agreement to agree and was not enforceable because it left material terms open for future negotiations, including the amount of K.P.'s profit. If the parties had not entered into a binding agreement, K.P. would not be able to show detrimental reliance, and summary judgment would have been proper as to K.P.'s fraud claims. Although the portion of the acceleration agreement relating to the promise to negotiate a future project was not a separately enforceable agreement, K.P.'s fraud claim addressed the misrepresentations made by La Quinta to induce K.P. to enter into the acceleration agreement, not simply the promise to negotiate a future project. K.P. alleged that by inducing it to enter into the acceleration agreement, K.P.'s principal officer was forced to suspend his bidding activities and supervision of other projects, and K.P. incurred additional expenses in meeting the commitment under the acceleration agreement, including personnel, labor, materials, and lost business opportunities. Accordingly, the trial court erred in granting summary judgment as to the fraud claims based on the absence of an enforceable agreement.

§ 6.04 THE OBLIGATION TO ACCELERATE

Page 6-11, add after sixth sentence in second full paragraph:

An owner cannot insist that a surety accelerate to complete at the original contract time if the surety elects to progress and pay liquidated damages. For example, in

St. Paul Fire & Marine Insurance Co. v. City of Green River, 2001 U.S. App. LEXIS 6269, 6 Fed. Appx. 828 (10th Cir. 2001), the Wyoming Joint Powers Water Board constructed a 32-million-gallon-per-day water treatment plant. St. Paul Fire and Marine issued a performance bond to guarantee the contractor's performance. The construction agreement required substantial completion of the project by December 1, 1998, and included a time is of the essence clause with a liquidated damages assessment after the stated December 1 completion date. The contractor fell behind in construction of the plant and the board issued a notice of termination for default. This triggered the surety's obligations under the performance bond. The surety had several options in the event of contractor default, and chose to complete the project by September 24, 1999, paying the daily penalty after December 1, 1998. The board, however, informed the surety that it had stepped into the shoes of the contractor for all purposes and thus was bound by the time is of the essence clause and the December 1, 1998, completion date. The board rejected the surety's proposed performance and declared the surety to be in breach. The surety sued to be discharged from further performance obligations on the ground that the board had breached the terms of the performance bond. Both parties sought summary judgment. The trial court granted summary judgment to the city. The surety appealed.

Although the surety essentially stepped into the shoes of the contractor upon the latter's default, it did not follow that the surety was bound by the original completion date. The board would continue to be entitled to $2,500 per day liquidated damages, but the surety under the bond need only proceed with reasonable promptness. The appellate court concluded that because the surety had selected one of the bond options available to it and was ready to proceed to completion, the board's rejection of its proposed performance breached the agreement. The surety's offer to substantially complete the project, by a date that the board conceded would have been acceptable, would have occasioned much greater expense. It was not unreasonable for the surety to seek compensation from the board for the cost of accelerating performance beyond the reasonable promptness standard the bond specified.

§6.05 CONSTRUCTIVE ACCELERATION

Page 6-18, add to note 51 in numbered list of second full paragraph:

The stated requirements for constructive acceleration have remained the same over the last decade. The requirements were restated in *Robust Construction, LLC,* ASBCA No. 54056, 05-2 B.C.A. (CCH) ¶33,019 (2005), where the contractor was awarded a fixed-price construction contract for lodgings at Altus Air Force Base in Oklahoma. Robust was required to submit updated CPM schedule on a regular basis, and encountered excusable delays (usually severe weather) and government-caused delays (delay in issuing utility dig permits) that impacted the critical path of the updates for 130 days. Robust gave timely notice of the delays and requested a time extension. Rather than grant a time extension for the full 130 days, the government only gave a 75-day time extension, and Robust attempted to meet the revised completion date by increasing its workforce. Robust filed a claim for constructive acceleration. The ASCBA in granting the claim reiterated the five requirements for constructive acceleration:

1. The contractor encountered a delay that is excusable under the contract
2. The contractor made a timely and sufficient request for an extension of the contract schedule
3. The government denied the contractor's request for an extension or failed to act on it within a reasonable time
4. The government insisted on completion of the contract within a period shorter than the period to which the contractor would be entitled
5. The contractor was required to expend extra resources to compensate for the lost time and remain of schedule

[A] Excusable Delay

Page 6-22, add at end of first full paragraph:

If there has been no excusable delay, there can be no acceleration. In *TLT Construction Corp. v. United States,* 60 Fed. Cl. 187, 2004 U.S. Claims LEXIS 67 (2004), TLT Construction Corporation and the Department of the Army contracted for construction, maintenance, and repair regarding the Smoke Bomb Hill Barracks Renovation Phase II at Fort Bragg, North Carolina. The contract was for an indefinite-delivery and indefinite-quantity with one base year and two option years. Prior to submitting its bid, TLT was provided minutes of the Pre-Bid conference, which provided that the first task order would be Bid Item No. 0001 and that Notice to Proceed would be issued for work to commence, not to purchase materials only. In addition, the meeting minutes stated that the construction durations were discussed and "the criticality of the schedule was emphasized." The Army and TLT disputed the completion date for the task order. The Notice to Proceed allowed TLT 94 days to complete Bid Item No. 0001, 34 days more than was estimated by TLT's bid. TLT provided a schedule at the Pre-Construction Conference that showed the work being completed in 32 weeks and were prepared to negotiate a time for completion. TLT confirmed its commitment to completion of Bid Item No. 0001 by the stated duration, but indicated that TLT might incur "acceleration costs" since there had been no negotiation of the delivery order's duration. The government maintained that the standard FAR Performance Provision incorporated into the contract by reference, which stated the duration was to be negotiated, was applicable only to the award of multiple buildings in a delivery order. Under those circumstances, negotiation would address the extent of construction overlap or concurrent construction and the availability of individual buildings from the start of construction. TLT submitted a formal claim to the CO seeking $263,126.00 for acceleration costs allegedly incurred as a result of the

Army's failure to negotiate the performance time for Delivery Order No. 0001. The CO did not issue a final decision on TLT's Acceleration Claim. TLT filed a complaint for breach of contract in the United States Court of Federal Claims.

The Solicitation Offer and Award prominently displayed the language of the standard NTP provision that the performance period was "mandatory." The performance period "negotiable" block was not checked. Therefore, the court interpreted the specific reference to the standard FAR NTP Provision requiring the contractor to commence within five days of the Notice to Proceed, together with the fact that the negotiable block was not checked to supercede and take preference over the subsequent general reference to the standard FAR Performance Period. The government's motion for summary judgment regarding TLT's "Acceleration Claim" was granted. TLT's reliance on constructive acceleration as a theory to support a claim for its "acceleration costs" was misplaced. TLT did not establish "any delays giving rise to the order were excusable" regarding Delivery Order No. 0001. In fact, TLT completed Delivery Order No. 0001 without delay. Also, the CO never issued an order that there was an excusable delay. The government's motion for summary judgment regarding TLT's "Acceleration Claim" was granted.

Page 6-23, add after note 75 in first full paragraph:

An examination of contract clauses other than the time extension clause may assist in determining whether a particular delay is excusable. For example, in *Unarco Material Handling,* USPS No. 4100, 00-1 B.C.A. (CCH) ¶ 30,682 (2000), the board considered a contract that allowed the owner to delay the start of the performance period. The U.S. Postal Service (USPS) awarded a contract to Unarco to install an integrated material and storage distribution system at a mail facility then being constructed. The contract allowed the USPS to delay the start of the work by 30 days without additional cost, and provided for a 60-day installation period with liquidated damages of $3,400 for each day that the contractor finished late. Because of delays and union problems affecting the contractor constructing the facility, Unarco's start date was postponed by the USPS for more than three months. Upon receipt of the notice to proceed, Unarco notified the USPS that it could not begin the installation work, because one of its subcontractors had scheduled other work. Because Unarco was unable to begin in a timely manner, it requested additional time for performance, but the request was denied by USPS. Unarco accelerated its performance to avoid liquidated damages, bringing in additional personnel and working overtime. The Postal Board of Contract Appeals examined the contract clause concerning notice to proceed and the facts relating to the claimed delays, and found in favor of Unarco on its acceleration claim. The USPS contract required the contractor to assume the risk of a late start, up to 30 days, but thereafter the risk of delayed notice to proceed was not the contractor's. Even though the contract performance time was adjusted for the duration of the late start, unavailability of the subcontractors to begin work was an excusable delay for Unarco, allowing for an additional time extension beyond the time extension to adjust for the late notice to proceed.

[C] Refusal to Grant a Time Extension

Page 6-30, add after note 96 in third full paragraph:

Not every untimely granting of a time extension will expose the owner to a claim of constructive acceleration. For example, in *Fraser Constr. Co. v. U.S.*, 2004 U.S. App. LEXIS 20338 (Fed. Cir. Sept. 27, 2004), the Army Corps of Engineers was engaged in a flood control project on the South Fork Zumbro River in Rochester, Minnesota. In connection with the project, the Corps contracted with Fraser Construction Company to excavate material from the bottom of Silver Lake, a shallow reservoir located along the river. Before the project began, the water level in the lake was to be lowered by approximately eight feet to facilitate excavation of the lake bottom. At that water elevation most of the lake would normally be dry, except for a small stream running through the lakebed.

Fraser intended to divert the stream into a trench along the edge of the dry lakebed and to construct an earthen dike to confine the water to the trench. The dike was originally designed to withstand a water flow rate of 800 cubic feet per second (cfs). A water flow in excess of the flow the dike was designed to handle could destroy the dike and flood the lakebed. According to statistics from the U.S. Geological Survey, a water flow of more than 800 cfs could be expected to occur, on average, approximately 2.4 times per year during the months of June, July, and August. In comments accompanying the acceptance of Fraser's construction progress schedule, in which Fraser described its plan for conducting the work, the Corps pointed out that the diversion system Fraser had selected "will be susceptible to damage by flow amounts which are anticipated to occur during the May to August time frame. Delays due to such flows are not justification for weather-related extension of the contract completion date."

Because of wet weather in the region, Silver Lake began to experience high water flows shortly after the project started. Before Fraser's diversion dike was completed, the flow rate reached 1,320 cfs, which damaged the dike and flooded the work site. Fraser spent much of the next two weeks repairing the dike, work that Fraser asserts was made more difficult than anticipated because high water flows continued through that period, rather than receding as Fraser had expected. Fraser requested a time extension based on weather conditions and high water flows. The Corps replied four days later, asking Fraser to provide further information regarding the request, such as what Fraser had anticipated about the site, what was different about the site, and how Fraser was damaged. That same day, the Corps sent Fraser a letter demanding that the company improve its progress and threatening to terminate the contract due to delays in the project. A copy of that letter was sent to Fraser's bonding company. Fraser sent the Corps a revised completion schedule in which it took the position that, up to that point, and assuming a 12-day extension, Fraser contended it was still on schedule to complete the work within the contract period as extended. The Corps denied the time extension request, but later granted a time extension. Fraser asserted a claim of constructive acceleration. Fraser contended that extensions granted after the fact were not sufficient to compensate the company for the extra expenses it had incurred in dealing with the high water flows, because upon being told that it would not receive time extensions for days of high flow or that those extensions would be dealt with at a later time, it was forced to continue operations at a substantial additional cost.

After completing the contract, Fraser submitted its claim of constructive acceleration to the contracting officer. Fraser contended the denial of reasonable extensions for high water flow conditions during the contract period entitled it to an additional $659,760 over the contract price. The contracting officer denied Fraser's claim. Fraser subsequently filed its action in the Court of Federal Claims. The trial court ruled that Fraser had not established the government's liability based on constructive acceleration. Fraser appealed the matter to the Federal Circuit.

Fraser argued that many of the time extensions that it was ultimately granted were of no use to it because the extensions were not granted on a timely basis. The trial court, however, found against Fraser on that issue. According to the court, the record "contains no indication that the Corps' granting of time extensions was tardy in any way or that the Corps deviated at all from the standard procedures used to grant time extensions." To the contrary, the court found that "Corps representatives participated in daily discussions with [Fraser] about the days to be granted," and that the days of each month for which the parties agreed to weather-related extensions were incorporated in contract modifications issued within a reasonable period after the conclusion of that month's work. As the trial court noted, and as the Court of Claims and the Boards of Contract Appeals have observed in other cases, it is not unusual for parties to negotiate after the fact as to the number of extra days that are justified under a contract and to incorporate the extensions in contract modifications issued weeks after the fact. *See Olin Mathieson Chem. Corp. v. United States*, 179 Ct. Cl. 368 (1967); *McNutt Constr. Co.*, 85-3 B.C.A. (CCH) ¶ 18,397, at 92,297 (Eng BCA 1985); *Fermont Div., Dynamics Corp. of Am.*, 75-1 B.C.A. (CCH) ¶ 11,139, at 52,997 (ASBCA 1975), *aff'd*, 578 F.2d 1389, 216 Ct. Cl. 448 (1978).

Any delay at all in determining whether a particular day will qualify for an extension under the contract will force the contractor to make a judgment as to whether to work in spite of the conditions or cease working in the expectation that an extension will later be granted. That dilemma is inherent in any case in which extensions are not granted on the spot, before work begins on a particular day. Although a trier of fact could have made a different finding on the issue of the timeliness of the extensions based on the evidence at trial, Fraser did not persuade the appellate court that the trial court's findings on that issue were clearly erroneous, and the matter was affirmed.

[D] Demand for Performance Without Extension

Page 6-34, add after carryover paragraph:

An owner's repeated emphasis on adherence to a contract schedule does not automatically give rise to a constructive notice to accelerate if time extensions have been granted. *Fraser Constr. Co. v. United States,* 57 Fed. Cl. 56, 2003 U.S. Claims LEXIS 166 (2003). The Army Corps of Engineers awarded a contract to Fraser Construction Company for flood control work. The contractor began work on the project on May 17, 1993, and the contract was scheduled to be completed by September 1, 1993. The contract's default clause absolved the contractor of responsibility for any delays in contract completion that arose "from unforeseeable causes beyond the control and without the fault or negligence of the Contractor." During the months of June, July, and August, however, the contractor experienced adverse weather conditions and high water flows that impeded its progress. In response to those conditions, the contractor repeatedly sought both time extensions and additional contract compensation. The Corps granted 33-day time extensions for unusually severe weather. With regard to the time extensions based on high water flows, however, the Corps denied all claims. The Corps additionally refused all requests for increased compensation. The contractor brought suit, seeking the additional days and acceleration costs.

The contractor contended that although it was permitted to cease operations at the disposal site on days for which it received time extensions, the Corps nonetheless specifically directed the contractor to continue operations at the excavation site, thereby forcing it to incur additional costs and essentially rendering the time extensions meaningless. The contractor pointed to various exhortations by the Corps that the contractor not fall behind its contract schedule. The court did not read the Corps' repeated emphasis on adherence to the contract schedule as a constructive notice to accelerate. The Corps specifically informed the contractor that there would be no acceleration unless it was formally ordered. Legitimate concerns by the Corps that the contractor adhere to its original schedule (or else be subject to default) was not an indication that the contractor's obligations had increased. The conditions inherent to the site were responsible for impeding the contractor's progress. With such a built-in hindrance to the timely accomplishment of the work, the Corps was correct to be concerned. The contractor had already received exactly what it sought: time extensions for the days in question. The contractor could only hope to extend the contract schedule to account for the time lost as a result of problems encountered at the site. The fact that the Corps granted time extensions on the basis of delays experienced at only one area was irrelevant: Having received an extension for a particular day, the contractor would have thereby exhausted the full extent of its relief under the contract. The court concluded that the contractor was not entitled to any additional compensation.

Page 6-34, add after note 115 in second full paragraph:

The fact that the granting of a time extension is delayed does not in itself constitute constructive acceleration. In *McKenzie Engineering Co.*, ASBCA No. 53374, 05-2 (CCH) B.C.A. ¶ 33,090 (2005), the contractor was awarded a fixed-price contract to perform concrete work on a spillway and powerhouse at Rock Arsenal in Illinois. The contract contained a winter exclusion period from November 15 through May 15 when work was not to be performed, and liquidated damages of $290 per day were set for late completion. Notice to proceed was given on October 28, 1996, and the contractor was not able to mobilize on May 15, 1997, because heavy rains caused high river water conditions. The contractor notified the government of the delay and requested a time extension. The government verbally agreed to grant a time extension but the modification granting the time extension was not issued until nine months later. The contractor submitted claims over $1 million for constructive acceleration, cumulative impact defective specifications, and differing site conditions, and argued that the nine-month delay in issuing the time extension constituted constructive acceleration because this in itself exposed the contractor to liquidated damages, forcing it to accelerate and work in the winter exclusion period. The Board did not find any threat by the government to assert liquidated damages, nor a directive to accelerate or work in the winter exclusion period, and held that the delayed issuance of a time extension does not meet the requirement for constructive acceleration.

§6.07 ACCELERATION COSTS

Page 6-46, add after first sentence of second full paragraph:

In the same way, contractors should record the overtime costs separately from regular time labor costs. Courts may not find it appropriate to use a percentage markup of the total labor cost to approximate overtime premiums (*J&K Plumbing and Heating Co., Inc. v. State of New York,* 235 A.D.2d 751, 652 N.Y.S.2d 369 (1997)) or the total cost method of proving overtime by comparing estimated labor costs with actual labor costs (*John F. Harkins Co. v. School Dist.,* 313 Pa. Super. 425, 460 A.2d 260 (1983)) because the actual overtime premium of the labor cost can be easily determined if proper records are kept.

CHAPTER 7

PROJECT DELIVERY METHODS AND DELAY CLAIMS

§ 7.02 DESIGN-BID-BUILD

Page 7-4, add at end of section:

Another variation of design-bid-build contracting is known as Job Order Contracting (JOC). JOC generally involves the use of a competitively bid, firm price contract that is awarded for a multiple-year period, based upon a comprehensive set of general specifications and a unit price catalog for the various types of work to be performed under the JOC procurement. In essence, JOC provides an umbrella contract for smaller construction, repair, and maintenance work. Typically, the types of work are repairs, maintenance items, or components of larger construction projects, like removing a wall, rebuilding interior partition walls, and relocating ductwork, rather than constructing an entirely new facility. The catalog of work items may contain as many as 30,000 line items, breaking down the price for the various work activities into labor rates, material and equipment costs, geographical factors, an overall pricing coefficient, a factor for work accomplished for work during other than normal work hours, and other adjustment factors. The bidder's pricing coefficient is to include indirect costs, design and engineering costs, general and administrative costs, insurance, wage rate changes, inflation, and profit. Individual work orders are issued to the contractor for specific tasks based upon a description of the task, unit of measurement, and a unit price.

JOC procurements are not merely minor contract awards, but a comprehensive procurement system that results in a master contract for many construction tasks and often spans a year or more. The JOC agreement is a "sole source" contract often designated as an indefinite quantity contract with aliases such as Delivery Order Construction, Simplified Acquisition of Base Engineer Requirements (SABER), Task Order Contracts, and Solution Order Concepts. The bounds of the contract amount may be set through a pre-established contract minimum and maximum dollar amount that can be awarded under the contract.

Typically, JOC contracting is used for facilities repair, maintenance, rehabilitation, modernization, tenant fit-out, and minor construction tasks by owners with large facilities or several locations, such as municipalities, school districts, colleges and universities, utility companies, public housing authorities, and large private owners. At the time of the JOC solicitation, the owner may not be able to give specific information concerning the types of projects or tasks anticipated for the JOC award, but may be able to provide information on its past JOC assignments or similar work. Precision in the standard or performance specifications and a fairly comprehensive list of work tasks will assist the bidders to understand the nature of the owner's intended procurement. The owner may also allow or encourage the bidders to visit its facilities to examine the conditions of the facilities, the types of work that may be required, and the working conditions.

The owner obtains bids and pricing from several bidders during the master JOC solicitation process. Each bidder is requested to provide a work task catalog with pricing for each work element. The bidders are to provide information on its past performance and qualifications. In many JOC procurements, the proposals are evaluated on several qualifications standards: management ability; subcontracting support; experience and capability of the technical staff; and fiscal management plan. The bid evaluation may be weighted between qualification factors and price considerations, often with price weighted less. On other JOC procurements, the bidders must first meet a qualifications threshold and then the pricing aspects become determinative for an award.

Once a JOC contract is awarded, task orders may be given to the JOC contractor based upon pricing in the work task catalog. The process for an individual task order may begin with a meeting where the contractor and owner's representative will review the scope of the individual task order. The scoping process may include an on-site inspection, a proposed written scope of work, the contractor's comments on the proposed scope, and the contractor's pricing proposal using the unit price data in the work task catalog. The details and design elements given to the contractor may vary depending upon the nature of the task, the difficulty of the work, and the owner's access to engineering and design resources. Some owners provide detailed designs, whereas other owners provide only programmatic, schematic, or general design criteria. The contractor may be required to submit a schedule and certain technical data with its proposal. The contractor's pricing may be compared with an estimate prepared by the owner, and reviewed for conformance with the scope of work elements and the pricing aspects of the work task catalog. Once the owner is satisfied with the contractor's proposal, a firm fixed price or lump-sum work order is given to the contractor as well as a notice to proceed for the individual task order.

Many owners find that there are several advantages of JOC. JOC allows the owner to effectively manage numerous minor construction and repair projects through a central control procedure, minimizing the cost and time for administering numerous bids, awards, and approvals. Owners find that JOC contractors have an ongoing incentive to perform quality and timely work even though the assignments are small because of the JOC contractor's belief that there may be many more tasks ahead that will make the overall JOC contract profitable. If the owner believes that quality work is not being performed, it does not have to award other task orders. Owners may stop task order awards after the minimum contract amount has been reached or may terminate the JOC contract for lack of performance. Some JOC contracts have partnering provisions to assist in quality and other concerns. Many owners find that using JOC contracting they are able to expeditiously complete small tasks, quickly mobilizing the designated JOC contractor. Because design specifications are not required for each new project, the leadtime for individual projects is also reduced. JOC contracts are reportedly producing responsive and dependable long-term relationships between owners and JOC contractors. Other owners indicate that they are able to maintain cost control and management, and overall the work is performed on a lower cost basis. Delay claims may be minimized due to similar reasons.

For a JOC procurement to be successful, there must be a measure of mutual trust. As one owner noted, the JOC process "is somewhat susceptible to being abused." This owner claimed that the JOC contractor fraudulently inflated the contract task order costs by misrepresenting the quantity, quality, and types of material used on over 200 projects. While such abuses may exist, overall the JOC process is found by many owners to be an acceptable method of procurement. (*Denying Guilt, Firm Pays Out Settlement to End Fraud Case*, Engineering News-Rec., 14 (Feb. 18, 2002); "General Services Agency Purchasing, Job Order Contracting," *www.ventura.org/gnd_jury*; "Job Order Contracting Overview," *www.ic.usa.edu/IC-Overview*).

§7.03 THE FAST TRACK

[A] The Potential for Delay on Fast-Track Projects

Page 7-5, add to note 5:

See also Wisch & Vaughan Constr. Co. v. Melrose Properties Corp., 21 S.W.3d 36 (Mo. Ct. App. 2000) ("fast tracked" meant that construction would begin prior to the completion of final architectural plans and specifications. On a "fast-tracked" project there was ample opportunity, if not almost a certainty, that disputes will arise between the general nature of the initial documents and the final architectural plans and specifications).

§7.07 TURNKEY CONSTRUCTION

Page 7-19, add to note 25:

A "turnkey" power plant is generally known in the construction industry as an arrangement whereby a contractor agrees to engineer, design, construct, commission, and start up the facility. Input from the owner is limited to establishing price, performance specifications, site and local conditions, and completion date. In contrast, an "owner-satisfaction" project assumes regular input from the owner and potential modification of design specifications. Fluor Daniel Intercontinental, Inc. v. General Elec. Co., 1999 U.S. Dist. LEXIS 12983 (S.D.N.Y. Aug. 20, 1999).

CHAPTER 8

RECOGNIZING AND DEALING WITH DELAYS AND OTHER DISRUPTIONS

§ 8.06 NOTICE

Page 8-13, add to note 18:

In Clifford R. Gray Inc. v. City Sch. Dist., 277 A.D.2d 843, 716 N.Y.S.2d 795 (2000), the contractor was not entitled to damages incurred before August 1995, the month that the contractor forwarded correspondence specifically complaining of delays. The plaintiff was, however, awarded damages totaling $175,583 encompassing the period from September 1995 to July 1996, for extended site overhead, labor rate increases, loss of labor productivity, expenses for small tools, and consumables and lost profit. It was denied reimbursement for extended home office overhead and warehouse worker and truck expenses. The owner argued that the contractor's request for damages should have been rejected in its entirety because the August 1995 correspondence was not sufficiently particularized to comply with the notice requirement. The contract required only the submission of "a claim in writing" and provided that one claim would be sufficient in the event of a continuing delay. Having previously notified the owner of its intent to file a claim for pecuniary losses due to the project delays, the contractor detailed the continuing delays in an August 2, 1995, letter.

Chapter 9
PROCESSING CLAIMS

§ 9.02 SUBMISSION OF THE CLAIM

[A] Subcontractor Claims

Page 9-5, add at end of first paragraph:

Under New York law, the general contractor is not responsible to a subcontractor for the delays that are the responsibility of the owner. For example, in *In re Regional Building Systems, Inc.*, 320 F.3d 482 (4th Cir. 2003), RBS (the general contractor) entered into a contract with Aspen Knolls (the owner) for the manufacture, delivery, and installation of 1,000 housing units in Staten Island, NY. RBS then entered into a subcontract with Bedford Construction (the subcontractor) to transport, erect, and complete the structures. The owner encountered financial difficulties and defaulted on payments to the general contractor. The general contractor in turn experienced severe cash flow problems, and was unable to deliver the manufactured housing units to the subcontractor to transport and erect. The subcontractor incurred the cost of an idled work force and labor. Later, the owner ceased paying the general contractor altogether, and the general contractor terminated its contract with the owner and its subcontract with Bedford. As the general contractor filed for relief under Chapter 11 of the Bankruptcy Code, the subcontractor submitted a claim which included delay damages. The bankruptcy court denied recovery for the subcontractor's delay damages, holding that New York law prevents a subcontractor from recovering delay damages from a general contractor when the owner, and not the general contractor, is responsible for the delay. Because the owner breached its payment obligation to the general contractor, the general contractor reasonably suspended its work under the general contract. Further, the general contractor terminated its contract with the owner when the owner further failed to make payments. Thus, the bankruptcy court found that the owner was wholly responsible for the delays in the project that also resulted in delay to the subcontractor. The general contractor was not guilty of any wrongdoing, nor responsible for the delays to the subcontractor, not liable to the subcontractor for delay damages.

Page 9-6, add at end of first full paragraph:

Under the law of certain states, a prime contractor may not be able a sponsor a subcontractor's claim merely on a "pass-through" basis. For example, in *Aetna*

Bridge Co. v. Rhode Island Department of Transportation, 795 A.2d 517 (R.I. 2002), Aetna Bridge was awarded a contract by the Rhode Island Department of Transportation for the reconstruction of a bridge. Aetna awarded the subcontract of the fabrication of the structural steel to Foster. In preparing the shop drawings for the structural steel, Foster alleged that deficiencies in the Department's design made the preparation of the shop drawings more difficult and costly, and filed a claim against Aetna. Aetna, later filed a claim in arbitration against the Department on behalf of Foster, noting at one point that the claim was a pass-through of Foster's claim to Aetna. The terms of the pass-through of the Foster claim to the Department were allegedly addressed in a liquidation agreement between Aetna and Foster, but because discovery of documents was not available under Rhode Island's Public Works Arbitration Act, the Department did not receive a copy of the liquidation agreement. After four days of arbitration, an award was entered against the Department, but there was no explanation of the findings of fact or the law. The Department moved to set aside the arbitration award on the basis that since there was no proof that Aetna was liable to Foster, a mere pass-through of Foster's claim to the Department was not an arbitrable issue under the Act. When the matter was considered on appeal to the Supreme Court of Rhode Island, the Court noted that because discovery was not allowed under the Act, and because the arbitration award did not provide the basis of the award, it was difficult to determine whether Aetna's claim against the Department was merely a pass-through of Foster's claim. However, if Aetna's claim was a pass-through, there would be no jurisdiction under the Act, as there would be no "concomitant liability to Aetna." The matter was remanded to a lower court to determine the basis of Aetna's claim against the Department.

Page 9-7, add after note 30 in first full paragraph:

In submitting claims to general contractors, subcontractors should act in accordance with the claim procedures spelled out in the subcontract. For example, in *Ferguson Electric Co. v. Kendal at Ithaca, Inc.*, 274 A.D.2d 890, 711 N.Y.S.2d 246 (2000), Ferguson was awarded a subcontract for the mechanical work at a nursing home by the general contractor. The dispute resolution provision required any claim to be submitted to the general contractor's designated representative for decision. Although that decision would be conclusive, final, and binding, it could be reviewed in court to determine whether it was "arbitrary, capricious or grossly erroneous to evidence bad faith." When the subcontractor encountered a payment dispute with the general contractor, it submitted a written demand but not a request for a final decision. The general contractor refused to make payment, and the subcontractor sued. In considering the matter, the New York Supreme Court, Appellate Division, noted that under New York law the parties must comply with a clear and unequivocal alternative dispute resolution agreement before taking the matter to court. The court determined that the contract clause was more than just an optional step in processing claims; it was an enforceable alternative dispute resolution (ADR) provision. Although the general contractor never appointed a representative to render a decision on the dispute, it was not necessary to do so until the subcontractor submitted its claim in accordance with the ADR clause.

§9.05 RELEASES

Page 9-14, add after third full paragraph:

A contractor's release of its claims against the owner did not release the contractor's claims against the designer. For example, in *Crown Corr, Inc. v. Wil-Freds Construction, Inc.*, 2000 U.S. Dist. LEXIS 17961 (N.D. Ill. Dec. 11, 2000), Bertrand Goldberg Associates, Inc., was hired as the architect and construction coordinator of the Wright Junior College Project by the City of Chicago, acting through its Public Building Commission. Wil-Freds was hired to construct the floors of the five buildings. The project was delayed. Wil-Freds claimed loss of productivity, wage and benefit increases, and extended general conditions. Wil-Freds and the city compromised on Wil-Freds's claims. The settlement agreement included a table that allocated dollar amounts of the settlement into various categories. Goldberg claimed that the "release" provision of the Wil-Freds/City settlement agreement precluded Wil-Freds from recovering any damages from Goldberg.

The "release" provided that "the Commission agrees to pay to Wil-Freds the sum of $720,000 as full and complete settlement of all amounts claimed against the Commission." The settlement agreement also provided that "specifically excluded from the forgoing release and waiver are any claims, demands or actions that Wil-Freds has or may have against any other entity not a party to this Settlement Agreement." Thus, the release in the Wil-Freds/City settlement agreement not only stated that it was solely releasing the city, it also made clear that it did not affect claims against any other possible defendants.

However, the designer was entitled to offset the contractor's claims by the amounts allocated by the owner in its settlement with the contractor. Goldberg argued that the allocation was binding on Wil-Freds. Wil-Freds asserted that the court could not use the allocation table for purposes of assessing whether Wil-Freds had already recovered money for its current claims against Goldberg. Wil-Freds asserted that it had received $720,000 under the settlement agreement, to allocate any way it wanted. Wil-Freds's position conflicted with the parol evidence rule. The settlement agreement provided that "the Commission agrees to pay to Wil-Freds the sum of $720,000 . . . Exhibit 'B' provides an allocation of the settlement amount." By signing the writing, Wil-Freds unambiguously accepted the allocations set forth in Exhibit B of the settlement agreement. Therefore, Wil-Freds was precluded from admitting evidence outside the four corners of the settlement agreement to contradict or explain its terms, and Wil-Freds was bound by the terms outlined in the agreement. The allocation table of the settlement agreement provided that $69,707 of the $720,000 went to the wage and benefit increases claim and $188,500 of the $720,000 went to the extended general conditions claim. Any recovery Wil-Freds would receive from Goldberg for wage rate increases was offset by $69,707. Any recovery Wil-Freds would receive from Goldberg for extended general damages must be offset by $188,500.

Page 9-15, add after note 48 in first paragraph:

For example, in *Eagle Asphalt & Oil, Inc.,* IBCA No. 4173-1999, 01-1 B.C.A. (CCH) ¶ 31,234 (2000), the contractor submitted two claims for suspension of the work against the federal government, but neither claim was properly certified nor did the submissions request a final decision as required by federal contracting claim procedures. The contracting officer indicated that the contractor should resubmit the claims in a proper manner, but the contractor failed to do so. When the contract work was completed, the contractor signed a general release, noting that it "reserved the right to submit a claim on the above-referenced contract" without providing any further specifics. The contractor later resubmitted its suspension of work claims and the matter came before the Board of Contract Appeals to determine whether the release precluded the contractor from proceeding on the claims. The Board found that the broad language in the release exception did not put the government on notice as to the specific nature and amount of the claims. The original submissions for suspension did not constitute "claims" under federal contract law, and the exception in the release was vague—a "naked intention to file an indeterminate future claim in an undetermined amount." The contractor was not allowed to proceed on the suspension of work claim.

Page 9-15, add after note 52 in second full paragraph:

General contractors should be careful to fully settle or include reservations of rights for subcontractor delay claims in the releases executed pursuant to change orders with the owner. For example, in *Trataros Construction, Inc. v. General Services Administration,* GSBCA No. 15344, 203-1 B.C.A. (CCH) ¶ 32,251 (2003), Trataros entered into a contract with GSA for the renovation of a U.S. Post Office and Courthouse. Trataros subcontracted with SPS to perform drilling and replacing mini-piles. The contract drawings indicated that certain obstructions would be encountered when drilling the mini-piles, but SPS allegedly encountered additional obstructions not noted on the drawings, and characterized these matters as differing site conditions. As a result of these differing site conditions, SPS experienced delays, unanticipated demobilizations and re-mobilizations, and disruption in the performance of the subcontract work, and filed a claim with Trataros requesting additional time and an increase in the subcontract amount. Trataros referred SPS's claim and other disputes to GSA, and requested payment for the SPS claim along with markups for profit and bond costs, but did not request an extension of time to the general contract. GSA and Trataros met to negotiate the SPS claim and other issues such as liquidated damages, and GSA indicated its willingness to pay for the actual added obstructions encountered during the installation of the piles. During the negotiations of the various claims the parties agreed to "global modification" rather than specific amount for each issue. The parties agreed to a contract modification adding $200,000 to the contract price and the waiver of liquidated damages in the amount of $1.4 million, and Trataros signed a release giving up all of its claims and its subcontractor claims, whether known or unknown, relating to delay, impact, and any other claims for time extension or costs through December 31, 1999. Later Trataros tried to negotiate the resolution of the specifics of the SPS claim, including the delay impact of the SPS differing site conditions, but GSA maintained that these issues were settled by the terms of the global modification, particularly the release language. Trataros, on the other hand, argued that the global modification did not specifically address the terms of SPS' claim, and because the SPS claim was not specifically negotiated during the discussions on the global settlement, there was no meeting of the minds. GSA's denial of the SPS claim was appealed by Trataros to the General Services Board of Contract Appeals. The Board considered each of Trataros' arguments, but found it implausible that Trataros could not have understood the significance of the global modification or would not have reserved its rights to pursue the SPS claim when executing the global modification and release language.

Page 9-17, add to note 54 in second full paragraph:

In *Larson Construction Services, Inc.*, ASBCA No. 53443, 02-1 B.C.A. (CCH) ¶31,730 (2002), Larson was awarded a contract by the U.S. Army Corps of Engineers for repairs of a dam. During the course of the work, Larson discovered an under-drain pipe that would interfere with the driving of sheet piling in the area. The resolution was to modify the design, which delayed Larson's work in the area. Larson and the government executed a bilateral agreement allowing an increase in the contract price and an extension in the contract time. The modification contained a general release to the effect that the government was released from all liability for "further equitable adjustments attributable to such facts or circumstances giving rise to the proposal for adjustment." Subsequently Larson submitted a claim for its alleged loss of productivity and added labor costs because Larson was required to drive the sheet piling during periods of unusually severe weather. Larson alleged that this claim was not precluded by the general release language in the modification because these costs were not incurred at the time the modification was signed. The claim was denied by the contracting officer and Larson appealed to the Board of Contract Appeals. The Board found against Larson, noting that Larson should have expressly reserved its rights to claim the loss of labor productivity when negotiating and executing the modification and release.

Page 9-17, add after note 55 in second carryover paragraph:

A contractor that does not explicitly or in a timely fashion reserve its right to assert an additional delay or impact claim may lose its right to assert the claim. In *Jackson Constr. Co. v. United States*, 2004 U.S. Claims LEXIS 245 (2004), Jackson Construction Company, Inc. constructed a building for the Department of the Army New York District Corps of Engineers. The contractor completed construction on the completion date mandated by the contract. Changes to the work were authorized by bilateral modifications. Although the contractor completed the work on time and was paid for all the changes required by the Corps, the contractor brought suit to recover additional compensation in the form of delay and impact damages. The contractor contended that it had to overcome a 120-day delay at the start of the project, during which it redesigned and relocated a waterline that ran through the footprint of the new building. The contractor also claimed additional damages were incurred due to the cumulative impact of the numerous changes required by the Corps during the course of performance. In its lawsuit, the contractor sought damages for an early completion delay claim, including additional home office overhead incurred during the delay period, and damages for the cumulative impact claim.

The Corps argued that, to the extent the project was delayed by the waterline relocation or by other problems, the release language found in all of the bilateral modifications barred the contractor from obtaining additional damages. While the contractor conceded that it executed bilateral modifications that covered the very same claims that it presented at trial, the contractor argued that the effect of the release language did not prevent it from bringing its damage claims. Arguing that it made a written reservation of rights to assert a claim at a later date, the contractor relied upon language that it inserted underneath the Corps' release in two of the modifications, stating that the contractor believed the time period was extended due to delays. Even if this notation had been sufficient to constitute a valid reservation of rights, the scope of the language utilized by the contractor only addressed the delay claim. The contractor's attempt to reserve its rights on either claim, however, was not valid because it was not made in a timely manner. There was no evidence that the contractor ever notified the Corps of its intent to assert a delay or impact claim during the course of performance, let alone evidence that the contractor reserved its rights to assert future claims when it executed bilateral modifications. The contractor could not have relied on the two reservations of rights that it made after-the-fact that were unrelated to the claims presented at trial. The contractor did not even attempt to assert an additional delay or impact claim until after the contract had already been completed. The contractor did not explicitly or in a timely fashion reserve its right to assert an additional delay or impact claim. The complaint was dismissed.

Page 9-17, add to note 56:

Centex Bateson, VABCA Nos. 4613, 5162, 5165, 99-1 B.C.A. (CCH) ¶ 30,153 (1999) (to succeed on a cumulative impact claim the claimant to provide a reasonable explanation of how the number of changes or other incidents caused the cost overruns. To demonstrate causation, a claimant might be required to show that there are no other reasons for the loss of productivity or cost overrun. This may amount to a requirement that the claimant prove a negative—that there were no other causes of the cost overrun. Keating & Burke, *Cumulative Impact Claims: Can They Still Succeed?,* 30 Constr. Law. (Apr. 2000)).

Page 9-18, add to note 57:

Uhle v. Tarlton Corp., 938 S.W.2d 594 (Mo. Ct. App. 1997).

Page 9-20, add after second full paragraph:

Release of lien in payment application does not bar a delay and acceleration claim. Consider *Battaglia Elec., Inc. v. Bancroft Construction Co.,* 2003 Del. Super. LEXIS 100 (2003). Bancroft Construction Co. subcontracted with Battaglia Electric, Inc. on a research center construction project. Both the contractor's steel and concrete subcontractors experienced delays in their performance, which the subcontractor averred significantly delayed the entire project schedule. Because of the impediments, the contractor then unilaterally decided to accelerate the project schedule. The subcontractor requested that it either be allowed some adjustment in its schedule or a provision be made to help it cover the costs of accelerating the schedule. The contractor responded that until the subcontractor demonstrated a meaningful effort to adhere to the construction schedule, no additional time or compensation would be given. The subcontractor executed two waiver and release forms. The first form, executed on April 26, 2000, was related to work the subcontractor performed before April 30, 2000. The second form, executed on May 24, 2000, was related to work the subcontractor performed before the date of execution of that release. Both forms were captioned "Waiver and Release of Lien Rights," and both contained language that upon receipt of payment as satisfaction in full for all labor, services, and materials furnished to the subcontractor, the document was effective to release pro tanto any mechanics' liens, stop notices, or bond rights the subcontractor had in connection with the project. The only apparent difference between the two forms was the typewritten assertion inserted by the subcontractor into the May 24, 2000 document that the release excluded all claims for delays, disruptions and inefficiency. The subcontractor then filed a complaint, alleging that it had incurred a significant increase in the cost to perform its work, and that it was entitled to damages. The contractor moved for summary judgment.

The contractor argued that the April 26, 2000 release covered all claims arising from work through April 30, 2000, and that the intent of the parties to include all claims was supported by the subcontractor's addition of language in the May 24, 2000 release. The contractor argued that the additional language would not have been necessary if the April 26, 2000 release did not include such claims for delays and disruptions. The subcontractor argued that the releases it executed applied only to claims for mechanics' liens. Each release executed by the subcontractor in the contractor's and the owner's favor was captioned "Waiver and Release of Lien Rights" and released any mechanics' liens, stop notices, or bond rights the subcontractor may have had in connection with the project. Thus, the scope of the subcontractor's April 26, 2000 release related solely to claims to secure payment for debt-type actions. The subcontractor's complaint was for breach of contract and not for any claim to a mechanics' lien it may have asserted. The contractor's argument that insertion of the additional language into the May 24, 2000 release showed that the subcontractor intended the April 26, 2000 release to preclude delay claims was not determinative. Even before execution of the May 24 release, the subcontractor was in the process of preparing a claim for the already-existing delays. The contractor was not entitled to summary judgment.

Page 9-21, add to end of section:

Summary judgment is inappropriate for an ambiguous change order referencing both a review and an audit. In *Kiewit Constr. Co. v. Capital Elec. Constr. Co.*, 2005 U.S. Dist. LEXIS 23621 (D. Neb. Oct. 12, 2005), the Omaha Metropolitan Entertainment & Convention Authority (owner) retained Kiewit Construction (construction manager) on a convention center and arena project. The construction manager agreed that it would substantially complete the work by a specified date. The construction manager negotiated contracts with its subcontractors, including Capital Electric (subcontractor). The construction manager directed the subcontractor to accelerate its work so that the project could be completed on time. The subcontractor complied, but the work was not completed on time. The subcontractor executed a change order because it had not been paid for what it billed to the construction manager. The construction manager brought suit against the subcontractor, who counterclaimed against the construction manager. The parties filed cross-motions for summary judgment.

The subcontractor sought additional compensation under the change order for the work it performed after the acceleration directive was given. The subcontractor characterized the change order as an agreement by the construction manager to make an interim partial payment that was due to the subcontractor as a result of the acceleration directive. The subcontractor asserted that the change order did not resolve all outstanding claims between the parties. The subcontractor further argued that the change order was not an accord and satisfaction. The construction manager argued that the subcontractor's claim was barred because the change order was an accord and satisfaction, as stated on the preprinted change order form. The construction manager argued that, even if the change order was not an accord and satisfaction, as a part of the contract documents the change order plainly and unambiguously bound the subcontractor to the results of the construction manager's audit of the acceleration costs. According to the construction manager, the audit determined that the construction manager overpaid the subcontractor. The change order was not intended to operate as a full discharge of amounts due to the subcontractor. Despite the preprinted form language, an attachment to the change order referenced interim payments. There was a genuine issue of material fact as to whether the change order was an accord and satisfaction. Although the parties agreed to be bound by the results of some type of future accounting, the change order was ambiguous, since it referenced both a review and an audit. The parties disputed whether an audit or a review of costs was conducted. Because there were genuine issues of material fact remaining on the subcontractor's claim, summary judgment for either party was not appropriate.

§9.06 PROCESSING THE DISPUTED CLAIM

[A] Federal Contract Claims Procedures

Page 9-21, add before first paragraph of subsection:

A contractor must show two things for the Court of Federal Claims to have jurisdiction over its claim: first, that it submitted a claim to the contracting officer (41 U.S.C. §605(a); *James M. Ellett Constr. Co., Inc. v. United States,* 93 F.3d 1537, 1541–42 (Fed. Cir. 1996)), and second, that the contracting officer rendered a final decision on the claim. *Bath Iron Works Corp. v. United States,* 20 F.3d 1567, 1578–79 (Fed. Cir. 1994); *Clearwater Constructors, Inc. v. United States,* 56 Fed. Cl. 303, 311 (2003). Because the Contract Disputes Act leaves the term "claim" undefined, the court looks to the FAR, the CDA's implementing regulations, to determine what is necessary for a contractor's submission to be deemed a claim. *Reflectone, Inc. v. Dalton,* 60 F.3d 1572, 1575 (Fed. Cir. 1995). The FAR defines a "claim" as:

> a written demand or written assertion by one of the contracting parties seeking, as a matter of right, the payment of money in a sum certain, the adjustment or interpretation of contract terms, or other relief arising from or relating to the contract. 48 C.F.R. §52.233-1(c) (2003).

Thus, a non-routine submission, seeking money damages, must meet three requirements to be a "valid" claim. It must be (1) a written demand or assertion, (2) seeking, as a matter of right, and (3) the payment of money in a sum certain. While a claim need not be "submitted in any particular form or use any particular wording," the contractor must provide "a clear and unequivocal statement that gives the contracting officer adequate notice of the basis and amount of the claim." Furthermore, the claim must request a final decision. *Contract Cleaning Maintenance, Inc. v. United States,* 811 F.2d 586, 592 (Fed. Cir. 1987). All that is necessary to satisfy this requirement is that contractor's submissions show an expression of interest in a final decision by the contracting officer (CO), which may be made expressly or implicitly. *Transamerica Ins. Corp. v. United States,* 973 F.2d 1572, 1577 (Fed. Cir. 1992). To determine whether the contractor has satisfied these requirements, the court uses a "common sense analysis" and views the contractor's submissions as a whole. *Kirkham Constructors, Inc. v. United States,* 30 Fed. Cl. 90, 93 (1993) (citing *Transamerica Ins. Corp., Inc.,* 973 F.2d at 1579).

To be adequate, the writings allegedly constituting a delay claim must sufficiently inform the CO of what is being claimed so that the CO can meaningfully review the claim. *CPS Mechanical Contractors, Inc. v. United States,* 59 Fed. Cl. 760, 2004 U.S. Claims LEXIS 50 (2004). The United States Army entered into a contract with CPS Mechanical Contractors, Inc. (CPS) for the renovation of sewer system components at Fort Jackson in South Carolina. According to the contract's statement of work provision, plaintiff's work included "installing 991 l.f. of

10-inch pvc gravity sewer, 1149 l.f. of 8-inch gravity sewer, 11 manholes, tieing [sic] to existing sewer system and installing 7 chemical injection units." A dispute arose when the Army directed CPS to install 988 l.f. of six-inch laterals. While CPS performed this work during the contract period, it nonetheless protested that the work was beyond that set out in the statement of work. CPS sent a series of letters to the CO expressing its view that the six-inch lateral installation work was beyond the scope of work under the contract and thus constituted additional work, which entitled CPS to an equitable adjustment under the contract. According to the CO, the statement of work is merely a general statement that provided an overall description of the work to be done. While certain work may not be included in a contract's statement of work, if it was included in the drawings, it was a part of the contract, and the contractor was responsible for completing such work. CPS filed a complaint seeking an equitable adjustment in the amount of $51,989.67 plus interest for the alleged additional work it performed in installing the six-inch laterals. In response, the Army filed a motion to dismiss for lack of subject matter jurisdiction, arguing that CPS failed to satisfy the requirements of the CDA by failing to submit to the CO a written demand seeking payment and further that the CO had not issued a final decision on the issue.

CPS argued that it informed the CO numerous times, in writing, that the installation of the six-inch pipes was extra-contractual work that required additional compensation, that the CO repeatedly stated that the installation was part of the work required by the contract, and that she would not authorize additional compensation for this work. Although CPS's letters gave the contracting officer adequate notice of the basis of its claim, CPS failed to seek a sum certain for the alleged additional work. CPS failed to present any evidence which showed that it submitted any type of written communication to the CO that mentioned a specific amount. CPS also failed to show that the amount it sought could have been easily determined by the CO through a simple mathematical calculation or from its submissions. If no sum certain was specified, the CO could not settle the claim by awarding a specific amount of money that precluded the contractor from filing suit seeking the difference between the amount awarded and some larger amount never specifically articulated to the contracting officer. Because plaintiff's submissions failed to satisfy the certain sum of money requirement, it was unnecessary to determine whether its submissions sought a final decision by the CO or whether the CO issued a final decision. The court granted the Army's motion to dismiss without prejudice.

Page 9-21, add after note 77 in first full paragraph:

The written submission must be a clear and unequivocal statement that gives the contracting officer adequate notice of the basis and amount of the claim. *Contract Cleaning and Maintenance v. United States,* 811 F.2d 586 (Fed. Cir. 1987). The written statement of the claim should be sufficient to inform the contracting officer of what is being claimed, providing enough information to allow a meaningful review. This may include factual allegations, the reasons that the claimant thinks the government is liable, supporting documentation, contract clauses, and references to specifications and drawings. Enough detail should be provided to allow the parties to begin negotiations. *Lee Ann Wyskiver,* PSBCA Nos. 3621, 94-3 B.C.A. (CCH) ¶ 27,118 (1994).

In some instances, the claim may not have to be submitted directly to the contracting officer, if the contractor gives the claim to its primary government contact, requesting a final decision with a reasonable expectation that the claim will be honored, and the claim is actually delivered to the contracting officer. *D.L. Braugher Co., Inc. v. West,* 127 F.3d 1467 (1997); or if the submission of the claim is reasonably calculated to ensure receipt by the contracting officer. *Roy McGinness & Co., Inc.,* ASBCA No. 40004, 91-1 B.C.A. (CCH) ¶ 23,395 (1991).

Page 9-21, add after note 79 in carryover paragraph:

In consideration of the $100,000 limit for the use of the accelerated appeals procedure, a contractor need not combine claims that are factually and legally distinct. For example, in *Linda Newman Constr. Co., Inc.,* VABCA Nos. 6307, 00-2 B.C.A. (CCH) ¶ 31,027 (2000), the claimant sought accelerated processing of its $42,533 claim for the impact of multiple change orders on unchanged work, and the normal processing of its $100,000 claim for late notice to proceed. The government unsuccessfully sought to remove both claims from the accelerated docket and consolidate the claims on the regular docket. The Board found the claims were factually and legally distinct, and further found that the contractor's actions were not an abuse of the accelerated appeals procedure. However, the Board partially consolidated the two cases for the purposes of conducting a hearing, but allowed them to proceed under the accelerated procedure.

Page 9-24, add after note 96 in second full paragraph:

Contractors including subcontractor claims as part of their own claims that have been properly certified must be sure that the subcontractor claim arises under the same set of operative facts. Otherwise the subcontractor claim may be determined to be a different claim that must be independently certified, and the court may conclude that it does not have jurisdiction to hear the subcontractor's claim. For example, in *AAB Joint Venture v. United States,* 2005 U.S. Claims LEXIS 321 (Nov. 1, 2005), AAB Joint Venture (contractor) contracted with the U.S. Army Engineer Corps–Europe to design and construct a storage and logistics base for use by the Israeli Defense Force. The contractor subcontracted for performance of the earthwork operations. The contractor performed exploratory borings, which revealed different subsurface conditions than indicated in the government's geotechnical report. The contractor submitted certified claims to the contracting officer for an equitable adjustment to cover added costs incurred by the contractor as a result of the differing site conditions. The contracting officer failed to issue final decisions within 60 days of receipt of the certified claims. Seeking additional compensation for itself and its subcontractors, the contractor brought suit against the government. The contractors claim requested compensation on behalf of the subcontractors for increased costs caused by delays resulting from the removal of unsuitable subsurface material. The government filed a motion to dismiss for lack of subject matter jurisdiction concerning the claim for one of the subcontractors, as that claim was never submitted to the contracting officer.

The government asserted that the contractor's claim on behalf of the subcontractor should have been dismissed because the claim was not certified and submitted in writing to the contracting officer for final decision as required by the Contract Disputes Act (statute) for claims exceeding $100,000. The contractor asserted that the subcontractor's claim arose from the same set of operative facts as its own certified claim to the contracting officer and thus did not constitute a new claim that had to be separately certified. Although the contractor maintained that the subcontractor's claim was merely a request for an increase in amount of the contractor's properly certified claim, the two claims did not arise from the same set of operative facts. The functions of the contractor and subcontractor were different, as well as the impact of the alleged differing site conditions on them. The claims were based on different facts. The contractor's claim was to recover the lost value of the excavated material, while the subcontractor's claim was to recover for delays in the earthwork operations. Although the subsurface conditions were the root of both claims, the claims involved proof of a distinct set of operative facts. Furthermore, there was no mention of the delay in the subcontractor's operations in the contractor's original certified claim, which would have alerted the contracting officer that the subcontractor's delay was part of the contractor's claim. By first raising the subcontractor's claim with the court, the contractor circumvented the contracting officer's role of rendering judgment on the delay costs incurred by the subcontractor. The contractor's failure to submit a certified claim on behalf of the subcontractor precluded jurisdiction under the statute.

Page 9-26, add after first sentence in first full paragraph:

The Contract Disputes Act itself does not give the definition of what is a claim; rather, *claim* is defined in the Federal Acquisition Regulations at § 33.201:

> Claim, as used in this subpart, means a written demand or written assertion . . . seeking as a matter of right, the payment of money in a sum certain, the adjustment or interpretation of contract terms. . . . [A] written demand or written assertion by the contractor seeking payment of money exceeding $100,000 is not a claim under the Contract Disputes Act of 1978 until certified by the Act and 33.207. A voucher, invoice, or other routine request for payment that is not in dispute when submitted is not a claim. The submission may be converted into a claim, by written notice to the contracting officer as provided in 33.206(a) if it is disputed either as to liability or amount or is not acted upon in a reasonable time.

There is a four-prong test to determine if a submission is truly a claim: (1) a written submission to the contracting officer claiming specific rights and requesting specific relief; (2) the amount requested is a sum certain; (3) a request is made to the contracting officer for a final decision; and (4) a claim over $100,000 must be certified. *Cubic Corp. v. United States,* 20 Cl. Ct. 610 (1990); *RSH Constructors, Inc. v. United States,* 14 Cl. Ct. 655 (1988).

Page 9-26, add after note 111 in first full paragraph:

An invoice certified in accordance with the requirements of the Contract Dispute Act and submitted after termination for default may be a claim, particularly when the government and the contractor dispute the percentage of completion. It may be an indication of the contractor's attempt to enforce its rights under the Contract Dispute Act. *American Service Supply, Inc.,* ASBCA No. 50606, 00-1 B.C.A. (CCH) ¶30,858 (2000).

Page 9-26, add after note 113 in first full paragraph:

The pricing of a claim based upon estimates may be adequate. *P.J. Dick Inc. v. General Serv. Admin.,* GSBCA Nos. 11783, 94-3 B.C.A. (CCH) ¶27,172 (1994); *Production Corp.,* DOTBCA Nos. 2424, 92-2 B.C.A. (CCH) ¶24,796 (1992); but a claim for an amount "in excess of" has been held to be inadequate. *Corbett Technology Co, Inc.,* ASBCA Nos. 47742, 95-1 B.C.A. (CCH) ¶27,587 (1995).

Page 9-27, add after note 118 in first full paragraph:

If the contractor files the appeal to the board of contract appeals prior to the date that the contracting officer designates for the decision, the appeal may be dismissed without prejudice, allowing the contractor to re-file the appeal upon receipt of the final decision. For example, in *Logicvision, Inc. v. United States*, 54 Fed. Cl. 549 (2002), Logicvision submitted a certified claim to the contracting officer alleging approximately $260,000 in delay damages. The contracting officer had 60 days to make a final decision or notify the claimant when the final decision would be issued. Fifty-three days after receiving Logicvision's claim the contracting officer wrote a letter to Logicvision stating that the decision would be issued on a certain date after the 60-day period. However, Logicvision did not receive the contracting officer's letter until 62 days after the certified claim was submitted. Logicvision nevertheless appealed its claim to the Court of Federal Claims. Logicvision maintained that because it had not received the letter from the contracting officer until after the 60-day period, it had not been "notified" within the 60-day period and that this should be deemed a denial of its claim. However, the court determined that because the contracting officer issued its letter (notifying the claimant when the final decision would be made) within the 60-day period, this complied with the notice requirement in the Contract Disputes Act. Logicvision's appeal was dismissed without prejudice, allowing Logicvision to submit an appeal upon receipt of the contracting officer's decision if Logicvision did not agree with it.

Page 9-27, add after first full paragraph:

The contracting officer's decision to delay making a final decision, which also delays the contractor's ability to appeal, is effective upon issuance of the decision to delay. Title 41 U.S.C. § 605(c)(2) provides that:

> A [contracting officer] shall, within sixty days of receipt of a submitted certified claim over $100,000—
>
> (A) issue a decision; or
> (B) notify the contractor of the time within which a decision will be issued;

Title 41 U.S.C. § 605(c)(5) provides that:

> Any failure by the [CO] to issue a decision on a contract claim within the period required will be deemed to be a decision by the [CO] denying the claim and will authorize the commencement of the appeal or suit on the claim as otherwise provided in this Act;

and 41 U.S.C. § 609(a) provides that:

> (1) . . . a contractor may bring an action directly on the claim in the United States Court of Federal Claims, notwithstanding any contract provision, regulation, or rule of law to the contrary; . . . .
>
> (3) Any action [on the claim] under paragraph (1) . . . shall be filed within twelve months from the date of the receipt by the contractor of the decision of the [contracting officer] concerning the claim, and shall proceed de novo in accordance with the rules of the appropriate court.

In *Logicvision, Inc. v. United States,* 54 Fed. Cl. 549 (Fed. Cl. 2002), Logicvision delivered to the United States Department of Commerce (DOC) a $259,177.82 claim for adjustment due to delay and disruption on October 12, 2001. The CO, by certified mail, sent plaintiff a letter, dated and issued on December 4, 2001, notifying plaintiff that a final decision would be issued by March 15, 2002. The CO's letter was not received by plaintiff until December 13, 2001. The 60-day deadline for issuance of a decision by the contracting officer under Section 5(c) of the Contract Disputes Act of 1978 expired two days earlier, on December 11, 2001. The complaint was filed on December 28, 2001. DOC argued that the court lacked subject matter jurisdiction to entertain plaintiff's complaint because it was filed before the CO issued a final decision and over two weeks after plaintiff received the CO's decision, issued within the 60-day period, that the final decision would be delayed.

Plaintiff contended that "notify" in 41 U.S.C. § 605(c)(2) meant actual receipt of a notice by the contractor. Reading the word "notify" to mean "the CO's making known or reporting" or even to turn on "issuance" rather than the contractor's receipt would not be unfair or adverse to a contractor, since it easily could call the contracting officer to inquire as to the status of its CDA claim prior to filing suit. Had plaintiff done so in this case, it would have learned, as it in fact did over two weeks before filing its complaint in this court, that the contracting officer had extended the 60-day period for issuing a final decision.

Dismissing the complaint on jurisdictional grounds also would not seriously prejudice plaintiff financially because interest under the CDA was paid "from the date the CO receives the claim . . . from the contractor until payment thereof." In addition, even if the claim was dismissed and subsequently denied by the contracting officer, plaintiff still would be able to file a complaint, and be entitled to otherwise payable interest from the date the contracting officer received the claim.

Plaintiff's suggested reading of the statute, on the other hand, may seriously prejudice defendant, and was inconsistent with the purposes of the CDA. The contracting officer in this case issued the notification seven days before the deadline for a final decision, which, generally, should provide a reasonable and adequate period for ensuring actual notice/receipt. Under plaintiff's reading, however, a notification letter mailed even several weeks before the 60-day deadline for a final decision could not be assured to arrive in time to extend the 60-day period. Requiring unduly early mailings to compensate for unknown mail delivery delays would force government agencies to decide large and complex monetary claims in significantly less time than the 60 days provided by statute, and thus, presumptively, in less time than intended by Congress.

Plaintiff's interpretation also would require the courts and Boards of Contract Appeals to act on claims that were not ripe and that might have been resolved more fairly, economically, and efficiently in the first instance, by those more familiar with the contract's administration, at the agency level. Moreover, if the date hinged on actual receipt, contractors would unilaterally determine the date and thus might have an incentive for manipulation. The contracting officer having issued a decision notifying plaintiff of an extension of the time for issuing a final decision on plaintiff's claim within the 60-day period provided by law, there was no "deemed denial" to trigger plaintiff's statutory right to immediately appeal its claim.

Page 9-29, add after note 132 in first full paragraph:

However, there are exceptions to the strict application of the timing requirement. In *Lawrence Harris Construction, Inc.*, VABCA No. 7219, 05-1 B.C.A. (CCH) ¶ 32,830 (2004), Harris was awarded a contract for the renovation of a mental health facility. During construction, Harris encountered excusable and compensable delays and filed a claim with the contracting officer. The contracting officer denied the claim in a letter to Harris, but failed to mention that Harris had 90 days to appeal the matter to the Department of Veterans Affairs Board of Contract Appeals. Harris filed its appeal 98 days after receiving the contracting officer's final decision. The VA moved to dismiss Harris' appeal for lack of jurisdiction because Harris failed to file within the 90-day period. The Board noted that although it didn't have authority to waive the 90-day appeal period, the contracting officer's final decision must substantially comply with the Federal Acquisition

Regulation requirements that include advising the claimant of the 90-day appeal period. Therefore the contracting officer's decision was defective, and the appeal period was not triggered. Harris' appeal, even though filed 98 days after the contracting officer's decision, was timely.

Page 9-30, add after note 138:

If the claimant desires to drop a specific claim, it must formally amend its complaint, and may not informally withdraw a claim and then attempt to pursue it later in another appeal after a decision has been made on the original claim. *Hitt Contracting, Inc.,* ASBCA No. 51594, 99-2 B.C.A. (CCH) ¶ 30,442 (1999).

Page 9-30, add at end of subsection:

Under the Contract Disputes Act, there is no presumption of delay. In *England v. Sherman R. Smoot Corp.*, 2004 U.S. App. LEXIS 22968 (Fed. Cir. Nov. 3, 2004), the Navy awarded Sherman R. Smoot Corporation a contract for renovation and construction work at the Washington Navy Yard. The contractor later notified the Navy project engineer of design and construction changes that would cumulatively delay the completion of the project by 51 days. The contractor asserted a claim to the contracting officer for an equitable adjustment for the time extension, seeking to recover extended overhead costs. The contracting officer issued modifications granting the contractor an extension of 51 days to complete the contract. When no final decision was forthcoming from the contracting officer regarding compensation, the contractor appealed to the Armed Services Board of Contract Appeals. The board awarded the contractor compensation for the delay, concluding that the Navy was responsible for the delay. Because of the contracting officer's extension of the contract completion date, the board invoked a rebuttable presumption that the Navy was responsible. The Navy appealed.

The contractor argued that the contracting officer's modification extending the contract completion date gave rise to a presumption that the Navy caused the delay. The Navy contended that the board erred in applying a rebuttable presumption because the presumption was at odds with the Contract Disputes Act (CDA). The Navy pointed out that under the CDA, any findings of fact by a contracting officer in a final decision was not binding in any subsequent proceeding. The Navy argued that a contracting officer's findings of fact should not be entitled to special deference on appeal. Although by its terms, the CDA only applied to contracting officers' final decisions, there was no basis for drawing a distinction between an interim and a final decision of a contracting officer. Apart from finality, the contracting officer's interim decision was indistinguishable from a final decision. There was no basis for distinguishing between the carefully crafted, written interim decisions by the contracting officer and a final decision. In addition, the presumption that the Navy caused the delay was unwarranted. The mere grant by the Navy of a contract extension did not indicate that the Navy was at fault; rather, other events external to the Navy could have been responsible. The board erred in using a rebuttable presumption based upon the contracting officer's extension of the contract completion date to determine the Navy was responsible for the delay. Vacated and remanded.

[B] Other Contract Claims Procedures

Page 9-31, add after note 146 in first carryover paragraph:

Other dispute resolution clauses require that claims first be submitted to a designated representative of the other party for a written decision that is final and conclusive, subject only to judicial review to determine whether the decision was fraudulent, arbitrary, or so grossly erroneous as to evidence bad faith. Such clauses are often upheld by courts, and failure to submit the claim as required may result in the claimant being barred from pursuing the matter in court. *Metropolitan Dade County v. Recchi America, Inc.*, 734 So. 2d 1123 (Fla. Dist. Ct. App. 1999); *Ferguson Electric Co. v. Kendal at Ithaca, Inc.*, 277 A.D.2d 843, 711 N.Y.S.2d 246 (2000).

Nebraska Revised Statute § 16-726 requires, as a condition precedent to maintaining an action against a first class city, that "the claimant . . . file such claim within ninety days of the accrual of the claim in the office of the city clerk." Neb. Rev. Stat. § 16-726. Section 16-726 functions as a statute of limitations, resulting in the time bar of suits not filed within the requisite 90 days. *Centric Jones Co. v. City of Kearney, Nebraska,* 324 F.3d 646 (8th Cir. 2003), *citing Sinn v. City of Seward,* 3 Neb. App. 59, 523 N.W.2d 39, 51 (1994).

In *Centric Jones Co. v. City of Kearney, Nebraska,* 324 F.3d 646 (8th Cir. 2003), Centric Jones Co. contracted with the City of Kearney, Nebraska for completion of the construction portion of the Kearney Wastewater Treatment Plant for $12,950,000. CH2M Hill, an engineering firm, served as the City's project engineer. The contract defined CH2M's duties: among other things, CH2M was to act as the City's representative during the construction period; receive payment applications from Centric and make recommendations to the city council regarding progress payments; and act as an impartial initial interpreter of the contract requirements and judge of the acceptability of the work, if and when disputes arose. If CH2M was unable to resolve an issue to the parties' satisfaction, either party could invoke the contract's dispute resolution procedures. If those dispute resolution procedures were unsuccessful, the contract provided that a party could institute suit in a court of competent jurisdiction. Centric submitted 30 payment applications to CH2M. CH2M reviewed the applications and forwarded them to the City's Utilities Director, along with either a certification that the application should be paid in full or a recommendation to pay a different amount. The applications were placed on the city council's agenda. The City adopted CH2M's certifications, resulting in full payment of 12 payment applications, refusal to pay in whole or in part 12 applications, and 6 instances of payment in excess of that actually requested on the application. As a result of the refusals to pay Centric the full amount requested, Centric requested that the parties mediate the dispute pursuant to the contract. The parties participated in an unsuccessful mediation. Centric sent a letter to the City's Utilities Director demanding payment of $1,266,486.53 within 30 days. The City rejected Centric's demand, proposing instead a final payment of $670,256.00. Centric filed suit against the City asserting causes of action for breach of contract and unjust enrichment. Centric sought damages in excess of $2,000,000. The district court granted the City's motion for summary judgment based on the affirmative defense that Centric had failed to comply with the claim statute governing claims against the City, Neb. Rev. Stat. § 16-726, and dismissed the whole case with prejudice.

On appeal, the Eighth Circuit held that substantial compliance was a limited concept under Nebraska law and inapplicable in this case. Moreover, under the existing case law, a clear requirement for substantial compliance was that the claim be filed with the statutorily mandated recipient, in this case the city clerk. Finally, there was no evidence that Centric ever filed a claim or payment application with anyone for the damages it sought in its complaint: namely, damages in excess of $2,000,000. Centric also argued that evidence of waiver of a contractual or statutory right by one party generally precluded summary judgment in favor of that party. The contract terms required compliance with the statutory notice-of-claim procedures, as well as all other applicable laws and regulations. There was not enough evidence of waiver to create a genuine issue of material fact. Thus, there were no genuine issues of material fact precluding summary judgment for the City.

Page 9-31, add after first full paragraph:

In *International Fidelity Insurance Co. v. County of Rockland,* 51 F. Supp. 2d 285 (S.D.N.Y. 1999), IFIC, the surety for NANCO (the original construction contractor), attempted to enjoin SWCF Architects Engineers Planners (SWCF) from adjudicating the parties' disputes over delay damages on two grounds: first, that the claims resolution clause requiring submission to the project architect did not include a dispute over post-completion delay damages; and second, that even if it did, the county's claim was not submitted to the architect for decision within 21 days after it arose, as required by the claims resolution clause.

The contract that NANCO signed contained a claims resolution clause which read:

> 4.4.2 Decision of Architect/Engineer. Claims, including those alleging an error or omission by the Architect/Engineer, shall be referred initially to the Architect/Engineer for action as provided in Paragraph 4.5.4, shall be required as a condition precedent to litigation of a Claim between the Contractor and Owner as to all such matters arising prior to the date final payment is due, regardless of (1) whether such matters relate to execution and progress of the Work or (2) the extent to which the Work has been completed. The decision by the Architect/Engineer in response to a Claim shall not be a condition precedent to litigation in the event (1) the position of Architect/Engineer is vacant, (2) the Architect/Engineer has failed to render a decision within agreed time limits, (3) the

> Architect/Engineer has failed to take action required under Subparagraph 4.5.4 within 60 days after the Claim is made, (4) 90 days have passed after the Claim has been referred to the Architect/Engineer or (5) the Claim relates to a mechanic's lien.

When it became clear that NANCO would be unable to complete the project in a timely manner, IFIC exercised its option to step into the contractor's shoes. Accordingly, it executed a Takeover Agreement, pursuant to which IFIC agreed to complete the contract in exchange for receipt of the payment of the remaining proceeds under the contract. The Takeover Agreement provided that all the terms of the underlying contract were incorporated by reference. It also provided that nothing in the Takeover Agreement affected any of IFIC's rights as surety under its payment and performance bonds.

IFIC engaged Harani Contracting Corporation to complete the project. Unfortunately, Harani proved unable to get the work done within the completion period. The county did not default IFIC and continued to pay under the contract through the Spring of 1995. Even after the county ceased payment in May 1995, IFIC and Harani continued work, and the project was accepted as substantially complete by SWCF, acting as the county's representative, on October 5, 1995. Harani thereafter ceased work on the project and filed for bankruptcy protection. The county defaulted IFIC on March 6, 1996, but ultimately permitted it to return and finally complete the project with another contractor.

Shortly before substantial completion, Rockland County served IFIC with a Notice of Claim for alleged delay damages totalling $4,554,735. The Notice of Claim stated that the county intended to submit the matter to SWCF, but the county did not submit the matter for decision to the architect at that time. By letter to counsel for the county, IFIC denied that delay damages were properly submissible to the architect, and further contended that the county had failed to submit the claim withing 21 days after it arose. Rockland County neither responded to these contentions nor sent the matter to SWCF until after final completion of the project, when it served IFIC with a more detailed analysis of claim and forwarded the matter to the architect for resolution.

IFIC then brought this proceeding, seeking compensatory damages against the county for money alleged to be due under the Takeover Agreement and a declaratory judgment that the claims resolution procedure was not applicable to it. IFIC also joined SWCF as a defendant and moved for a permanent injunction against resolution of the delay damages issue by the architect.

IFIC first asserted that the question of delay damages could not be submitted to the architect because it was not bound in any manner, shape, or form by the claims resolution clause contained in § 4.4.2. The court stated that when it signed the Takeover Agreement, which incorporated all the terms of the underlying construction contract by reference, IFIC bound itself to send all disputes falling under §§ 4.4.1 and 4.4.2 of that contract to the architect for its review.

IFIC also argued that delay damages could not be decided by the architect, even if it fell within the clause, because Rockland County did not submit the dispute to SWCF within 21 days of the time when the dispute arose or when it first became aware of the matter giving rise to the dispute (whichever was later), as required by § 4.4.3. IFIC further contended that, even if the issue was timely submitted, the failure of the architect to take action on the matter within 60 days after the claim was first made removed an obligation on IFIC's part to wait on a decision by SWCF as a condition precedent to litigation.

Rockland County took the position that it made the claim when it served IFIC (but not the architect) with a notice of claim for over $4 million in delay damages. The contract stated that pursuant to § 4.4.2 of the contract, the decision by the architect in response to a claim was not a condition precedent to litigation in the event that the architect failed to take action required under Subparagraph 4.5.4 within 60 days after the claim was made. There could be no dispute that the architect had not taken action on the claim within 60 days. Therefore, by the express terms of the contract, IFIC was freed from any restriction that may have limited its right to have a court determine whether the county is owed delay damages.

If, on the other hand, the claim was not actually made until the dispute over delay damages was submitted to SWCF for resolution, then the conclusion that it was made too late was inescapable. The county asserted in a brief to the court that it first became aware that it had a claim for delay damages on September 12, 1995. Section 4.4.3 of the contract gave it 21 days from that date to submit the claim to the architect for resolution under the alternative dispute resolution clause. There was no dispute that the information needed to resolve the dispute was not sent to SWCF until the Spring of 1997, well beyond the brief limitations period agreed to by the original parties to the contract.

According to the court, New York courts examining such clauses have rejected the notion that architects are empowered to resolve delay damages disputes. Citing *Liebhafsky v. ComStruct Assocs., Inc.,* 62 N.Y.2d 439, 440, 466 N.E.2d 844, 844, 478 N.Y.S.2d 252, 252 (1984) (authority of architect under resolution clause was centered on the operational phases of construction and a claim for delay damages, asserted two years after substantial completion need not be submitted to the architect); *County of Rockland v. Primiano Construction Co.,* 51 N.Y.2d 1, 11, 409 N.E.2d 951, 956, 431 N.Y.S.2d 478, 484 (1980) (claims, disputes and other matters in question between the contractor and the owner that must first be referred to the architect are those relating to the execution or progress of the work, but claims asserted after substantial completion of the work do not fall within the scope of the duties of the architect); *Tsombikos v. Grager,* 147 Misc. 2d 995, 999, 559 N.Y.S.2d 460, 462 (Sup. Ct. 1990) (where claims were made subsequent to the termination of contractor's role as a contractor, architect's responsibility to supervise the contractor's performance and, by extension, initially mediate the dispute was at an end). Delay damages by their nature arise after the construction project is complete—a time when the architect has ceased to be responsible for supervising the contractor's performance. IFIC's claim for a declaratory judgment was granted, and SWCF was enjoined from entertaining the county's claim.

In *County Commissioners of Caroline County v. J. Roland Dashiell & Sons, Inc.,* 747 A.D.2d 600 (Md. 2000), the contractor's suit for unjust enrichment was dismissed because the contractor failed to submit its claim in a timely manner as required by the contract. The County awarded a contract to Dashiell for the renovation and expansion of a correctional facility. The contract included the American Institute of Architect's General Conditions (Doc. A201-1987) that required the submission of a claim to the other party within 21 days of the event giving rise to the claim or 21 days after the claimant recognizes the conditions giving rise to the claim. Dashiell allegedly experienced an increase in its performance costs due to design errors and differing site conditions that were claimed to be the responsibility of the County, but Dashiell did not submit a claim until after substantial completion of the contract. The County denied the claim for various reasons, and Dashiell brought suit under two theories—breach of contract and unjust enrichment. The trial court dismissed the suit under both theories, and an appeal reached the Court of Appeals of Maryland. The court of appeals ruled against the contractor on both theories. The court reasoned that the contractor failed to pursue its contractual claim in a timely manner. Further when a claim was covered by the express terms of the contract, a party may not recover under a quasi-contractual theory of unjust enrichment.

Page 9-33, add after note 160 in first full paragraph:

The question of whether a contractor may change its claim, add different components to its claim, or increase the amount of its claim is often not addressed in the contract provisions relating to claims procedures, so courts may look to judicial procedures to resolve these issues. For example, in Oklahoma a contractor may increase its claim for delay damages in subsequent judicial proceedings over the amount originally claimed before the state transportation authority. In *M.J. Lee Constr. Co. v. Okla. Transp. Auth.*, 2005 Okla. LEXIS 99 (Dec. 6, 2005), the Oklahoma Transportation Authority (authority) awarded a bridge project contract to M.J. Lee (contractor). The contract included the authority's specifications for turnpike construction. After the authority's chief engineer suspended and delayed the contractor's work, the contractor filed a claim for additional compensation for delay costs due to suspension of work for an unreasonable period of time. The chief engineer and deputy director denied the claim. The contractor filed suit against the authority in district court. At pretrial, the contractor increased the alleged damages above the amount originally claimed. The district court ruled that the contractor's claims as presented to the authority could not be changed in a subsequent judicial proceeding. The district court certified an interlocutory order construing the authority's specifications for turnpike construction.

The contractor argued that it was entitled to recover additional compensation for suspension and delays of work, as well as changes in the character of work. The contractor asserted that the increased amount of its claim in district court reflected costs that could not be calculated until it received documentation from the authority. The contractor argued that the specifications did not limit the damages that could be recovered in a judicial proceeding for breach of contract. The authority asserted that because the contractor's notice of claim and claim for additional compensation were conditions precedent to its liability for any delay damages, the contractor could not increase its claim in district court. The specifications' language that provided the procedure for a multi-tiered review of the contractor's claim was not ambiguous. There was no condition precedent to the authority's liability since the authority's potential liability was not dependent on the contractor's notice of claim or claim for additional compensation. The authority's specifications for turnpike construction did not limit the damages the contractor could recover in district court to the amount originally claimed before the authority. Except for suspension of work, the contractor did not preserve any other claims against the authority because it did not provide separate notices of claim. Therefore the interlocutory order was vacated, and the matter was remanded to the district court for further proceedings.

Page 9-34, add after note 171 in last paragraph:

However, the architect's decision under the American Institute of Architects agreements is not an arbitrator's award. For example, in *Martel v. Bulotti*, 65 P.3d 192 (Idaho 2003), the owner (Martel) and the contractor (Bulotti) entered a contract for the construction of an addition to a house and used the American Institute of Architect (AIA) contract documents. The AIA documents provided that any disputes between the parties to be submitted to the architect for a decision, that would become final and binding unless within 30 days one of the parties demanded the arbitration of the disputes under the rules of the American Arbitration Association (AAA). When a performance-based dispute arose between the owner and contractor, the matter was referred to the architect who awarded damages to the owner. The contractor faxed a demand for arbitration to the architect within 30 days of the decision, but did not send the demand to the owner or the AAA as required by the contract. The owner filed in the state district court to confirm the architect's decision as an arbitration award and the district court entered judgment against the contractor. The contractor appealed the judgment, and the matter was taken to the Supreme Court of Idaho. Upon consideration, the Court held that the architect's decision could not be an arbitration award because the architect was not neutral. Since the architect was engaged by the owner there was a potential for bias such that the architect may "discourage a fair investigation of the facts." Even though the architect's decision was not an arbitration award, the architect's decision in this matter was final and binding since the contractor failed to demand arbitration in the proper manner as required by the terms of the contract. The contract required that the demand for arbitration be served upon the owner and the AAA, not the architect. Notice to the architect would not "trigger the arbitration process," and under the terms of the AIA contract, the architect's decision was final and binding upon the contractor.

Chapter 11
USING THE SCHEDULE TO PROVE DELAY AND DISRUPTIONS

§ 11.01 WHY USE A SCHEDULE?

Page 11-3, add after note 2 in first full paragraph:

In order to be awarded delay damages, one multiple prime contractor must demonstrate that the delays by another multi prime contractor specifically affected its work performance. In *Biemann & Rowell Co. v. Donohoe Cos., Inc.*, 147 N.C. App. 239, 556 S.E. 2d 1 (2001), Biemann & Rowell Co. and Donohoe Cos., entered into separate contracts on June 1, 2002, with the State of North Carolina to build a multimillion-dollar neuropsychiatric hospital. Delays occurred throughout the project and the contractors were frequently forced to complete work out of the anticipated sequence. To accelerate completion of the project, it was the understanding of the contractors, architect, and schedule coordinator that Donohoe would "dry-in" the building by installing a moisture seal. Donohoe never installed the building seal. The architect indicated that Donohoe failed to complete work according to the project schedule and to fulfill its duties as project expediter. Meeting minutes and observation reports indicated that Donohoe also contributed to overall project delay. Biemann and Rowell maintained that the contractor's failure to install the building seal, and its failure to supervise and properly schedule its subcontractors, caused delays to Biemann and Rowell's work. The installation of the temporary seal first appeared as an activity on the critical path long after the project began experiencing delays. Biemann and Rowell merely presented a chart of instances of delay allegedly attributable to Donohoe and relied on anecdotal testimony about the delays. The trial court rendered judgment in favor of Donohoe. Biemann and Rowell argued on appeal that it was not necessary to prove that Donohoe proximately caused injury to it because an injured contractor may recover delay damages by merely demonstrating that such damages were within the contemplation of the parties at the time the contract was entered. Donohoe had a duty to the other prime contractors for the full performance of all duties and obligations due under the terms of the separate contracts and in accordance with the plans and specifications. To recover damages, Biemann and Rowell had to show that Donohoe breached the contract and that the breach caused Biemann and Rowell's damages. Although there was evidence that Donohoe may have contributed to overall project delay, Biemann and Rowell failed to show how delays specifically caused by Donohoe affected its work performance. Biemann and Rowell also failed to take into account delays attributable to other causes.

Page 11-3, add after first paragraph:

However, in one instance, it was held that the claimant has no need for a delay analysis if promises in the contract are breached. In *Missouri Department of Transportation v. Safeco Insurance Company of America*, 2002 Mo. App. LEXIS 1761 (Aug. 27, 2002), Robertson Contractors, Inc. (C) was the general contractor, and PR Developers, Inc. (PRD) was the subcontractor for the base rock and paving for the construction of embankments, ramps, and roadways at the interchange of an interstate highway. The Missouri Highway and Transportation Commission (o) bid materials informed bidders that above-ground high voltage electrical utility lines were located in the northwest portion of the project which needed to be raised by AmerenUE for embankment and ramp work in that area. MHTC had already informed AmerenUE of the need to do this for the project. The contract provided that Robertson would have 130 working days to complete the project, starting from March 15, 1999. The specifications stated that if the contractor experienced delays due to the owners of utility facilities, the delay would be figured into the count of working days, and that would be the contractor's sole compensation from MHTC. The subcontract incorporated by reference the contract and the specifications. The subcontract stated that time was of the essence. PRD was told that Robertson wanted PRD prepared to begin its work on March 15, 1999. PRD indicated that it would like to start work on the west side of the project. PRD was not able to begin its work until the end of May, 1999, and had to begin work on the east side of the project, a delay of approximately 50 working days. PRD encountered problems with the tacks and the grades on the east side ramps which caused several days of delay, which required restaking the tack lines and grades by MHTC. PRD eventually finished the paving on the ramps on the east side of the project by July 14, and was ready to begin working on the west side of the project by July 19. Robertson had not yet completed the embankments and subgrade on the west side at that time, in part due to further staking problems by MHTC, and it was not until August 16 that the subgrades on the west side ramps were ready for PRD to work, a delay of 20 working days, during which PRD could not use its paving equipment. PRD also claimed that throughout the course of the project, it did various extra tasks which were not part of its duties under the subcontract pursuant to oral agreements with Robertson. PRD filed suit against Robertson. The jury returned verdicts in favor of PRD on its claims, and awarded: damages for delay of $1,581,192.50 against Robertson; and $74.810.16 against Robertson for non-delay damages.

On appeal, Robertson asserted that the trial court erred in failing to grant a new trial because PRD failed to submit any evidence establishing that actions of Robertson were a substantial contributory factor to the delay and the delay was not due to the acts or omissions of third persons or other events. It was irrelevant whether or not Robertson's actions or inaction caused the delays and consequent damages to PRD or were even a substantial contributory factor. When a party in a contract charges itself with a duty possible to performed, it must perform it, unless the performance was made impossible by the act of God, by the law, or the other party. It appeared that much of the delay was due to the time required for AmerenUE to relocate the utility pole on the west side of the Project, which prevented PRD from working on the west side for a considerable period of time, and delayed Robertson in completing the embankments on the west side. Robertson did not choose the sandy subgrade material that deformed under PRD's paver, nor did Robertson prevent MHTC from agreeing on the need to add a stabilizer to firm up the subgrade. Nevertheless, the subgrade did fail, causing delays and other costs. The evidence showed that throughout the course of the Project that MHTC's staking and grades were inaccurate, which delayed PRD from performing its work until MHTC performed the staking and grading properly. By entering into the subcontract, Robertson implicitly promised to have a site ready for PRD to work thereafter, and promised explicitly to maintain the subgrade so that it would support construction traffic, which would include PRD's equipment, to have an on-site superintendent, and to give reasonable notice to the MHTC engineer of intent to perform work to permit proper staking. Nothing appeared to limit these promises or make them conditional, and there was sufficient evidence to support the jury's finding that these promises were made. The trial court did not err in denying the motion for a new trial.

§11.03 LINEAR SCHEDULING METHOD

Page 11-16, add new subsection:

[A] Other Types of Schedule

Another type of scheduling is evolving—4D scheduling. A 4D schedule model combines three-dimensional computer assisted design (3D CAD) graphics with the computerized schedules and can be used to develop videos of construction sequences over time. Construction activities are displayed in three dimensions in time-sequenced scenarios such that the construction activities can be viewed from any point of view during any stage of the progress. Visualizing a project schedule may serve to facilitate analysis of a variety of issues such as design, construction, problem resolution, and even delays, as various project participants such as owners, contractors, engineers, subcontractors, and others are able to evaluate the consequences of the time dimension on a project. For example, coordination of different trade contractors may be enhanced to minimize conflicts, facilitate work in congested areas, expedite material handling, and improve communications. Commentators have discussed the possibility that 4D facilitates more accurate assessment of the impact of pending changes and potential delays and disruptions. For example, time and space conflicts between various work activities, changes, and crews can be visualized from various physical perspectives by professionals from different disciplines. 4D schedules can be used to develop two-week look-ahead schedules to assist in the coordination of materials, crew, trades, and equipment, minimizing the disruptions on a tight project site or complex project. Time intervals can be illustrated with the activities that may be affected by changes, equipment problems, site access restriction, suspensions on portions of the work, and other delay-causing events. Finders of fact such as juries and arbitration panels may value such computerized graphics as being understandable and persuasive in determining the impact, responsibility and causation of project delays. S. Chong, et al., *4D Schedule Modeling—What Are the Advantages*, Construction Superconference (Dec. 11, 2003).

§ 11.05 DEMONSTRATING SCHEDULE RELIABILITY

[B] Reliable Schedule Is Updated

Page 11-29, add to note 40:

Blinderman Constr. Co. v. United States, 39 Fed. Cl. 529 (1997), *aff'd mem.*, 178 F.3d 1307 (Fed. Cir. 1998).

§ 11.06 MANIPULATION OF THE SCHEDULE

Page 11-38, add after carryover paragraph:

One common problem is the failure to submit a schedule at the beginning of construction. The failure to submit the schedule prior to the start of construction provides a claimant with the opportunity to develop an "after-the-fact as-planned" schedule that may depict its intended sequence of construction in ways that would reflect delays and problems of others. For example, an after-the-fact schedule could be developed with the critical path running through the activities later discovered to be affected by the delays of the owner. To prevent the development of after-the-fact schedules, many owners have developed detailed scheduling specifications requiring the submittal and acceptance of the project schedule as a prerequisite to receiving payment for mobilization or for the entire first progress payment. Some owners require the contractor to submit a preliminary schedule prior to receiving notice to proceed.

Schedule manipulation may also occur in updates. Despite the initial accuracy of the schedule logic, some activities may actually begin prior to the completion of activities that are depicted as "precedessor" activities—follow-on activities start out of sequence. Some scheduling software allows the scheduler to employ two different approaches to automatically deal with this common situation. One technique is known as "retained logic" where if an activity starts out of sequence but is not completed in the update period, the remaining duration of the activity is shown starting after the predecessor activities are completed. Another technique is "progress override" that allows the scheduler in essence to ignore the activity relationships and depict the activity start without being tied into the previously established logic. Rather than allow the software to use either "retained logic" or "progress override" as the "default" technique in updating schedules, a better approach may be to require the scheduler to redefine the logic relationships between activities to reflect the current plan for completing the activities.

Software may also automatically assign start and completion dates for activities, rather than reflect its actual start and finish dates. This is often referred to as "percent complete updating." For example, when a schedule is updated on the first day of every month, if an activity started on May 25 and was worked on for five days, the June update would indicate that the activity started on June 1 and was 50 percent complete with five days remaining duration. In reality, the activity began on May 25 and may require more than five days to complete. This schedule updating technique may also distort the project schedule. Therefore, many owners require the scheduler to use actual dates in reporting on updates.

Page 11-44, add after second full paragraph:

It is possible to improperly manipulate the schedule by using and misusing the feature on scheduling software that allows the use of multiple calendars. Using different calendars for different types of work activities is often a reality on large and complex construction projects. For example, a 40-hour week schedule is the most commonly used schedule for work and installation activities in the United States—one shift per day, five days a week, with non-working Saturdays and Sundays with certain designated union holidays. For testing of equipment or systems, contractors may use a six-day work week, given the fact that many testing crews work on Saturdays. Also in order to accelerate the completion of the project, certain trades may work on overtime on Saturdays, or a trade contractor may decide to add a second or third shift. This may be reflected on the project schedule by a different calendar for the trade contractor, but the other contractors' work is scheduled using the 40-hour week calendar. Certain contractors may have different holiday policies, and hence a different calendar is used in the project schedule for this contractor's work. For projects not allowing roadway lane closures on weekdays, portions of the work may involve only weekend working day calendars. A seven-day/week calendar may be used for submittal and long lead-time procurement activities. The seven-day week calendar is often used to calculate contract completion dates, as contract durations are often stated as a certain number of days after Notice to Proceed.

The use of different calendars for different activities may result in complications and confusion when calculating activity completion dates, as well as the critical path for the schedule dates. For example, if an activity has a stated duration of 30 days using a seven-day/week calendar starting on January 1, it will have a completion date of January 30. The same activity duration of 30 days starting on January 1 using a five-day/week calendar will have a completion date of February 11. Changing calendars for activities as part of an update or schedule revision may result in different completion dates, critical path, and activity float calculations. The use of multiple calendars is something that should be carefully done, evaluated, and monitored. Scavino, *Effect of Multiple Calendars on Total Float and Critical Path*, 45 Cost Eng'g 11 (June 2003).

Page 11-45, add at end of section:

Features and functions available on software programs have facilitated the manipulation of schedules. Some commentators have opined that the problems are so widespread that it is "an industry crisis" that has its origin in the failure of scheduling software to require the users to employ discipline and realism in planning and scheduling. Wickwire & Ockman, *Use of Critical Path Method on Contract Claims—2000*, 19 Const. Law. 12 (Oct. 1999).

Several scheduling software features or techniques that may distort schedules include:

- Scheduling reports and updates that do not list predecessor and successor relationships, listing only activities that the scheduler deems to be on the critical path;
- Program options that allow the scheduler to hide flawed logic, such as restraints that interrupt the calculation of the critical path;
- Software manuals that instruct schedulers on how to fix certain dates as driving the schedule rather than the logic and duration of activities;
- Software manuals that encourage schedulers to misrepresent the status of the project and distort the impact of events and problems on the project schedule;
- Software tools that allow the scheduler to mask or delete key information from scheduling reports;
- Linking cost-loaded schedules developed for project billing purposes with schedule activities in such a way as to distort updates;
- Extensive use of leads and lags in such a way that the critical path is no longer sequential;
- Scheduling reports that do not identify which part of a critical "overlapping" activity is truly critical and which part has float;
- Scheduling reports that do not identify how much of any given activity must be in place before its immediate successor can start;
- Use of multiple calendars that allow the scheduler to apply different calendars for various parts of the schedule;
- Reporting activity status by its percentage of completion rather than recording actual start and completion dates—for example, automatically assigning an

activity's start date as the schedule status date when the activity progress was first reported as more than 0 percent, rather than its actual start date;

- Misusing the "retained logic" function in updating schedules. If a follow-on activity has started before its predecessor activities are complete, the remaining duration is shown as starting only after all of the predecessor activities are complete. This may have the affect of distorting the relationships between the activities;
- Misusing the "progress override" function in updating schedules. If a follow-on activity has started before its predecessor activities are complete, the relationships are ignored or "overridden" to allow activities to start out of sequence. *Id.* at 17–19.

The authors suggest several procedural and software changes that will help resolve some of these problems: audit trails programs that identify changes in the schedules; monthly reports for predecessor-successor relationships; program requirements for the reporting of actual start and completion dates of activities; restricted use of arbitrarily assigned milestones or other constraints; reports that highlight schedule changes made in each update or revision; restrictions on the automatic use of "progress override" or "retained logic" features in schedule updating; and restrictions on the use of inappropriately calculated start and completion dates for activities. *Id.* at 20.

§11.07 METHODS OF SCHEDULE ANALYSIS

Page 11-46, add at end of first full paragraph:

Construction law practitioners and programme (schedule) analysts in the United Kingdom are also addressing the issues raised by delay claims and the various methods of evaluating construction schedules. In October 2002, the Society of Construction Law published the *Delay and Disruption Protocol*, to guide parties to the construction process in dealing with construction delay in both dispute prevention and dispute resolution. One of the goals of the *Protocol* was to provide answers to common questions arising from contractors' requests for extension of time and requests for compensation for additional time and resources to complete projects. The *Protocol* addresses many issues: the kinds of programmes and records that ought to be kept during the course of construction; entitlement to extensions of time; effects of delay; float; concurrent delay; mitigation of delay and losses; calculation of compensation for prolongation (extended performance); acceleration; disruption; and claim preparation and analysis. As part of the discussion on delay analysis, the *Protocol* addresses the methods that have been used to evaluate programmes and delays. The *Protocol* recommends that the programme should be prepared as a critical path network in sufficient detail to enable the determination of changes and delays on the dates and resources, and the updated programme should be the primary tool used to determine entitlement to extensions of time for delays. During the course of the project, if a contractor requests an extension of time, the contractor should submit a sub-network (group of activities or durations linked logically) to reflect the effect of the claimed delays, and insert the sub-network into the updated programme to demonstrate the actual or anticipated effect of the delay, along with the documents and records to demonstrate entitlement. The impacts of delays should be "mapped out" at the point it time at which the delays occur, as this will allow the determination of the discrete effect of individual events and delays. This methodology is deemed "time impact analysis."

If the analysis of the impact of events is not carried out during the course of construction, the analysis can be performed retrospectively. The *Protocol* discusses different types of retrospective programme analysis. The collapsed as-built method is where the effect of events is subtracted from the as-built programme to determine what would have occurred but for the event. The *Protocol* notes that this method of analysis is simple to perform but is often laborious and subjective. The impacted as-planned method shows the effect of delays on the planned programme, and is a relatively simple form of analysis but is restrictive due to the theoretical nature of the analysis. The as-planned versus as-built is the least expensive and simplest method but is restricted by its ability to identify concurrency, re-sequencing, mitigation, or acceleration. The *Protocol* notes that the preferred method of retrospective programme and delay assessment is the time impact analysis where the impacts of delays are to be determined out at the point in time at which they occur. However, this method can be time consuming and costly to use. Knowles & Hall, *Society of Law: Protocol for Determining Extensions of Time & Compensation for Delay & Disruption* (2002).

Commentators have noted that the *Protocol* is broadly in line with recent case law in the United Kingdom, such as *Ascon Contracting v. Alfred McAlpine Construction Isle of Man,* 66 Con. L.R. 119 (1999) and *McAlpine Humberoak v. McDermott International,* 58 B.L.R. 1 (1992). Burr & Lane, *The SCL Delay and Disruption Protocol: Hunting Snarks*, 19 Constr. L.J. 135 (2003). However, the commentators noted that it remains to be seen to what extent the judiciary in the United Kingdom would follow the *Protocol*, rather than the "somewhat incremental and individualistic approach to delay and disruption." Burr & Lane at 136.

Other commentators have noted that there are problems with the retrospective time impact analysis preferred by the *Protocol*. For example, two programmes are generally produced for each impact—one that reflects the progress before the delays are impacted, and one after. Given a two-year construction project with monthly updates, there may be 48 programme iterations. The quality of the base programme (as-planned schedule) must be verified as being achievable, and at the end, what actually happened may have no resemblance to the plan. The lack of adequate and consistent progress information may also affect the quality of a time impact analysis since it is dependent upon this information. Further, there may be "on-going tinkering" with the programme logic at each update to reflect the re-sequencing of the work activities. Palles-Clark, *Problems With Using Critical Path Analysis for Proving Delay Claims,* Civil Eng'g Surveyor Constr. L. Rev. 47 (2003).

[A] As-Built Schedules

Page 11-47, add after carryover paragraph:

Courts may also find the as-built schedule methodology to be unreliable if only a single activity or portion of the schedule is analyzed without consideration of the progress on other portions of the schedule. In *Galaxy Builders*, ASBCA No. 50018, 00-1 B.C.A. (CCH) ¶31,040 (2000), the contractor was required to develop a critical path method schedule for its work in building an Air Force support facility. Once it was approved by the government, the schedule was to be updated monthly, and be the basis for determining time extensions and delays. The contractor submitted shop drawings for the roofing system, and the government took longer than the 60-day period for response. The contractor submitted a delay claim for the late approval, using an as-built schedule that isolated the roofing activities only. This was "proof" that the submittal response problem delayed the other activities and overall project completion date. The contractor's analysis did not track the progress of other activities. The government submitted an alternative analysis indicating that there were concurrent contractor delays in the HVAC and masonry activities, such that the late roofing submittal and the roofing activities were never on the critical path. The board found that the government had rebutted the contractor's as-built schedule analysis. The contractor's focus on a single line of activities, excluding the actual progress information of all other activities, was a distortion of the schedule and not a proper delay analysis.

In *Morganti National, Inc. v. United States*, 49 Fed. Cl. 110 (2001), an as-built delay analysis was rejected and a time impact analysis substituted. Morganti was required to prepare a schedule, using critical path method scheduling, for the Federal Bureau of Prisons. The contractor's progress was to be tracked in monthly schedule updates. The contract also required that Morganti submit a time impact analysis (TIA) in support of any time extension requests. The contract did not allow the contractor to add time to the contract schedule until the time was approved by a bilateral modification. The contractor brought in a scheduling expert to produce a project schedule. The schedule expert prepared a schedule that eventually became known as "Schedule B." The parties agreed that the activity durations and schedule logic in one Schedule B update (WC04) would serve as the baseline schedule for measuring any impacts to the schedule.

The contractor was required to redesign 387 cell panels of the 5,800 cell panels required for the project. Because of the impact these changes had on cell-panel production, installation of cell panels and follow-on work had to be resequenced. The WC04 update took into account the additional delay caused by the disruption to the sequencing of cell-panel installation activities. Contemporaneous with the first panels arriving on site, the contractor submitted a TIA prepared by the scheduling expert that indicated a time extension of 213 work days.

The contractor abandoned the TIA analysis at trial. Another expert provided a different scheduling analysis on behalf of the contractor at trial, based on an as-built analysis. The other expert testified that the as-built critical path of the project ran through cell-panel installation and explained that the "total time" until project completion represented the duration of the government-caused delay, which entitled the contractor to a time extension of 511 calendar days.

The court found that the cell-panel change caused some delay to the critical path, and thus caused delay to the project as a whole. The contractor's trial expert's as-built delay analysis never referenced the durations or logic in the contractor's schedule updates. Rather, the other expert determined the number of days of excusable delay by comparing the as-built schedule with the initial schedule. According to the court, the contractor's trial expert's as-built analysis was a "total time" approach, which was of virtually no value. The court stated that the "total time" theory of proving delay was insufficient to prove that government-caused delay actually delayed overall completion of the project. The court used the TIA prepared by the contractor's first expert to measure the delay. According to the court, the contractor's trial expert's "total time" analysis was "wholly lacking" in the face of contemporaneous schedule updates showing other causes for the contractor's remaining delay.

Page 11-50, add to note 97:

According to one proponent of the collapsed as-built method of schedule analysis, it is a comparative method of quantifying the critical path delay by using actual progress data rather than theoretical progress. It avoids debates as to the reasonableness of the as-planned schedule, and focuses the analysis upon the actual progress that can be determined by an independent review of the project record. Purportedly the delay analysis can be reduced to factual verification and contractual entitlement analysis. The impact of critical delays is measured by comparison of the actual completion date with the completion date calculated by the "collapsed as-built CPM." According to this commentator, the effects of concurrent delays are factored into the collapsed as-built assessment because if any claimant-caused concurrent delays are not extracted from the as-built, they will remain in the schedule analysis and attributed to the claimant. The disadvantages of the collapsed as-built method are that it is time-consuming to perform and therefore expensive. Hoshino, *Collapsed As-Built CPM Analysis,* Project Management Institute 26 Annual Seminar/Symposium (Oct. 16–18, 1995); Hoshino, *Collapsed As-Built: The Right Way*, AACE International 44th Annual Meeting (June 25–28, 2000).

The key to the collapsed as-built approach is the development of the accurate "as-built critical path." There are two types—the controlling path and the analogous critical path. The controlling critical path is a path that actually determines the completion of the project. The analogous critical path is the "calculated" critical path logic imposed upon the as-built schedule. This allows the schedule analyst to determine the total number of compensable delays by "subtracting" the collapsed completion date from the as-built completion date. The analogous critical path is what remains after the compensable delays have been extracted from the "as-built critical path."

There are seven key steps to a proper collapsed as-built schedule analysis. The first step is to develop an as-built bar chart from the project record—daily diaries, meeting minutes, inspection reports, schedule updates, etc. A detailed as-built schedule should be performed on a daily basis. The second step is to determine the actual duration of the activities in the schedule, as well as lead and lag relationships. A seven-day week calendar, rather than a five-day work day per week calendar should be used because it is usually the case that construction crews occasionally work on non-working days. The third step is to develop actual logic relationships between the schedule activities. This is an important step because the collapsed as-built method seeks to simulate the "actual" activity sequences and relationships similar to a computerized, forward-looking CPM schedule. This step is the development of the "as-built CPM model" and is often a "multi-iterative process" of identifying faulty or incomplete logic by performing various "trial collapses" to "de-bug" the model. This may be a weakness of the approach; unless the schedule analysts is careful, accurate and complete, logic flaws may distort the analysis. The fourth step is to identify all significant delays without regard to responsibility or liability. The delays may be of different types: (1) suspensions where there is no work on one or more activity; (2) delayed start of activities that could have begun but for some problem; (3) extended duration delays where an activity exceeds a reasonable duration; and (4) pacing delays where an activity duration is extended due to concurrent delay in a predecessor activity (direct pacing) or where an activity duration is extended merely to fill in time waiting for other activities. The fifth step is to "extract" various delays from the as-built schedule to create the collapsed as-built. For example, only owner-caused or compensable delays are subtracted from the as-built schedule. Any concurrent delays by the contractor would remain in the schedule, having the effect of automatically compensating for any contractor-caused delay. Thus the collapsed as-built method of schedule analysis does not require a special concurrency assessment. The sixth step is to transfer the calculated critical path of the collapsed as-built onto the analogous logic of the as-built schedule. A comparison of the collapsed as-built with the as-built will result in a list of delays that add up to the net difference in the overall duration between the two schedules. The final step is to review the collapsed schedules, adjust the model for any illogical, impractical, or inappropriate logic such as unreasonably short activity durations, overburdened resources, excessive concurrency of activities, violation or procurement restraints, and performing work in periods of adverse weather. According to this proponent of the collapsed as-built schedule method, the approach is the most sound, fact-based, and most effective communication approach to performing complex schedule analysis. *Id.*

[C] As-Planned Method

Page 11-61, add after note 110 in first full paragraph:

For example, it may be a distortion of the schedule merely to insert additional activity days to reflect change-order work without considering the impact of other delays or the actual effect of the change orders on the progress of the work. Additional activity days for change orders inserted into a schedule do not necessarily equate to day-for-day delays to overall project completion. *J.T. Jones Construction Co.*, ASBCA No. 48303, 98-2 B.C.A. (CCH) ¶ 29,892 (1997).

[E] Other Delay Analysis Methods

Page 11-66, add to note 123:

See C&D Lumber, Inc., VABCA No. 2877, 91-1 B.C.A. (CCH) ¶ 23,544 (1991); Norair Eng'g Corp., ENG BCA No. 3804, 90-1 B.C.A. (CCH) ¶ 22,327 (1990); Blackhawk Heating & Plumbing Co., GSBCA No. 2432, 75-1 B.C.A. (CCH) ¶ 11,261 (1975).

Page 11-72, add after note 130 in third full paragraph:

In *Kinetic Builders, Inc. v. Secretary of Air Force*, 226 F.3d 1307 (Fed. Cir. 2000), the Air Force awarded Kinetic the contract for alteration of a building. The government took beneficial occupancy of the building 37 days after the scheduled completion date and withheld liquidated damages. The contractor submitted multiple claims to the contracting officer (CO) for government-caused delays. The CO determined that the contractor was permitted 30 days of delay. The contractor appealed to the Board. The contractor contended that progress in completing the contract had been delayed because of the Air Force's delay in providing a solution to a defective restroom design. The contractor contended that the progress schedule indicated that the defective design and the Air Force's slow response caused it to go from 4.2 percent ahead of schedule to 5.8 percent behind schedule, clearly indicating that its progress had been negatively affected by the defective design. Moreover, the contractor contended that the progress reports for drywall work indicated that it went from 60 percent completion of the work, at the time the Air Force was notified of the defect, to 14 percent behind schedule when the Air Force solved that defect, also evidencing that the defect caused delay. The contractor further contended that the evidence of progress on the drywall work at least entitled it to a contract extension to restart and complete the drywall work. The Board noted that the mere fact that the progress rate declined, between the time the contractor notified the Air Force about the design defect to the time the Air Force corrected that defect, did not establish a causal connection between the defect and the slowing of the progress. The Board noted indications in the record that other drywall installation and other work could have been completed, but was not, during the time the defective design was being resolved. Moreover, the Board noted that 0.7 percent of the drywall work was not completed until after resolution of the design defect. The Board also found that the contractor failed to show that the design defect itself delayed or prevented any drywall work. The Board determined that the contractor was entitled to 17 days of excusable delay because of the Air Force's delay, and referred the quantum to the parties. The contractor appealed the Board's decision to the Federal Circuit court. The Federal Circuit court concluded that the contractor's method of delay analysis did not establish that the Air Force's delay in correcting the defective design had caused the contractor's overall slowing of progress or slowing of the progress on drywall installation.

§11.08 CHOOSING A SCHEDULE ANALYSIS METHOD

Page 11-77, add after first full paragraph:

The use of an expert for schedule and delay analysis may be subject to judicial "gate-keeping" scrutiny to determine whether the expert testimony is reliable and should be allowed into evidence according to the general principles set forth in *Kumho Tire Co. v. Carmichael*, 526 U.S. 137 (1999) and *Daubert v. Merrell Dow Pharmaceuticals*, 509 U.S. 579 (1993). The essence of the *Daubert* requirement is that the methods used by experts must be reliable, and courts seem to be imposing the *Daubert* evaluation to any expert. Simpson, *Experts: A Little Daubert'll Do Ya,* GPSolo 30 (Oct./Nov. 2002). There are several aspects of the analysis of a scheduling and delay expert that may be appropriate for the judicial gate keeping function before allowing the scheduling expert's testimony into evidence.

- Whether the method of schedule analysis used by the expert is recognized, tested and accepted for use in analyzing construction delay and establishing causation.
- Whether the as-planned schedule or other baseline schedule used by the expert to evaluate delays was a reasonable and workable plan for construction.
- Whether the records used to develop the as-built schedule were reliable and accurate.
- Whether the delay analysis ignores available scheduling data and relevant progress information recorded contemporaneously during construction.
- Whether the analysis was based upon limited scheduling and project information provided by the party retaining the expert.
- Whether the network logic used in the expert's schedules accurately depicted the sequence of the work and the interdependencies of the activities.

- Whether the expert identified all relevant causes of delay and properly accounted for them in the delay evaluation.
- Whether the conclusions reached by the expert are objectively justified and not merely subjective assessments.

Many of these considerations may also provide the basis for the impeachment of the scheduling expert's opinions even if it allowed into the record. Bell & Heady, *Expertise That Is "Fausse" and Science That Is Junky: Challenging a Scheduling Expert*, Procurement Law. 1 (Fall 1999).

One of the difficult parts of schedule analysis, whether preparing a delay analysis or evaluating one that has already been prepared, is choosing what particular method to use to measure the delay. Besides the confusing choices among the standard Time Impact Analysis, As-Built, As-Planned, or one of the many variations of a "window" approach, practitioners of delay measurement enjoy putting their own individual creativity on their products, providing even more choices.

One certainty in this environment of choice is the contract. Many contracts will identify the method to be used to measure time extensions or delay in either the schedule or disputes specification. Often the contract schedule is identified as the medium for the measurement, and the critical path on the update closest to the delaying event the measure for determining the validity and length of the time extension.

It is surprising how often and passionately those who chose a method that is not the contract-required method defend their choice. It is most often seen when one party disputes the length or entitlement to an extension. Consultants challenging an extension request seem to prefer a variation of an as-built method that sets off every hiccup in a contractor's progress as a concurrent delay. This does not conform to defining delay from the critical path on a particular update, but it is hard to accept a method that ignores late activities that are not critical and show an extension is due, if there is another view that says the extension is not warranted.

However, when the contract specifies a particular method, there is no choice: the contract method must be used. Consider *P.J. Dick, Inc. v. Principi*, 324 F.3d 1364 (Fed. Cir. 2003). The Department of Veterans Affairs (VA) selected P.J. Dick, Inc. (PJD) to construct the Clinical Addition to a VA Medical Center. Under the contract, PJD was due to complete the work by January 12, 1998. The specifications required:

> The Contracting Officer's determination as to the total number of days of contract extension will be based upon the current computer-produced calendar-dated schedule for the time period in question and all other relevant information.

During the contract, the government issued over 400 orders changing the contract and causing various delays to different aspects of the project. These modifications increased the contract price by over 5 percent and caused the VA to grant 107 days of additional contract performance time. PJD completed the contract 260 days after the original contract completion date and 153 days after the revised date. PJD presented to the government contracting officer claims for additional relief as a result of the delays to the contract. All the claims were denied by the contracting officer (CO). PJD timely appealed the CO's denials to the Board. PJD claimed it was entitled to a time extension for all 260 days of delayed performance and sought field and home office overhead damages for most of these delays. A large portion of the extension (201 days) granted by the Board resulted from PJD's claim for delays due to the "combined directives." The "combined directives" were six separate contract change orders issued by the DVA over a 10-month period, all relating to installation of certain equipment in the Clinical Addition. The Board analyzed the effect of each of the changes as of the date of the earliest change, which gave PJD a larger delay than it would have received were the effects of the change orders analyzed separately on the dates the DVA issued them. The Secretary and PJD each appealed different aspects of the Board's decision. The Secretary argued that the Board erred in its determination that the "time period in question" for all six of the changes included in the "combined directives" was November 1995—the date the DVA issued the first of those change orders.

The language of the contract required the Board to analyze the changes of the "combined directives" separately utilizing the most recent monthly update of the "computer-produced calendar-dated schedule." The express language of the contract established that the "current" schedule "will be" the basis for determining the extent of the delay. By using "will," the contract indicated that this was the sole method of calculating the delay. The Board circumvented this language by crediting litigation arguments that the six unrelated changes issued over ten months had a unitary effect, making the "time period in question" the date the DVA issued the first change order. This was error because the language of the contract required the use of the "current" computer schedule as the basis for making such a determination. In other words, regardless of what the testimony showed, under the contract, the only way of determining the effect of the changes was to analyze each of them using the current computer schedule. Thus, if the changes did have a unitary effect, it had to be demonstrated by the computer model, not the testimony PJD presented. The Board's determination was reversed that the "combined directives" should all be analyzed utilizing the October 1995 schedule update—the update most current as of November 1995—and remanded for a damages analysis in accordance with the contract. On remand, the Board should determine whether PJD's delay claims that it found were not controlling because of the "combined directive" delay (the underground conduit and the radiology and cardiology claims) become controlling as a result of any reduction in the "combined directive" delay and analyze them accordingly.

An interesting circumstance in *P.J. Dick* was that *all the parties acquiesced to another method and the BCA accepted the parties' agreement.* Still, the Federal Claim Court said the contract method was to be employed.

Using the method specified in the contract also may be inconvenient for the claimant. Alternate methods for delay measurement may be faster and cheaper than the contract method, and contractors—being generally optimistic—may believe that the other, less expensive way may just work for a negotiation. However, be sure you know that a contractor's delay claim that does not use the contract method also does not satisfy the contract requirements for a change order request or a claim if the contract identifies the method to measure delay. The parties in *P.J. Dick* learned that lesson, but only after a considerable time and money investment.

Page 11-77, add after fifth full paragraph:

The study in the *Journal of Construction Engineering and Management* indicated that different delay analyses methods provide different results with the same facts. Three accepted methods were considered. The As-Planned Method; the As-Built Method; and the Modified As-Built Method. Two analyses were performed using the As-Planned Method. First, after representing the delay as activities, the delays were input one at a time and the schedule was calculated after each input. The resulting project completion date and total float available were recorded. Second, the six delays were input to the As-Planned Schedule, then the schedule was calculated. The results were the same for both variations. A. Bubshait & M. Cunningham, *Comparison of Delay Analysis Methodologies*, 124 J. Constr. Eng'g & Mgmt. 315, 322 (No. 4, July/Aug. 1998).

The As-Planned Method considered all delays at once, therefore, the completion date was adjusted to the largest delay on the most critical path. Of the six delays, the results showed that the project completion date had been delayed by 56 days caused by delay No. 2-B, which was the late delivery of electrical duct heaters. After the delay, the total float was reduced to 56 days, delaying project completion by 56 days to 8/18/93. Delay No. 6-A caused two-days' delay to two activities; however, these two days were included into the larger delay caused by delay No. 2-B. Bubshait & Cunningham, at 319–320.

The As-Built procedure compared original float to float available after the delay. Delay No. 2 delayed two activities on the work plan by a total of 127 days. The first activity initially had 169 days positive float, and the second had 71 days positive float. After the delay, activity was reduced to 256 days total float, thus delaying project completion by that amount. The result was 22 days to the project completion date. The As-Built Method calculation showed that the project completion date had been delayed by 58 days. Bubshait & Cunningham, at 320.

Using the Modified As-Built Method, impacts were incorporated as they occurred, the network recalculated, and the results immediately evaluated. After delay No. 2, there was no positive float available. Negative float was 70 days and the revised completion date was 9/1/93. Delay 5 had an impact on the completion date by increasing negative float from 217 to 268 days and resulted in a revised completion date of 8/31/93. The Modified As-Built Method resulted in delays of 70, 51, and 87 days, for delay Nos. 2, 5, and 6, respectively. Bubshait & Cunningham, at 320–321.

The As-Planned Method accessed a penalty duration of 56 days. The As-Built Method penalized the project by 58 days. Under the Modified As-Built Method, the project completion date was delayed by 87 days. Bubshait & Cunningham, at 321.

Page 11-78, add after first full paragraph:

The importance of the skill of the scheduler and the methodology used is underscored by the case of *Cogefar-Impresit U.S.A., Inc.*, DOTCAB No. 2721, 97-2 B.C.A. (CCH) ¶ 29,188 (1997). In the *Cogefar* case, both the government and the contractor retained scheduling experts to perform analyses of the delays and to provide expert witness testimony at the board hearings. The parties' experts employed different methods of schedule analysis. The contractor's expert measured delays by comparing the as-planned schedule against the as-built schedule. The contractor's expert took great pains to document the as-built schedule, reviewing daily reports, monthly progress reports, field reports, cost reports, payroll records, submittal log, and depositions. The contractor's experts also viewed videos and interviewed 20 contractor employees. The government's expert used a "contemporaneous time frame analysis," using information available at the time of the delay and evaluating the delay in a "window" of time using two successive schedule updates contemporaneous with the timeframe in which the delay occurred. The court noted that the conclusions of the experts were "heavily slanted" towards the position of the party that retained them, but found the contractor's expert to be more reliable because of its "thorough" review and attempt to determine the "as-built critical path." The government's expert relied on the contractor's updates without changing the update logic to reflect the "actual logic on the job." The board determined that the approach used by the government's expert did not use a current CPM schedule, and yielded a less precise result.

Page 11-78, add at end of section:

Some commentators have classified the various schedule analysis methods into three fundamental groups—prospective, retrospective, and contemporaneous. A prospective method is any schedule analysis approach that purports to demonstrate the likely impact of the delaying events. A prospective approach or foresight method will use a model of reality—essentially theoretical in nature. The model used is frequently the as-planned schedule, and the model is impacted with delaying events to demonstrate or predict the effect upon the overall project completion. A retrospective approach attempts to show what actually happened on the project. A retrospective analysis or hindsight method will often focus on when activities actually occurred, what the actual delays were, how the actual progress differed from what was planned, and reaches conclusions about what caused the project to be late. This method often emphasizes the "as-built schedule" over the "as-planned" model. The contemporaneous method attempts to establish the status of the project at the time the delay-causing events occurred, and then examine the impact upon the planned and actual completion of the project. A contemporaneous approach will often update the schedule at the time of the delay from the project record, and then incorporate appropriate changes into the schedule that may be necessary to reflect the proposed planning to complete the project. The contemporaneous method may use both prospective and retrospective aspects, as long as it is documented from the contemporaneous project record. Whatever schedule analysis method is chosen, it may be argued that it is preferable to take a fact-based approach by tying the schedule analysis to facts that can be gleaned from the project record. This makes the analysis more credible from the perspective of the trier of fact, whether it is a jury, judge, or arbitrator. References to the project records such as daily reports, job meeting minutes, inspection reports, correspondence, progress photos, and other contemporaneous documents generated during the course of construction often provides a credible link. While not necessarily proof of "cause and effect," tying the schedule analysis to a detailed analysis of the project record will at least provide solid basis for recovery. Gibson, *Analysis of Project Delay—Theoretical or Interrogation of the Facts?,* TECBAR Review 7 (July 2003); Raper & Cohen, *Delay Claims in the Construction Industry*, In-House Lawyer 38 (Feb. 2003); Brennan & D'Onofrio, *Techniques and Methods for Assessing Delays*, Cause Effect Newsletter (Spring 2002).

In selecting a scheduling method to analyze the impact of project delays, one commentator developed guidelines to ensure proper schedule analysis based on recent case law. The guidelines were stated as the "Ten Commandments" of schedule analysis:

1. Do not rely upon a schedule that was not followed during the project.
2. Consider actual performance.
3. Avoid "as-planned plus impact" analyses.
4. Establish a correlation between the plan, changes, actual performance, and contemporaneous records.
5. Consider your own delays in a delay analysis.
6. Keep schedules concurrent and reflect delays as they occur.
7. Involve the right people in the delay analysis.
8. Be objective and avoid adversarial interests that damage credibility.
9. Recognize the right to finish early.
10. Recognize reasonable resource leveling.

Wickwire, *The Scheduling Expert's Commandments*, The Millenium Construction Superconference (Dec. 9, 1999).

§ 11.09 APPORTIONING CONCURRENT DELAYS

Page 11-80, add after note 144 in second full paragraph:

In the United Kingdom, there are competing approaches that are being considered to define and apportion concurrent delays. One approach is to attempt to differentiate the various causes of delay in terms of their "causative potency." One delay may not have the same potential impact as the other, only being a minor cause, in which case, it is treated as if it did not cause the delay at all. Only the "effective cause" of delay would be considered in the delay analysis.

The responsibility to make decisions under the "dominant cause approach" could be given to the architect to determine which delay was "dominant." The architect may already have the contractual authority, as it has the responsibility under many standard form agreements to decide claims and time extensions. This approach may help sort out concurrent delays in some situations, but there are instances where more than one event is of "approximate equal causative potency." This is in essence a winner-take-all approach that actually treats the delays as not being truly concurrent in equal value. Martin, *Concurrent Delay*, 18 Constr. L.J. 6, 436–48, at 437, 445–46 (2002).

Another approach is to apply a "but-for" test. For example, in the event of two concurrent delays, one caused by extra work ordered by the architect and one caused by the contractor is delaying the performance of the extra work, the contractor may argue that the delay would not have occurred at all but for the extra work. The but-for approach may assist in sorting out such delays; however, the but-for test does not seem to have a basis in English case law for determining contract claims. *Id.* at 443.

Another approach is known as the "Malmaison approach," based upon the rationale of the case, *Henry Boot Construction (U.K.) Limited v. Malmaison Hotel (Manchester) Limited*, 70 Con. L.R. 32, QBD (TCC) (1999). In *Malmaison*, the court distinguished between "relevant events" and events that were not relevant, both of which were concurrent delays. Relevancy of an event is determined through consideration of the contractual allocation of risk between the parties. For example, do the terms of the contract specifically allocate the risk of certain delays, such as "exceptionally inclement weather," but remain silent concerning delays due to labor shortage? The contract would be examined to determine if one of the concurrent delays were specifically stated to be grounds for relief under the contract. If so, such a delay would be deemed to be relevant under the terms of the contract, and be more important to the parties. *Id.* at 447–48.

Canadian law also permits the apportionment of concurrent delay. Rather than simply permit a "winner-take-all" approach, a contractor proceeding on a breach of contract theory for delay may be able to sort out its own delays and yet still recover for the delay of the employer (owner). Martin, *Concurrent Delay*, 18 Const. L.J. No. 6 at 439 (2002), *citing Tompkins Hardware Limited v. North Western Flying Services Limited*, 139 D.L.R.3d 329 (1982); *Ribic v. Weinstein*, 140 D.L.R. (3d) 258 (1982); and *Doiron v. Caisse Populaire D'Inkereman Ltee*, 17 D.L.R. 4th 660 (1985), and *citing Convert-A-Wall Limited v. Brampton Hydro-Electric Commission*, 65 Or. 2d 385, 441 (1988).

Page 11-80, add after note 146 in second full paragraph:

The inability of a court or trier of fact to apportion concurrent delays may be due to the fact that the delays are so intertwined, the complexity of the scheduling issues, or that the parties simply failed to present credible evidence that would allow the fact finder to make a proper determination. For example, in *PCL Construction Services, Inc. v. United States*, 53 Fed. Cl. 479 (2002), PCL was awarded a contract by the United States Bureau of Reclamation (USBR) for the construction of the Visitor's Center and Parking Structure at the Hoover Dam. During the course of construction, PCL was allegedly delayed by defective design drawings and specifications, differing site conditions, and change order work, and submitted a request for equitable adjustment in the amount of $23 million. USBR invited PCL to participate in a joint fact-finding session to prepare an as-built schedule that would be used to agree upon responsibility for critical delays to the project completion. PCL declined the offer, and interpreted the USBR's response as a denial of its claim. The project was over a year late, and the government withheld $1.3 million for liquidated damages and incomplete contract and punch list work. PCL disputed its obligation to perform the work related to the retainage allegedly because it was outside of the scope of the contract. The government terminated the contract for default, and the contract disputes went to a trial before the United States Court of Federal Claims. The court's first post-trial opinion addressed most of the legal and factual issues, and the second opinion addressed the issues of the PCL's entitlement to the retainage for liquidated damages and uncompleted punch list work. In considering the liquidated damages claim, the court noted that neither party submitted sufficient credible evidence on how the changes and delays alleged affected the other activities so that the court could clearly apportion the delays between the parties. PCL's claim for delay and disruption was a total cost claim, and did not present an in-depth critical path analysis. In regards to PCL's claim for return of the liquidated damages, PCL presented a "summary level as-planned and as-built retrospective analysis." This analysis used the as-planned schedule and made adjustments to it to develop a baseline schedule. "Summary roll-up" activities were developed from the project records, and the as-planned baseline schedule was compared to the summary level as-built roll-up schedule. From this as-planned/as-built comparison, PCL's scheduling analysis concluded that USBR was responsible for 18.8 months of delay, and that three months of "unconceded" project delay was concurrent with government caused delay, so the three months was non-compensable, excusable delay. The government's expert also used a comparison of the as-planned and as-built schedules, although in more detail, and the government's expert concluded that USBR was responsible for only 16 percent of the extended performance period.

However, the court was not impressed with either party's presentation or analysis of the project delays. The court noted several flaws in the facts contained in both PCL's and USBR's expert reports. PCL's expert did not even rely upon the "summary level as-planned and as-built retrospective analysis" prepared by his firm in giving expert opinion on the breach of contract claims. PCL's expert acknowledged that several project documents were either missing or overlooked during the preparation of the schedule analysis. The court noted that USBR's expert did not analyze exactly how many days were caused by either party or were concurrent. The court criticized the fact that the USBR expert focused only on government directed change work because of his assumption that if delay was not identified resulting from an actual change order, the delay must be attributable to PCL. The court concluded that the parties did not provide the court with sufficient evidence for a clear apportionment of the delays that occurred on the project.

Page 11-80, add after note 147 in second full paragraph:

When critical path delays in a project are caused by both the government and a contractor and cannot be apportioned with any certainty, an assessment of liquidated damages against the contractor is not valid. For example, in *George Sollitt Constr. Co. v. United States*, 2005 U.S. Claims LEXIS 52 (Feb. 23, 2005), the United States Department of the Navy awarded George Sollitt Construction a contract for the renovation and new construction of various buildings at a naval training center. The work was to be conducted in three phases, but the phases were not completed on time. The contractor had been required by the contract to deliver a baseline construction schedule and to provide monthly updates to that schedule throughout the duration of the project. The contractor was late in providing its baseline Critical Path Method (CPM) and failed to provide monthly CPM updates for some months. The absence of CPM schedule updates made the proof of delay damages difficult. The contractor filed suit against the government alleging numerous claims, including compensable delay.

The contractor alleged that circumstances chargeable to the government delayed the completion of Phase I construction. The government's post-award revisions to a ship's trainer, changes to a building's electrical service, and changes and unforeseen conditions encountered when constructing new buildings all allegedly caused delays in Phase I. The government argued that the contractor did not prove that the three alleged circumstances caused the delayed completion of Phase I construction. The government also argued that the contractor delayed the completion of Phase I construction primarily because of its tardy procurement of structural steel and other items. Although the contractor was able to prove that 28 calendar days of delay were chargeable to the government due to its revisions to the ship's trainer, some or all of those days were also chargeable to the contractor for its concurrent delays in steel procurement. In addition, the delay in the substantial completion of the new buildings was not proved to be excusable. Therefore the contractor was also responsible for its delay for that portion of Phase I construction. The delays were intertwined and could not be apportioned with any certainty. No expert for either party attempted to trace one critical path through Phase I construction or to apportion delays attributed to each party along such a critical path. The substantial completion date for Phase I construction would have been highly speculative. The court could not compare and apportion delays to portions of Phase I construction. Apportionment of liquidated damages based on the overall concurrent delays to all of Phase I construction would have been even more speculative, as the record did not offer the tools to accomplish that task. Because the concurrent critical path delays of the two parties could not be apportioned with any certainty, the assessment of liquidated damages for Phase I construction was not valid and had to be returned to the contractor. Final judgment entered for contractor.

Page 11-81, add after carryover paragraph:

In *Clark Construction Group, Inc.*, No. VABCA-5674, 00-1 B.C.A. (CCH) ¶ 30,870 (2000), Clark originally planned to construct a VA hospital using a "horizontal" construction sequence. Clark intended to complete the structure of each floor of all three wings before constructing the next floor. Clark developed an "interim" schedule that reflected scheduling logic using a horizontal construction sequence. As a consequence of a stop order, Clark re-sequenced the job from the planned horizontal to vertical construction. Consequently, Clark commenced vertical construction of the West Tower where the foundation was complete while concurrently, on a limited basis, continuing East and Center Wing foundation work. The interim CPM schedule required by the contract, still under development at the time the first stop work order was issued, was abandoned.

However, the change to vertical construction took place well before the major part of the subcontractors' work began. Apparently, the subcontractors took no actions to adjust their material delivery or packaging schedules to mitigate the disruption. The fact that Clark did not seek nor did the subcontractors attempt any input or apparent influence on the development of the vertical construction CPM schedule provided some explanation for the subcontractors' failure to adjust their material deliveries. Also providing some reason for the subcontractors' material handling difficulties, the record showed that Clark did not progress the job according to the schedule, and the subcontractors were reduced to looking out the window to see where Clark was working in order to assign its forces. The subcontractors' claims did not consider all the other delays identified in the record, but blamed all the delay on the government. The board concluded that the subcontractors failed to adequately prove that the change in working conditions resulting from the sequence change was the sole cause of its material handling inefficiencies.

Similarly, in *Par Painting, Inc. v. Greenhorne & O'Mara Inc.*, 61 Conn. App. 317, 763 A.2d 1078 (2001), the Connecticut appellate court refused to permit a subcontractor to collect its delay damages when the delays were concurrent with the subcontractor's own delays. A subcontractor that sandblasted and repainted steel structures brought an action against a bridge inspector on the repainting projects, alleging that it had inspected the subcontractor's work too slowly and in bad faith, resulting in the subcontractor's inability to perform under the contract to repaint a bridge. The subcontractor claimed tortious interference with business relationships. A jury found in favor of the subcontractor. However, the trial court granted the inspector's motion to set aside the verdict.

According to the appellate court, the jury could not properly have concluded that the inspector's action was the cause of the subcontractor's delayed completion. Despite evidence of the subcontractor's very late start and other difficulties performing its work, employees' high blood lead levels that prevented work, and lapse of the subcontractor's certification, the jury found that the conduct of the inspector was the factor that kept the subcontractor from finishing its work on time. Even when viewing the evidence most favorably to the subcontractor and assuming that duplicate two-hour inspections in fact occurred, the jury could not with reasonable certainty have concluded that such inspections were the reason the subcontractor had completed only 33 percent of the work well after the contract period had expired. The reviewing court agreed with the trial court's determination that it was impossible for the jury to find (without speculating) that the inspector's tortious conduct, as opposed to its legitimate inspections or the subcontractor's own delays, caused the subcontractor's losses, considering the many nonactionable delays that occurred on the project. The court held that there was no showing of a causal connection between the inspector's conduct and the subcontractor's delays. The inspector presented ample evidence that the subcontractor's work was inadequate and that the inspectors were merely doing what they were hired to do. The trial court judgment was affirmed.

In *Net Construction, Inc. v. C & C Rehab and Construction, Inc.*, 2003 U.S. Dist. LEXIS 6141 (E.D. Pa. 2003), C & C Rehab and Construction, Inc. served as the general contractor for the Chester Housing Authority. From February 3, 1997 through March 25, 1999, when C & C was removed from the project by a federally appointed receiver, Net Construction, Inc. performed installation of concrete site work services. Net's work was delayed and its performance costs increased. Net filed suit, claiming $383,366.29 in consequential damages as a result of C & C's breach of its contract with Net.

A default judgment hearing determined the amount of damages suffered by Net. Net's only witness was its president who testified that the project proceeded in two "phases" each of which involved work on 10 to 12 homes that were basically identical. As the concrete site work contractor, Net performed work directly related to the construction of the buildings, and involved pouring concrete "footers" and "base slabs." Net also performed miscellaneous site-work, including the construction of curbs and sidewalks. Work on Phase I was supposed to be completed in July 1997, but it did not end until April of 1998. Net lost $103,188 on Phase I of the project. The whole project, including the completion of Phase II, was supposed to end in April 1998, but the project was not complete until April 1999. Net lost $139,000 on Phase II. The delay in finishing the project was due primarily to C & C's poor management, especially during of Phase I of the project. However, there were other problems with the construction site that led to delay, including the existence of asbestos at the site, unsuitable soils, and weather-related issues. Further more, Net itself contributed to the delay of the project's completion because Net experienced problems with its concrete supplier and with pouring a floor.

The court found that Net failed to establish a reasonable allocation of its extra costs as a result of particular delays caused by C & C. While Net established that there were delays in completing the construction project, it failed to demonstrate that its damages for lost productivity were the result of delays for which C & C was responsible. "No basis appears for even an educated guess" as to the increased costs suffered by Net as a result of delays for which C & C was responsible. Net failed to distinguish losses suffered as a result of delays by C & C from losses that it might have suffered because of Net's own performance problems, weather, unsuitable soils, or the existence of asbestos at the site. Net was not entitled to recover damages on its claim of lost productivity and extended home office overhead.

Updated schedules submitted during the performance period that do not apportion concurrent delay due to change orders do not satisfy the obligation to prove how project delays are the responsibility of the government. *CEMS, Inc. v. United States*, 59 Fed. Cl. 168, 2003 U.S. Claims LEXIS 324 (2003). The Department of Transportation, Federal Highway Administration (FHA) awarded a contract to CEMS for the construction of the bicycle path on a portion of the Historic Columbia River Highway. On April 14, 1998, the FHA received a CEMS schedule for the completion of the bicycle path. The April 14, 1998 schedule indicated the critical path of the bicycle path project, the activities on the critical path, and the float time in the activities that could affect the critical path. During the course of the project, CEMS submitted 21 revised construction schedules to the FHA. The construction schedules included two-week narratives for upcoming work, preliminary schedules, or critical path schedules. The revised schedules noted changes on the project affecting the CEMS schedule. In an October 8, 1998 transmittal letter for CEMS's 21st revised construction schedule, tabulation, and written narrative, CEMS notes "that several Government changes and delays are extending scheduled work beyond the current completion date. These scheduled work items, changes and delays include, but are not limited to, the following. . . ." The FHA assessed $32,800.00 in liquidated damages at $800.00 a day for 41 days. The plaintiff alleged changes, delays, and other compensable acts or omissions of the FHA and requested an equitable adjustment under the terms of the contract. Following the denial of many of the plaintiff's claims by the contracting officer, CEMS filed its complaint in court for $1,241,203.70.

CEMS stated that the FHA was constantly provided with updated construction schedules, which showed CEMS's critical path delay, due to "ongoing FHA changes and delays." The schedules provided by the plaintiff, however, did not provide the court with the ability to determine whether the alleged delays claimed by the plaintiff were on the critical path, or whether concurrent delays associated with other problems also contributed to the delay of the project. At trial, CEMS testified that the delayed work was on the critical path, but that other unidentifiable activities also were on the critical path. The court was not able to determine the degree that CEMS was delayed due to the changes ordered by the defendant. The court was unable to base an accurate award on the delay calculations made by the contractor. The contractor's delay analysis identified delay from August 15, 1998 through May 22, 1999, but had not shown that the days of delay asserted under the specific claims were based on much more than CEMS's own estimates and speculation provided at trial. Nor could CEMS explain how the continuously changing days of delay were determined, or why it took until the day of trial to determine the number of delay days. For example, CEMS testified regarding one aspect of the delay claim as follows: "I haven't had a chance to review all the documents on this particular one, but I wrote a number down here of 10 to 14 days." The court found that CEMS had not carried its burden of establishing compensable delays. The plaintiff was not entitled to recover liquidated damages withheld by the agency.

In *City of Westminster v. Centric-Jones Constructors*, 2003 Colo. App. LEXIS 1432 (2003), the City of Westminster and Centric-Jones Constructors contracted to expand the City's water treatment plant. Jones began construction using designs and specifications prepared by the engineering firms. The contract between the City and Jones provided $1,000 per day in liquidated damages for each calendar day following the designated completion date that the project was not operational. Work continued beyond the scheduled completion date. Disagreements arose over water leakage and structural problems in the walls and foundation. After most of the work was done, the City terminated Centric-Jones. The City then hired new engineers. They recommended that the City demolish and rebuild, using a significantly different foundation design, and demolish and rebuild the masonry walls to new specifications. The City did so, although expenses substantially exceeded the original construction costs and the new specifications added features. The City sued Jones for breach of contract and negligence, claiming as damages the entire cost of removing, redesigning, and rebuilding the walls, and modifying the foundation. Jones counterclaimed for lost profits from unfinished work based on wrongful termination of the contract and for nonpayment for work performed to the termination date. The court entered summary judgments in favor of Jones on the City's negligence claim. The trial court concluded that the City failed to provide any basis on which the jury could apportion damages, either between the benefit of the City's bargain with Jones and additional benefit to the City from rebuilding the structures to new specifications, or between particular breaches by Jones and design errors of others for which Jones was not responsible.

The City presented testimony that it was entitled to recover $1,994,500 for approximately six years of delay. However, the evidence also showed that the delay was caused in part by the City's decision to redesign the structures, including correction of errors by the initial designers. The City's liquidated damages claim failed for two reasons. First, because the City's redesign went beyond correcting breaches by Jones, the City was at least partially responsible for the delays. Second, the City failed to apportion the total delay between optional redesign and necessary correction of Jones's defective construction. Consequently, the jury would have been required to speculate as to what portion of the delay resulted from Jones's construction defects and the portion of the delay damages for which Jones was responsible. The trial court properly directed a verdict as to liquidated delay damages.

Page 11-82, add to carryover paragraph:

If the owner admits that it delayed the project, it will not be able to assess any liquidated damages. In order for the owner to assess liquidated damages, it must be able to apportion responsibility for delays between the owner and contractor with sufficient credibility that the court will be able to clearly apportion the responsibility for delay. *PCL Constr. Services, Inc. v. United States*, 53 Fed. Cl. 479 (2002).

[A] Apportioning Consecutive Concurrent Delay

Page 11-82, add to Abrahamson cite at note 154:

at 139 (4th ed. 1979).

Page 11-83, add after fourth full paragraph:

In *Essex Electro Engineers, Inc. v. Danzig*, 224 F.3d 1283 (Fed. Cir. 2000), the federal court found that shop drawing submittal and approval are inherently apportionable, but may distort the overall measurement of delay. In *Essex*, the Marine Corps Research, Development and Acquisition Command awarded Essex a contract to manufacture, test, and deliver a number of skid-mounted floodlights. The contract required Essex to produce and test two prototypes. Essex notified the government that the drawings contained errors and omissions that had to be corrected before Essex could assemble the prototypes. Essex stopped work because of the errors. The government's technical project officer contacted Essex by telephone on May 10 and provided additional details and revised dimensions that addressed some of Essex's concerns. Then, on May 21, the contracting officer sent Essex a letter advising that changes to the drawings appeared to be indicated and directing that the proposed changes to the drawings be submitted as engineering change proposals (ECPs) pursuant to the contract. On November 26, Essex submitted the ECPs on the correct form, providing the total proposed cost for each ECP, but no cost details. The government, on February 11, 1992, notified Essex that it had approved some of the ECPs but had rejected others. Two of the disapproved ECPs concerned the dimensional errors and detail omissions that were the subjects of the government's May 10, 1991, telephone call to Essex. The government informed Essex that Essex needed to submit additional cost details and a revised delivery schedule to complete the approved ECPs. Essex submitted the required cost details on May 5, 1992.

The Board found that the government had provided Essex with defective drawings and was liable for the resulting harm. The Board awarded Essex recovery of the engineering and production costs to incorporate the disapproved ECPs into the design. In addition, the Board found the government liable for Essex's costs associated with the 25-day delay between Essex's notice to the government of the drawing defects and the government's May 10 telephone call providing additional detail and revised dimensional information, because that delay affected the production of the prototypes. For the period after May 10, however, the Board concluded that Essex had acted unreasonably in stopping work pending the government's acceptance of the ECPs. Any government-caused delays after that date were concurrent with Essexcaused delays and precluded Essex from further recovery based on delays associated with the erroneous drawings.

On appeal to the Federal Circuit court, Essex argued that the government's liability for the defective drawings did not end with the May 10 telephone call in which the government allegedly provided Essex with correct dimensions and details. The telephone call from the government required Essex to choose whether to produce sets that did not conform to the official drawings, and thus could be rejected by the government; or to defer further prototype development pending formal ECP approval by the contracting officer. Neither the contract nor case law required Essex to proceed at its peril following the government's telephone call. The Federal Circuit court did not, however, accept Essex's contention that the government's liability extended to all the delays associated with approval of the ECPs. Essex caused delay in approval of the ECPs by failing to follow the applicable standards prescribed by the contract to provide cost information. Both Essex and the government may have delayed the approval of the ECPs in other ways as well. The government requested ECPs for the defective drawings on May 21, 1991, and Essex submitted them on July 30. The government was liable for whatever period of time Essex reasonably needed to respond to the government's written request for the ECPs, but not for any longer. The Board on remand had to determine whether Essex acted reasonably in taking 70 days to submit the original ECPs, and if not, what constituted a reasonable period for preparing the ECPs. Even though Essex submitted the ECPs on the wrong form and may have taken an unreasonable amount of time to submit the original ECPs, Essex's conduct, in turn, did not relieve the government of its obligation to proceed reasonably thereafter. After Essex's initial ECP submission on July 30, 1991, the government took 48 days to respond simply that Essex had not followed the applicable standard. Within 15 days of the government's response, Essex requested a meeting with government officials concerning the ECPs. Thirty-nine more days elapsed before the government met with Essex on November 8 and stated that the ECPs required cost data. Essex then submitted ECPs with estimates of total costs. The record suggested that the government did not respond again for another 77 days, when it notified Essex that some of the ECPs had been approved, subject to the submission of cost details. On remand, the Board had to determine whether the government acted reasonably in the substance and pace of its responses to Essex's various submissions, keeping in mind that the need for the ECPs arose from the government's defective specifications in the first place. A finding that the government did not act reasonably in responding to Essex's ECP submissions may affect the Board's conclusion that most government-caused delays were concurrent with Essex-caused delays. The sequential nature of Essex's submissions and the government's responses rendered each party's delays inherently apportionable, at least in the case of ECP-related delays.

The amount of any overall delay chargeable to the government should be equal to the period between the time the submissions would have been approved absent culpable delay by the government and the time Essex actually began testing. This inquiry required the Board to focus on the overall effect that government-caused delay had on the beginning of prototype testing, and not to focus on each discrete period of delay and then automatically treat as concurrent delay any period of government-caused delay during which Essex was also causing unrelated delay. That type of instance-by-instance analysis of the delays could result in distortion of the proper measure of overall delay. In the absence of any government-caused delay, Essex's unrelated delays might have been concurrent with each other (rather than concurrent with government-caused delays), so that the overall delay in contract completion would not have been as great.

Page 11-84, add after first paragraph:

In *Alcan Electrical & Engineering Co. v. Samaritan Hospital*, 2002 Wash. App. LEXIS 33 (2002), the court concluded that subcontractors that each contribute to project delay were not jointly and severally liable for the entire project delay if their delays can be apportioned. Gilbert H. Moen Company was the general contractor for the renovation and expansion of Samaritan Hospital. Alcan Electrical & Engineering Co. was a subcontractor on the project. Alcan sued Moen for payment. Moen filed a counterclaim against Alcan, arguing that any payment was subject to an offset for liquidated damages resulting from Alcan's delay of the project. The trial court found a 201-day delay was the result of another subcontractor's unauthorized installation of nonconforming pipe, as well as Alcan's failure to submit fire alarm shop drawings to the Fire Marshall in a timely manner. The court determined that the other subcontractor's installation of the wrong pipe delayed the project 170 days. The court found Alcan's untimely submission of fire alarm shop drawings caused a 31-day delay. Alcan's award was reduced for its 31-day delay. Claiming Alcan was responsible for more than 31 days of the delay, Moen appealed.

Moen asserted that the two subcontractors caused a concurrent delay and were therefore both liable for liquidated damages for the same period of time. Moen claimed that both Alcan and the other subcontractor should have been held to be jointly and severally liable for the entire delay. However, the appellate court declined to so hold. Moen also took issue with the court's apportionment of the delay between the two subcontractors. It argued that Alcan was responsible for more than 31 days of the delay. During November and December 1991, Alcan was working with the Fire Marshall to prepare satisfactory drawings. In December 1991, Alcan wrote to Moen, asking when it would get a response from the architect regarding the fire alarm system and when it would be able to complete its work. During this time, tests of the other subcontractor's nonconforming pipe were being done. These tests required the removal of walls and ceilings. The pipe testing process went almost into January 1992. It could be inferred that, until the pipe testing was complete, Alcan could not have installed the fire alarm system. Therefore, it was reasonable for the court to find that Alcan was only responsible for the 31-day delay from January 4 to February 4, 1992.

§ 11.10 USING THE SCHEDULE TO PROVE OUT-OF-SEQUENCE WORK AND OTHER DISRUPTIONS

Page 11-97, add at end of section:

Contractors' efforts to protect their productivity rates also are hampered by the prevailing view that total float is "owned by the project" and is freely available on a first-come-first-serve basis to both owners and contractors. When attempting to analyze the disruptive impacts of delays, however, there are two arguments in favor of using free float rather than total float. First, while total float determines delays to overall project completion, free float is a more relevant schedule characteristic when looking for delays to activities and increases in activity overlaps. Since disruption is largely caused by activity overlaps, it may be reasonable to analyze disruptive results using free float and leave total float to delay analyses. Second, total float indicates the amount of float available to a chain or path of activities. Total float is shared by all activities along a given chain, and does not necessarily represent the float available to a specific activity. M. Finke, *Schedule Density as a Tool for Pricing Compensable Float Consumption*, 42 Cost Eng'g 34, at 35 (No. 6, June 2000).

Changes in schedule density are practical indicators of the disruptive effects of free float consumption. Schedule density adds the total of all activities' free float to determine the proportion of activity durations to free float. Reducing the free float of activities that have high labor costs should be more disruptive than if the activities had relatively low labor costs. Disruption may be measured by adding labor costs to the density determination and then measuring the effect of any reduction of free float. If such information were provided for activities in the schedule, weighted schedule densities could be calculated for specific locations, trades, and/or supervisors to more narrowly focus on specific causal mechanisms. Finke, at 35.

Increases in schedule density should never reduce a contractor's costs, instead, they should tend to increase activity-specific costs for the remainder of the work. Changes in (remaining) weighted schedule density should be considered in light of the schedule's total remaining labor costs. A given change in weighted schedule density may have a greater effect if the schedule's remaining labor costs total $500,000 rather than if they total only $100,000. Conversely, a given schedule change also will tend to cause smaller changes in weighted schedule density if it occurs earlier in a schedule. Without any revisions to logic or activity descriptions, a noncritical delay should always cause a potentially disruptive increase in weighted schedule density because the sum of activity durations will stay the same while the sum of activity free floats will decrease. On the other hand, a critical delay will decrease a schedule's weighted density because the sum of activity durations will stay the same while the sum of activity free floats will stay the same or increase. As a result, a critical delay could very well be less disruptive than a noncritical delay. Finke, at 36–37.

CHAPTER 12
DELAY DAMAGES

§ 12.01 THE IMPORTANCE OF PROVING DAMAGES

Page 12-3, add after first paragraph:

In *Clark Construction Group, Inc.*, No. VABCA-5674, 2000 VA B.C.A. LEXIS 4 (Apr. 5, 2000), the VA did not contest that it took extended periods to respond to some RFIs, primarily because of the failure of its A/E to act promptly. The average response time to RFIs relating to the 10 systems or areas impacting Clark's subcontractors ranged from 38–235 days, which appeared, according to the board, to raise questions. However, the reasonableness of RFI response time, in terms of disruption of installation labor, must be gauged from the context of when the RFI was submitted and when the work was to be performed. Most of the RFIs at issue were submitted beginning in the last quarter of 1991 through 1992. The VA's failure to respond until mid-1992 to an RFI submitted in December 1991 relating to an installation scheduled for December 1992 was not necessarily unreasonable and, by itself, did not result in VA's liability for a labor disruption experienced in that installation.

In addition, proof that the late RFI response caused the disruption of its labor required corroboration by reference to contemporaneous documentation. Disruption should leave a trail through the project documentation, including daily logs, CPM fragments, correspondence, and other contemporaneous contract documentation to support that the late RFI responses changed the expected working conditions and that the change to working conditions disrupted labor and the extent of the disruption. The board inferred, in light of the subcontractor's failure to cite any of the extensive contemporaneous documentation in the record to support late RFI response caused disruption, that the documentation in the record did not support the disruption. The board concluded that the subcontractors failed to prove by a preponderance of the evidence that the VA's response to any particular RFI was unreasonably late or that a late RFI response by the VA caused a loss of labor productivity.

Causation between the delays and the claimed costs must also be demonstrated. In *PCL Construction Services, Inc. v. United States,* 2004 U.S. App. LEXIS 6705 (Fed. Cir. 2004), PCL was awarded a $33.8 million contract by the U.S. Bureau of Reclamation (USBR) in 1991 to build a visitors' center and parking structure for the Hoover Dam site. During the course of construction, the contractor alleged various design deficiencies, delays, and differing site conditions. The parking structure was over one year late—the contract required completion of the parking structure on February 15, 1994, but it was not substantially complete until May 12, 1995. The USBR terminated the construction contract for default on March 6, 1996, and PCL filed suit for $33.5 million on a quantum meruit basis. The trial in the U.S. Court of Federal Claims resulted in a rejection of the PCL claim because PCL did not show a "cause and effect relationship" between USBR changes, design deficiencies, or other USBR actions and PCL's claimed costs. There was no critical path analysis to quantify the amount of delay to specific design errors, changes, or other USBR actions. The trial court stated that PCL only presented broad assertions by its experts that USBR was responsible for some 22 months of delay but made no allocation of the delay between USBR and PCL, and that PCL did not present evidence that the delay increased its costs. U.S. Court of Appeals for the Federal Circuit agreed that PCL did not meet its burden of proof on the issue of causation of damages, that is, the burden to show the nature and extent of the various delays for which damages are claimed and connect them to some act or omission of the USBR.

Page 12-4, add after first carryover paragraph:

A contractor that is not fully cooperative in sharing financial reports and uses an unsophisticated accounting system is still able to prove that it incurred increased costs caused by a change in site conditions by submitting its handwritten payroll records. For example, in *Appeal of Environmental Safety Consultants, Inc.*, 2005 ASBCA LEXIS 21 (Mar. 8, 2005), the U.S. Navy awarded a contract to Environmental Safety Consultants (contractor) to collect, transport, and dispose of sludge from two lagoons. The contractor made a claim to the contracting officer for increased costs. The contracting officer denied most of the contractor's claim by final decision. The Armed Services Board of Contract Appeals held that the contractor was entitled to an equitable adjustment for additional costs incurred in performance of its contract. Pursuant to the board's order, the company filed its statement of costs detailing its increased costs. Among other items, the statement included costs incurred due to higher solids content than expected in the lagoons, which increased the cost to break down the solids to dispose of the sludge.

According to the contractor, the government received benefit from the contractor's work but did not provide compensation. The government argued that the contractor had not met its burden of proof and there was an absence of any reliable or credible records. According to the government, the contractor failed to show in its accounting records that it incurred increased costs solely because of the changed conditions. The government objected to the contractor's resistance to producing specific records. The government argued that the contractor did not properly allocate and segregate its costs for base work and changed work. Since the contractor did not accumulate cost data and could not identify its actual costs attributable to changes, estimates were permissible to quantify the increased costs the contractor incurred. The contractor was entitled to recover the reasonable additional costs of removing and disposing of the sludge that resulted from the change in the solids content of the sludge. The government relied on the absence of complete records that would generally have been found in an established corporate government contractor accounting system to dispute the contractor's recovery for those costs, but the lack of sophistication in the contractor's accounting system did not affect the credibility of the documentation provided in support of the costs claimed. Although the contractor was not fully cooperative in permitting the inspection and copying of all financial records, the contractor met its burden of proof by submitting handwritten payroll records that demonstrated the increased costs caused by the change in conditions. The government's argument that the contractor failed to meet its burden of proving causation was without merit. The court accepted the contractor's proof of increased costs concerning the lagoons. The appeal was sustained in part and denied in part.

§12.04 DAMAGE CALCULATION THEORIES

[B] Total Cost Method

Page 12-13, add after second sentence in first paragraph:

Federal courts hearing contract disputes also take a conservative approach to the use of the total cost method of calculating damages. The U.S. Court of Federal Claims stated, "Trial courts are advised to use this method with caution because bidding inaccuracies can create an unrealistically low estimate of the contractor's costs, and performance inefficiencies can increase the costs incurred." *Baldi Brothers Constructors v. United States*, 50 Fed. Cl. 74, 79–80 (2001).

[1] Permissible Uses of the Total Cost Method

Page 12-17, add after second full paragraph:

Ordinarily, recovery under the total cost method requires all elements must be proved. Where the action is maintained under the Miller Act, the rules are somewhat different because whether or not the contractor was at fault is not at issue. Recovery under the Miller Act does not require proof of the contractor's fault. Without the fault allocation between the contractor and the subcontractor, the total cost method is a good tool for subcontractor's recovery under the Miller Act. *Lighting & Power Servs. v. Roberts*, 354 F.3d 817, 2004 U.S. App. LEXIS 364 (8th Cir.). See also **§15.02**.

Page 12-20, add after carryover paragraph:

The case of *Mid-Atlantic Constructors, Inc. v. Stone & Webster Constr., Inc.*, 2005 U.S. Dist. LEXIS 26624 (E.D. Pa. Nov. 3, 2005) indicates that in Pennsylvania, the total cost method may be used to calculate a subcontractor's damages. In *Mid-Atlantic Constructors,* Stone & Webster Construction (contractor) subcontracted with Mid-Atlantic Constructors (subcontractor) in connection with the construction of a combined cycle power plant. The subcontract required the subcontractor to set equipment and install piping and pipe racks. The contractor later terminated the subcontractor. The subcontractor alleged that the contractor caused delays, accelerated conditions, and inefficiencies in the subcontractor's performance of its duties, causing it to incur substantial cost increases and delays. The subcontractor brought a breach of contract action against the contractor. The parties filed cross motions for summary judgment.

The contractor argued that the subcontractor's claim had to fail because the total cost method used to calculate the subcontractor's damages was invalid in Pennsylvania and because the subcontractor did not account for delays and productivity losses it caused. As evidence of the subcontractor's role in some of the delay and lost productivity, the contractor cited deposition testimony of the subcontractor's employees who admitted that the subcontractor could have done more to increase productivity. The contractor claimed that the subcontractor used the total cost method because it could not isolate the causes of the breach. The subcontractor countered that it could justify its use of the total cost method at trial. The subcontractor asserted that it did not include delays for which it was responsible in the damages it sought. The subcontractor cited its expert reports as evidence of delays for which the contractor was directly responsible. The total cost method was viable in Pennsylvania as a measure of damages. To demonstrate that the total cost method should be used, the subcontractor had to show that the nature of its losses made it impossible or impracticable to determine damages by another method, that its bid was realistic, that its actual costs for the project were reasonable, and that it was not responsible for the added costs of the project. Because there was a genuine issue of material fact as to whether the total cost method applied and whether damages from delay were due, summary judgment as to the subcontractor's breach of contract claim was not warranted. The contractor's motion was granted in part and denied in part, and the subcontractor's motion was denied.

Page 12-20, add at end of subsection:

In *McKie v. Huntley*, 620 N.W.2d 599 (S.D. 2000), the concrete contractor priced its claims for delay and extra costs upon the actual labor and material costs for amount of concrete placed, and then subtracted the amount of the payments it received. The trial court granted summary judgment to the defendant because the damages as calculated by the claimant were vague and speculative. On appeal, the South Dakota Supreme Court reversed the summary judgment because the lower court failed to recognize that the total cost method has legitimacy under certain circumstances. If the claimant was able to prove the prerequisite elements or modify the total cost calculation for its own problems, the court might entertain use of the total cost method. However, to reject the total cost methodology as a matter of law was inappropriate.

In *Baldi Brothers Constructors v. United States*, 50 Fed. Cl. 74, 79–80 (2001), the contractor was awarded a contract by the government to construct a tank training range where extensive earthwork was required. However, the project was located on federally protected wetlands, a fact that was not indicated in the contract documents. After award, the government provided the contractor with a wetlands designation map indicating that 80 percent of the site was affected. The government admitted that its failure to disclose the wetlands was a Type II differing site condition but rejected the contractor's pricing of the claim on a total cost basis. The court found the modified total cost method of calculation to be acceptable because nearly all of the work was impacted—this was not a situation where only discrete work activities were subject to increased performance costs. The court did modify the total costs claimed by the contractor based upon what it found to be overly optimistic equipment production rates.

[2] Challenging the Total Cost Method

Page 12-21, add to end of section:

An effective challenge to the use of the total cost method of calculating damages is to point out that the claimant using the total cost method was responsible for some of the problems or that there were problems that were not the responsibility of the defendant that may have resulted in damages that are included in the total costs claimed. In *Net Construction, Inc. v. C&C Rehab and Construction, Inc.*, 256 F. Supp. 2d 350 (E.D. Pa. 2003), C&C was awarded contract for work on a public housing project. C&C subcontracted the concrete work for the foundations, curb, and pavements to Net. There were many problems on the site that delayed Net's subcontract work and the overall project finished late. Net sued C&C in federal district court, alleging that C&C's mismanagement of the project caused the delays and further required Net to change its planned method of performing the concrete work, reducing Net's productivity and increasing Net's costs of performance. In proving its damages against C&C, Net compared the actual costs of its performance of the subcontract work with the originally estimated costs, and sought the difference from C&C.

The U.S. District Court rejected C&C's damages because they were based upon a total cost method of calculating damages. The problem was that Net could not distinguish between the damages resulting from C&C's alleged mismanagement and the damages that may have been caused by other problems that included the discovery of asbestos on the project site, unstable soils conditions, bad weather, Net's own defective concrete work, and Net's problems with its concrete supplier. The court reasoned that it would be unfair to hold C&C liable for the cost of problems that were not C&C's responsibility, and Net could not distinguish between the various causes of its losses.

In *Biemann & Rowell Co. v. The Donohoe Companies, Inc.*, 556 S.E.2d 1 (N.C. App. 2001), the defendant successfully challenged the plaintiff's total cost claim on numerous grounds. The parties were separate prime contractors for the construction of a neuropsychiatric hospital at the University of North Carolina Chapel Hill campus. The plaintiff, Biemann and Rowell (B&R), was the HVAC prime contractor, and Donohoe was the prime contractor for general construction. Under North Carolina state law, a separate prime contractor may sue another prime contractor for economic loss for failing to perform its duties under the terms of the separate prime contract. Donohoe, as the general contractor, was assigned the role of project expediter and scheduler. Donohoe hired a schedule coordinator to develop and update a CPM schedule that indicated completion of the project in 1,004 days, but the project was delayed by 369 days. According to the bench trial, the delays were due to bad weather, logistical problems in coordinating the contractors on a congested site, structural defects, and owner changes. B&R was required to work out-of-sequence and accelerate its work pace, and incurred delay and disruption costs for which it sued Donohoe, alleging that Donohoe failed to expedite the project, failed to properly perform its work, supervise the work of its subcontractors, and properly schedule the work on the project. Rather than use a direct or actual cost method of quantifying its losses stemming from Donohoe's actions, B&R relied upon a modified total cost method of calculating damages. The court characterized B&R's modified total cost method as allocating "only a narrow set of costs to itself, and then attributed the remainder of the cost overrun entirely to the defendant." The trial court determined that B&R failed to meet most of the requirements necessary to recover under the total cost method. First, B&R failed to establish that it was impractical to prove the losses directly. Even though B&R kept a daily log of its labor overruns, it made no attempt to tie the extra labor costs to Donohoe's delays. Second, B&R did not establish that its bid was reasonable. B&R's own employees testified that its bid was "aggressive" and B&R did not produce any other HVAC bids for comparison. Third, B&R failed to prove that its costs were reasonable, or that its damages could be ascertained and measured with reasonable certainty. The Court of Appeals of North Carolina reviewed the findings of the trial court, and applied the traditional four-part legal test for the recovery under the total cost method, and affirmed the trial court's decision.

In the seminal case of *Amelco Electric v. City of Thousand Oaks*, 38 P.3d 1120 (Cal. 2002), the court found that the contractor failed to meet the fourth requirement of not being responsible for any of the added costs that it might be claiming. Amelco was awarded a $6.1 million lump sum, prime contract for the electrical work on an office building, city council chamber, forum theatre, and performing arts theater. The city selected a construction manager and bid the project using multiple prime contracts. During the course of performance, Amelco claimed that it was delayed and disrupted by hundreds of design changes that made it impossible to keep track of the impact of any one change on the project. There were 1,018 sketches that clarified or changed the contract drawings, 248 that affected Amelco's electrical work, Amelco requested 221 change orders, and 32 change orders were negotiated for over $1 million—a net increase of nearly 17 percent. By the time the work was complete, there were serious disputes between Amelco and the city, and Amelco submitted a total cost claim for $2.2 million. Amelco filed suit against the city for breach of contract and abandonment of the contract by the city, alleging that the city failed to follow the change order procedures in the contract and that the final project differed substantially from the original design. The jury found that the contract had been abandoned by the city, and awarded Amelco $2.1 million using a modified total cost approach. The California Court of Appeals upheld the verdict. The city appealed the matter to the California Supreme Court.

The California Supreme Court in considering the total cost damages issue, set forth the four-part test as the standard for California law. However, the court determined that Amelco failed to produce evidence that it was not responsible for the added expenses of the project. The court noted that Amelco made no effort during construction to distinguish between its own inefficiencies and those caused by the city. In the claim, Amelco acknowledged that some of the costs could have been caused by its own inefficiencies, and reduced its claim by 5 percent to account for its own problems. The court reversed the judgment and remanded the case for a retrial on the breach claim on the issue of damages.

The court's ruling on the abandonment count was profound in that the crux of its holding was that abandonment of the contract cannot be the basis of recovery against a public agency because it circumvents public contract bidding laws. The court expressed public policy concerns that required contractors to bid fixed price contracts for public work and then allowed them to be set aside or "abandoned" when the contract experienced excessive change orders. This created a risk that fixed price public contracts could be converted into cost-reimbursement contracts with no ceiling price.

§12.05 EXTENDED GENERAL CONDITIONS

Page 12-23, add after note 79 in first paragraph:

Courts have allowed extended general conditions to be paid on a daily rate basis. For example, in *Techdyne Systems Corp. v. Whittaker Corp.*, 427 S.E.2d 334 (Va. 1993), a prime contractor calculated its extended general conditions as a daily rate of $2,068 and was able to recover these costs against a subcontractor that delayed the critical work activities.

When the government grants a time extension for change order work, there is a rebuttable presumption that the time extension is compensable and the contractor may be entitled to extended field overhead costs. For example, in *Gottfried Corp.*, ASBCA No. 51041, 98-2 B.C.A. (CCH) ¶30,063 (1998), the contractor was awarded a contract for the construction of a classroom building at an Air Force base. During the course of construction, the government issued a contract modification to the contractor extending the contract time by 16 days to deal with a structural steel design defect. Later, when the contractor sought to recover its delay costs for extended field overhead, the government argued that contractor-caused delays were concurrent with the design-related delay such that the contractor was not entitled to compensation for the time extension. The Board held that the granting of a time extension created a rebuttable presumption that the delay was solely due to government-caused delay, was not concurrent with contractor-caused delay, and was critical (delayed the overall completion of the contract). A similar result was reached in *Sherman R. Smoot Corp.*, ASBCA No. 53115, 2003 ASBCA LEXIS 25 (2003), where the government made several changes to the contract work, and even though the changes were not on the critical path of the schedule, the contractor alleged that administering the modifications delayed the project. The government issued a contract modification extending the contract time, but did not provide for any time-related compensation, such as extended field overhead costs. In the hearing before the board on the claim for its delay costs, the contractor did not provide any critical path analysis to prove that the change orders caused any delay, but relied upon the time extension granted that related to the changes. The board agreed that in granting a time extension associated with changed work, the government creates a rebuttable presumption that the delays resulting from the changes were critical and that the government was responsible for the delay-related costs. The government may be able to rebut this presumption with a schedule and entitlement analysis, but did not do so in the instant case.

Even if the change order is issued unilaterally by the owner and a time extension is given, there may be a rebuttable presumption that the owner is responsible for the delay costs associated with the time extension. In *Gavosto Associates, Inc.*, PSBCA No. 4058, 2001-01 B.C.A. (CCH) ¶ 31,389 (2001), Gavosto was awarded a fixed-price construction contract for a truck wash facility to be completed within 150 days. During the course of construction, the Postal Service issued numerous unilateral change orders, some of which extended the performance period, but did not include an overhead markup or extended field overhead costs. Gavosto brought its claim for extended field overhead to the Board, which in finding for Gavosto noted that when an owner directs additional work by unilateral modifications and grants time extensions for the work, it creates a rebuttable presumption that the owner is responsible for the delays and the contractor is entitled to compensation for time-related costs, such as extended field overhead costs.

Page 12-24, add at end of third full paragraph:

Extended general conditions may also include additional subsistence costs for on-site supervisory personnel, transportation expenses and sanitary facilities. Such costs should be demonstrated to be time-related. For example, the time-related rental costs for a dumpster may be properly recovered, but the roll-off charge disallowed as it would have been incurred whether or not the project time was extended. *Adventure Group, Inc.*, ASBCA No. 50188, 97-2 B.C.A. (CCH) ¶ 29,081 (1997).

Page 12-26, add after note 86 in carryover paragraph:

In seeking to recover its general conditions cost on change orders, contractors often use the percentage markup on the change order cost. If the change order involves a time extension, a contractor may also seek to recover a per diem for the extended general conditions. Using both methods on a given contract may be problematic. Two key cases on this issue are *M.A. Mortenson Co.*, ASBCA No. 40750, 97-1 B.C.A. (CCH) ¶28,623 (1997), *aff'd on recons.* 98-1 B.C.A. (CCH) ¶29,658 (1998) and *Caddell Construction Co., Inc.*, ASBCA No. 49333, 00-1 B.C.A. (CCH), ¶30,702, *aff'd on recons.*, 00-1 B.C.A. (CCH) ¶30,859 (2000). In *Mortenson*, the contractor added a percentage markup for general conditions for change order work that did not extend the performance time and used a daily rate for extended general conditions costs when the change order involved a time extension. The government allowed a 3½ percent markup for home office overhead but did not allow a markup for field overhead costs without an extension in the performance time. For example, if the project already required a full-time supervisor in the field, the performance of change order work would not increase its supervision costs, unless the change order involved a time extension requiring the supervisor to be on-site for a longer duration. The Board found in favor of the government in the original appeal and upon reconsideration. Switching between a per diem rate and a percentage markup on the same contract would allow the contractor to include the field general conditions costs in two different overhead pools. According to the Board's reading of Federal Acquisition Regulation (FAR) § 31.203 (b) this is not allowed. The *Mortenson* rule was confirmed in *Caddell Construction.* The contractor building a wastewater treatment plant for the Corps of Engineers used a percentage markup for general conditions costs on change order proposals that did not involve time extensions and a per diem rate for general conditions on change order proposals involving a request for compensable extension of time. After an audit by the Defense Contract Audit Agency, the Corps gave the contractor the option of choosing either, but not both methods. In considering the matter on appeal, the Board ruled that the contractor must use one method in order to comply with the requirement of FAR § 31.203 (b) for a single distribution base for allocating an overhead pool. The "single cost pool" requirement was the basis for the decision.

However, in *The Sherman R. Smoot Corp*, ASBCA No. 52261, 01-1 B.C.A. (CCH) ¶31,267 (2001), the Board did not preclude the practice of using both a percentage markup and a per diem to compensate the contractor for general conditions costs in limited situations. The fixed-price contract for renovation of buildings stipulated overhead and profit percentage markups for change orders up to 10 percent of the original contract value, and that field overhead was not to be a direct cost to change orders. The contractor received unilateral modifications granting time extensions providing for extended field overhead on a per diem basis, in addition to the amount of overhead paid on previous change orders. The contractor submitted a claim for extended field overhead calculated on a per diem basis, arguing that the percentage markup limitation did not apply to delays *not* associated with change orders. The board denied the government's motion for summary judgment because a fact issue existed as to whether the government waived the percentage limitation by issuing the unilateral modifications authorizing a daily rate for extended field general conditions.

The issue of paying general conditions as a daily rate or a percentage of the cost of change orders was also raised in *W.M. Schlosser, Inc. v. United States*, 50 Fed. Cl. 147 (2001), where the contractor argued that the government's approval of the pricing of extended field overhead on a per diem basis for three change orders established an implied agreement for future change orders to alter the contract provisions that extended field overhead was to be priced as a percentage of the cost of a change order. The contractor claimed that the extended field overhead also should be priced as a daily rate for other changes. Only 3 of the 95 change orders issued by the government priced extended field overhead on a per diem basis, and the government argued that this was a mistake. The court determined that an implied agreement to alter the terms of the contract requires proof of mutual intent and found the three change orders pricing field overhead on a daily rate were inadvertent. Further an explicit provision like the percentage markup pricing method cannot be changed by an implied agreement.

§ 12.06 HOME OFFICE OVERHEAD

Page 12-29, add after second full paragraph:

Broward County v. Russell, Inc., 589 So. 2d 983 (Fla. App. 1991); *Department on Transportation v. Herbert R. Imbt, Inc.*, 630 A.2d 550 (Pa. Cmmw. Ct. 1993); *Chilton Ins. Co. v. Pate & Pate Enterprises, Inc.*, 930 S.W.2d 877 (Tex. Ct. App. 1996); *Fairfax County Redevelopment and Hous. Auth. v. Worcester Brothers Co., Inc.*, 257 Va. 382, 514 S.E.2d 147 (1999).

[A] *Eichleay* Formula

Page 12-32, add after second sentence in first paragraph:

The use of the *Eichleay* formula to calculate home office overhead damages for delay has been accepted by various state jurisdictions. For example, Florida's Court of Appeals, Fourth District first adopted the *Eichleay* formula in *Broward County v. Russell*, 589 So. 2d 983 (Fla. Dist. Ct. App. 1991), and later the use of the *Eichleay* formula was affirmed in *Triple R Paving v. Broward County*, 774 So. 2d 50 (Fla. Dist. Ct. App. 2000). When the federal courts adopted the more restrictive requirements for the recovery of *Eichleay* damages, Florida's Court of Appeals Fourth District followed suit in *Broward County v. Brooks Builders, Inc.*, 908 So. 2d 536 (Fla. Dist. Ct. App. 2005). The case arose from a contract for the construction of a fire station adjacent to a runway at the Ft. Lauderdale airport that was awarded to Brooks Builders for $5.4 million that was to be completed in October 2001. The location near an active runway was challenging enough, but the aftermath of the terrorists attack in New York City on September 11, 2001, and revisions in the architectural plans and phasing caused the project to be completed eight months late. Brooks Builders sued the County for unpaid changes and delay damages which included home office overhead costs calculated by the use of the *Eichleay* formula. Brooks Builders was awarded over $1 million by the trial court, but the judgment did not itemize the damages. In considering the County's appeal of *Eichleay* damages, the Court of Appeals Fourth District noted that the federal court standards for awarding *Eichleay* damages became more stringent. The court considered the case of *P.J. Dick v. Principi*, 324 F.3d 1364 (Fed. Cir. 2003), where, in order to recover home office overhead according to the *Eichleay* formula, the contractor was required to show "effective suspension of much, if not all of the work." Even though Brook Builder's principal testified that it faced delays at "virtually every turn," he also testified "we kept working the best we could" but not in a "sufficient manner." The appellate court noted that this did not demonstrate "effective suspension of much, if not all of the work on the contract," and that the trial court, therefore, erred in awarding *Eichleay* damages. The case was remanded to the trial court to calculate the exact amount of the *Eichleay* damages that were awarded and subtract that from the final judgment.

The recovery of *Eichleay* damages may also be recoverable in Maryland. In *Gladwynne Construction Company v. Mayor and City Council of Baltimore*, 807 A.2d 1141 (Ct. Sp. App. Md. 2002), Gladwynne was awarded a contract to renovate several laboratories at the Polytechnic High School for the City of Baltimore. The work was to be completed within 180 calendar days, but construction was completed 309 days late. The contract was substantially completed on September 11, 1998. Nevertheless, change orders continued until December 1998, and the work was not fully completed until January 31, 1999. Gladwynne encountered asbestos which had to be removed, taking 170 days of the 180-day contract. Gladwynne also was delayed by design changes and repair work performed by another contractor. Gladwynne alleged that the unforeseen problems and changes extended the job and caused it to incur additional field and home office expenses. After the city failed to make full payment under the contract, Gladwynne filed suit. Gladwynne's claim included extended field costs and home office overhead costs. The trial court awarded Gladwynne $66,083.55 for extended field costs for a period of 195 days, rather than the 309 days claimed. The trial court declined to award any home office overhead expenses because it said that Gladwynne failed to satisfy the requirements for the application of the *Eichleay* formula.

On appeal, Gladwynne argued that the court erred in failing to award any damages for extended home office overhead. Gladwynne asserted there was no way that Gladwynne could guess when the city would complete the redesign process, and thus Gladwynne had to remain ready to perform changed and additional work until the final change order was completed in January 1999. The City maintained that Gladwynne failed to meet its burden to satisfy the *Eichleay* criteria because Gladwynne failed to present evidence as to the inability to assume other work during the delay. The City contended that a significant portion of Gladwynne's alleged *Eichleay* damages occurred after Gladwynne started work on a replacement project in November 1998. Gladwynne countered that it could not stop working until the City stopped redesigning. The change order work slackened in Fall 1998 to the point where Gladwynne was able to take on a replacement job in November 1998. But, prior to that time, Gladwynne was working on a contract which was supposed to have expired in February 1998 and was receiving no additional compensation from the City. Gladwynne also insisted that it was unable to take on additional replacement work due to bonding restrictions, which were not "eased" until November 1998. The City's project was almost finished in November 1998 when Gladwynne began work on another project; nevertheless, Gladwynne continued to work on the City project through January 1999. Gladwynne argued that even if it submitted bids for other work during its performance of the project, this would not establish that Gladwynne was able to take on other work during the delay period because its bonding restrictions were not eased until November 1998. Gladwynne presented evidence to satisfy the *Eichleay* formula for at least some portion of the total delay. The trial court declined to award any delay damages for overhead expenses, on the ground that Gladwynne did not meet all of the elements of *Eichleay*, but it failed to identify which element was not satisfied. Because it was not apparent why the court determined that the *Eichleay* formula was not applicable for any period of time, the appellate court vacated the judgment. On remand, the trial court was to reconsider whether the *Eichleay* formula was satisfied, and, if so, for what period of time and at what daily rate.

In Ohio, the *Eichleay* formula is accepted with some modifications as one, but not the exclusive method of calculating unabsorbed home office overhead for owner-caused delay. This was set forth in *Complete General Construction Co. v. Ohio Department of Transportation*, 760 N.E.2d 364 (Ohio 2002). In this case Complete General was awarded four of the five sections of I-670 between Columbus and the Port Columbus Airport. On one of the projects, Complete General was to construct a section of the highway, erect three new bridges, and widen another bridge. Design errors caused a seven-month delay and this in turn triggered another five-month delay because of a mandatory winter shutdown period. Another two-month delay was caused by additional work. The Ohio Department of Transportation (ODOT) and Complete General attempted to negotiate the additional compensation, but ultimately Complete General sued ODOT to recover unabsorbed home office overhead (calculated by the *Eichleay* formula), idle equipment costs, extended equipment costs, additional bond costs, and interest. The trial court (Ohio Court of Claims) awarded Complete General unabsorbed overhead, idle equipment, additional bond costs, and interest, but denied the claim for extended equipment costs. Both parties appealed to the Franklin County Court of Appeals which affirmed in part, reversed in part, and remanded for further proceedings. The matter was brought to the Supreme Court of Ohio on a discretionary appeal.

On appeal, ODOT challenged the use of the *Eichleay* formula to calculate the home office overhead during the delay period. The Court noted that the *Eichleay* formula was the most prominent method to calculate overhead costs allocable to delay, and determined that federal case law properly required two elements before *Eichleay* recovery was allowed: (1) The contractor must demonstrate that it was on "standby," that is, the work was suspended for an uncertain period of time, and the contractor remained bound to immediately return to the project at any time; and

(2) the contractor must prove that it was unable to take on other work while on standby, that is, the uncertainty of the duration of the delay made it unable to commit to replacement work on another project. A third requirement was also noted: The suspension of the work had to affect the project completion date. These elements establish the *prima facie* case for recovery of unabsorbed overhead damages, but this can be rebutted if the government can demonstrate either (1) that it was not impractical for the contractor to obtain replacement work or (2) that the contractor's inability to obtain replacement work was not caused by the government's suspension. The court noted differences between federal contracting and Ohio highway contracting ("like comparing apples and orange barrels"), and modified the use of the *Eichleay* formula in Ohio, such that in Ohio the courts must allow owners the opportunity to dispute particular items in the contractor's overhead cost pool as unallowable costs, whereas in federal contracting this is determined by the Federal Acquisition Regulations. Further, in Ohio, the *Eichleay* formula is only one way of determining home office overhead, not the exclusive method. Ohio courts could apply a direct cost formula: A comparison of the direct costs actually attributed to a project as a portion of all of the direct costs incurred by a business over a particular period; the result is a ratio by which the percentage of indirect costs can be calculated, including home office applicable to a particular project. In conclusion, the Court found that the trial court acted within its discretion in applying the *Eichleay* formula, but erred not allowing ODOT to dispute the overhead cost items that may not have bestowed any benefit on the project.

Other states also have accepted the *Eichleay* formula, although with some restrictions. *See California Department of Transportation v. Heritage Engineering Construction, Inc.*, 2004 Cal. App. Unpub. LEXIS 10631 (Nov. 22, 2004); *Fairfax County Redevelopment and Housing Authority v. Worcester Brothers Co.*, 514 S.E.2d 147 (Va. 1999); *PDM Plumbing & Heating, Inc. v. Findlen,* 431 594 (Mass. App. 1982); *Southern New England Contracting Co. v. State*, 345 A.2d 550 (Conn. 1974).

Page 12-35, add to end of section:

The *Eichleay* formula does not apply where the delays are incurred before Notice to Proceed (NTP) and the start of contract performance. In *Nicon, Inc. v. United States*, 331 F.3d 878 (Fed. Cir. 2003), Nicon was awarded a $1.4 million contract to repair a dormitory at MacDill Air Force Base, but before NTP a bid protest was filed which suspended action on the contract before the repair work began or before Nicon secured the payment or performance bonds. Even though the bid protest was dismissed, the government never directed Nicon to proceed with the work, and eventually terminated the contract for convenience some 288 days after Nicon was notified of the contract award. Nicon submitted a termination settlement proposal to the government in which it sought, among other things, unabsorbed home office overhead for the period of time between the award of the contract and the termination for convenience, using a modified version of the *Eichleay* formula. The contracting officer denied recovery of the unabsorbed home office overhead, and

Nicon filed suit in the Court of Federal Claims. The Court of Federal Claims granted the government's motion for summary judgment concluding that it was inappropriate to apply an alternate to the accepted *Eichleay* formula and it could not modify the *Eichleay* formula as Nicon requested. On appeal to the U.S. Court of Appeals for the Federal Circuit, the issue was considered as to whether the *Eichleay* formula could be modified to apply to situations where there was no performance on the contract. In order to make an *Eichleay* calculation, the court would have to use "constructive figures," that is, substitute the original contract price for contract billings and substitute the anticipated period of performance for the actual days of performance. The court would have to assume that Nicon would have taken the entire time period allowed by contract and that the extended period of performance would be the period of time between contract award and contract termination. The court reasoned that this "asks us to make too many assumptions and stretches the Eichleay formula beyond the circumstances for which it was intended." The court indicated in situations where contract performance had not begun, the contractor could recover some of its unabsorbed overhead costs as part of a termination of convenience settlement but must use a reasonable method of allocation, but not a modified *Eichleay* formula. The court reaffirmed that the *Eichleay* formula is the exclusive formula for the calculation of unabsorbed overhead due to government-caused delay in situations where contract performance had begun, and the formula must be strictly applied. The Federal Circuit vacated the summary judgment and remanded to the Court of Federal Claims for further proceedings to determine if unabsorbed overhead could be allocated by some other method.

[B] Establishing That Home Office Overhead Expenses Were Affected

Page 12-36, add last sentence of first paragraph:

, and (4) the contractor's performance is extended by change order work. Courts reason that the contractor's home office overhead has been absorbed by the value of the change order overhead markup. (See **§ 12.06 [E]**.)

Page 12-38, add after second full paragraph:

The requirement to show the uncertain duration of a delay gives rise to interesting applications in situations where the project has not started. For example, the Federal Circuit held that if notice to proceed has not been given and the construction contract has been terminated for convenience, the contractor may be entitled to recover for unabsorbed overhead, but not according to the *Eichleay* formula. The contractor should use a method to allocate its fixed home office expenses during the suspension period between the suspended contract and its other contracts. *Nicon, Inc. v. United States*, 331 F.3d 878 (Fed. Cir. 2003).

Page 12-39, add after note 149 in first full paragraph:

In *Interstate General Government Contractors v. West*, 12 F.3d 1053 (Fed. Cir. 1993), the Federal Circuit stated that a contractor proving that it is on "standby" does not need to prove that it has workers or equipment idly standing around waiting for productive work tests. The standby requirement can be met by the contractor demonstrating that the delay or suspended performance is for an uncertain duration during which period the contractor remains ready to perform. However, if the contractor continues to progress the work in any substantial manner, there has been no standby sufficient to support home office overhead claims.

The Federal Circuit in *P.J. Dick, Inc. v. Principi*, 324 F.3d 1364 (Fed. Cir. 2003), explained in much greater detail how a public owner should make a standby analysis. In making the standby inquiry, the court should first determine whether the contracting officer (CO) has issued a written order that suspended all the work on the contract for an uncertain duration and required the contractor to remain ready to resume work immediately or on short notice. *P.J. Dick, Inc. v. Principi, citing Interstate*, 12 F.3d at 1055, 1057 n.4. In such a case, the contractor need not offer further proof of standby. In the cases where the CO does not issue such a written order, the contractor must then prove standby by indirect evidence. To do so, the contractor must show three things.

First, the contractor must show that the government-caused delay was not only substantial but was of an indefinite duration. *P.J. Dick, Inc. v. Principi, citing id.* at 1058. For example, where the government suspends all work on the contract, but tells the contractor work will begin again on a date certain, the contractor could not be on standby. *P.J. Dick, Inc. v. Principi, citing Melka*, 187 F.3d at 1376.

Second, the contractor must show that during that delay it was required to be ready to resume work on the contract, at full speed as well as immediately. Once the suspension period is over, the contractor must be required to be ready to "resume full work immediately." *P.J. Dick, Inc. v. Principi, citing id.* at 1375; *All State Boiler*, 146 F.3d at 1373. Thus, where the government gives the contractor a reasonable amount of time to remobilize its workforce once the suspension is lifted, the contractor cannot be on standby. *P.J. Dick, Inc. v. Principi, citing Mech-Con Corp. v. West*, 61 F.3d 883, 887 (Fed. Cir. 1995) (holding the contractor could not be on standby where the government gave the contractor three months to remobilize its workforce on site). Presumably, the same result would follow if the government required immediate resumption of the work, but only with a reduced workforce and allowed the contractor to gradually increase its workforce over some reasonable amount of time. *P.J. Dick, Inc. v. Principi, citing Melka*, 187 F.3d at 1375. In addition, satisfaction of this element of standby clearly requires something more than an uncertain delay. The implication is that the contractor must be required to keep at least some of its workers and necessary equipment at the site, even if idle, ready to resume work on the contract (i.e., doing nothing or working on something elsewhere that allows them to get back to the contract site on short notice). *P.J. Dick, Inc. v. Principi, citing Sergent Mech. Sys. v. United States*, 54 Fed. Cl. 47, 49-50, 54 Fed. Cl. 636 (2002).

Third, the contractor must show effective suspension of much, if not all, of the work on the contract. *P.J. Dick, Inc. v. Principi, citing Melka*, 187 F.3d at 1375. Early decisions contained some statements that supported the notion that suspension of the work and idleness are not prerequisites to a determination that the contractor is on standby. *P.J. Dick, Inc. v. Principi, citing Altmayer v. Johnson*, 79 F.3d 1129, 1134 (Fed. Cir. 1996) ("There is no requirement that a contract be suspended before a contractor is entitled to recover under *Eichleay.*"); *Interstate*, 12 F.3d at 1057 n.4 ("Although idleness of workers is evidence that a contractor is on standby, i.e., performance has been suspended, it is neither conclusive nor required."). At no time, however, has the Federal Circuit court held that a contractor has been placed on standby merely because a government-caused delay of uncertain duration occurred, at the end of which the contractor must be ready to resume work. According to the Federal Circuit, *Altmayer*, oft cited for that proposition, held no such thing. *Altmayer* merely held that a contractor's performance of "minor tasks" during a suspension does not prevent it from recovering *Eichleay* damages. *P.J. Dick, Inc. v. Principi, citing Altmayer*, 79 F.3d at 1134. Nor does *Interstate*'s statement that idleness is not a prerequisite to standby support the idea that a contractor can be deemed on standby where there is no delay or suspension of the work. *P.J. Dick, Inc. v. Principi, citing Interstate*, 12 F.3d at 1057 n.4. *Interstate* indicates that its reference to "idleness" simply means the workers need not be "physically standing by idly." *P.J. Dick, Inc. v. Principi, citing id.* at 1057 n.5 (referencing a quote in a previous opinion and stating "these two phrases ["stand by idly and suspend its work"] clearly refer to standing by in the sense that no work is being performed on the contract, not that there must be workers physically standing by idly"); *see also id.* at 1057 n.4 ("If the test were whether the contractor's work force assigned to the contract in issue was standing by, the contractor would be penalized for, and thus deterred from, mitigating its damages for direct costs by reassigning its employees to other jobs or laying them off during the period of delay."). Every case where the Federal Circuit has held a contractor to be placed on standby has involved a complete suspension or delay of all the work or at most continued performance of only insubstantial work on the contract. *P.J. Dick, Inc. v. Principi, citing E.R. Mitchell Constr. Co. v. Danzig*, 175 F.3d 1369, 1372, 1374 (Fed. Cir. 1999) (holding subcontractor was entitled to *Eichleay* damages where it performed "some work" on the contract, but where most "work could not proceed until the faults [causing the suspension] were cured"); *All State Boiler, Inc.*, 146 F.3d at 1370, 1373 (holding contractor was entitled to *Eichleay* damages where the government suspended all work on the contract); *Satellite Elec. Co. v. Dalton*, 105 F.3d 1418, 1421 (Fed. Cir. 1997) (holding contractor was on standby where all the work on the contract was stopped, but denying recovery of *Eichleay* damages for other reasons); *Altmayer*, 79 F.3d at 1134; *Mech-Con*, 61 F.3d at 887 (holding contractor was entitled to *Eichleay* damages where work on the contract was completely suspended); *see also Interstate*, 12 F.3d at 1058 (noting that the record "could support" a conclusion that a contractor was on standby where all work on the contract was suspended).

Later decisions by the Federal Circuit explicitly state that such a suspension or delay of the work on the contract is a prerequisite to a finding that the government placed the contractor on standby. *P.J. Dick, Inc. v. Principi, citing Melka*, 187 F.3d at 1375–76; *see also Interstate*, 12 F.3d at 1057 (discussing standby as requiring "suspension of work on the contract"). In *Melka*, the Circuit Court held that a contractor was not on standby where it "was working on the contract and the government had not suspended all contract work." *P.J. Dick, Inc. v. Principi, citing Melka*, 187 F.3d at 1375. There, the government stopped work on one type of work, but, by resequencing the work under the contract, the contractor was able to perform substantial work on another type of work with comparable direct cost billings. *P.J. Dick, Inc. v. Principi, citing id.* at 1375–76. Subsequent decisions by the Court of Federal Claims and the Contract Boards also followed this view of the law. *P.J. Dick, Inc. v. Principi, citing Pete Vicari, Gen. Contractor, Inc. v. United States*, 53 Fed. Cl. 357, 368–70 (2002); *Carousel Dev., Inc.*, ASBCA No. 50719, 2001-1 B.C.A. (CCH) ¶ 31,262 (Jan. 23, 2001) (concluding contractor was not on standby where it continued to perform substantial amounts of work on the contract—here "approximately one quarter of the entire scope of the contract work"). So too does a major treatise on government contracts. *P.J. Dick, Inc. v. Principi, citing* John Cibinic, Jr. & Ralph C. Nash, Jr., Administration of Government Contracts 737 (3d ed. 1995) ("The standby test precludes a contractor from receiving extended or unabsorbed overhead unless there has been some period when the work was significantly slowed or stopped."). Thus, even though it is the typical scenario, formal suspension is not an absolute prerequisite. Contract performance also could be stopped or significantly slowed by government inaction, such as failure to vacate spaces in which the contract was to be performed.

The Court of Appeals for the Federal Circuit confirmed that the "standby" requirement truly means that the contractor could not continue performing any work on the contract in *Charles G. Williams Construction, Inc. v. White*, 326 F.3d 1376 (Fed. Cir. 2003). In this case, the contractor was awarded a fixed price contract to improve and repair a government building in two phases. There were 93 days of delay in Phase One of the work: Some of the delays were attributable to the government (deficiencies in the specifications, failure to vacate portions of the building in a timely manner, and change orders); and some of the delays were attributable to the contractor (deficiencies in the performance of the contractor and its subcontractors). Phase Two of the contract was terminated for convenience. The contractor submitted claims, some of which were accepted by the Armed Services Board of Contract Appeals (the Board), but the Board rejected the contractor's claim for home office overhead recovery under the *Eichleay* formula. At first the Board rejected the *Eichleay* claim based upon a finding by the Defense Contract Audit Agency (DCAA) report that stated the overhead for the extended period of contract performance was fully absorbed by the base contract, change orders, and other project. In considering the contractor's first appeal from the Board's denial, the Federal Circuit remanded the matter to the Board (*Williams Construction, Inc. v. White*, 227 F.3d 1055 (Fed. Cir. 2001)), stating that it was the Board's function to determine whether the contractor met the *Eichleay* requirements, not to determine whether the DCAA's auditor's report was supported by the record before the Board. On remand, Board determined that "based upon the auditor's report and testimony, the daily reports and other data" the contractor did not prove that "the performance of the work was suspended or significantly interrupted" during the period claimed in the contractor's *Eichleay* calculation. Indeed the daily reports and other project records indicated that "the site was manned without significant interruption" by the contractor and its subcontractors throughout the period. In the second round of appeals to the Federal Circuit, the court again considered the standby prerequisites to the recovery of *Eichleay* damages: The contractor must be on "standby" which is a "delay or suspension of contract performance for an uncertain duration, during which a contractor is required to remain ready to perform." The Federal Circuit underscored the point that the recovery of *Eichleay* damages is an extraordinary remedy, and it is not enough to show that the government "significantly interfered with the efficient performance of the contract" in order to benefit from this extraordinary remedy. The Federal Circuit reasoned that "as long as the contractor is able to continue performing the contract, although not in the same way or as efficiently or effectively as it had anticipated it could do so, it can allocate a portion of its indirect costs to that contract." The Federal Circuit affirmed the decision of the Board.

Another example of where the contractor failed the "standby" test was in *Sergeant Mechanical Systems, Inc. v. United States*, 54 Fed. Cl. 47 (2002), where the Department of Veterans Affairs (VA) awarded Sergeant a contract for the improvements to a boiler plant. When the work was substantially complete, the VA inspected the project and prepared a punch list that included the calibration of a sensor on the hydronic bridge system and training of VA employees on the operation of the system. Sergeant then learned through its subcontractor that there was problem with the pressure in the chilled water system that prevented the sensor from being calibrated. It took several months for VA to resolve the problem even though Sergeant asked the VA to close out the contract. Sergeant submitted a claim to unabsorbed home office overhead using the *Eichleay* formula, but the claim was denied by the VA. In considering Sergeant's appeal, the Court of Federal Claims determined that Sergeant failed to meet the "standby" test. The court noted that Sergeant was not required to keep any of its own workforce ready for continued performance. It was Sergeant's subcontractor who was responsible for having workers ready when the VA required, and there were no subcontractor damages in Sergeant's claim. Sergeant only monitored the contract and had to notify the subcontractor when the hydronic bridge work was ready to be completed. Further, there was no "short notice" given to Sergeant that required immediate completion of the hydronic bridge work. The facts indicated that Sergeant was never actually required to perform work on short notice; for example, it took over a month for Sergeant to notify the VA that there were chilled water problems, and it was nearly a month after the VA notified Sergeant that it was ready for training before the training was completed by Sergeant. Finally there was no evidence in the record that Sergeant had an expectation that it would be required to complete the contract work on short notice.

An example of the standby was found in *Huff Seal*, ASBCA No. 53587, 2002-1 B.C.A. (CCH) ¶ 31,855 (2002), where the contractor was awarded a contract to repair runways at a Naval Air Station in Puerto Rico. The government stopped work of four occasions for brief periods of time for training exercises. Although the government argued that these work stoppages were unforeseeable and reasonable under the terms of the Suspensions Clause, the Board noted that the training exercises were already planned at the time of contract award and therefore were not unforeseen circumstances. The Board allowed the contractor to recover its unabsorbed home office overhead using the *Eichleay* formula.

In short, a court evaluating a contractor's claim for *Eichleay* damages should ask the following questions: (1) was there a government-caused delay that was not concurrent with another delay caused by some other source; (2) did the contractor demonstrate that it incurred additional overhead (was the original time frame for completion extended or did the contractor satisfy the *Interstate* three-part test); (3) did the government CO issue a suspension or other order expressly putting the contractor on standby; (4) if not, can the contractor prove there was a delay of indefinite duration during which it could not bill substantial amounts of work on the contract and at the end of which it was required to be able to return to work on the contract at full speed and immediately; (5) can the government satisfy its burden of production showing that it was not impractical for the contractor to take on replacement work (i.e., a new contract) and thereby mitigate its damages; and (6) if the government meets its burden of production, can the contractor satisfy its burden of persuasion that it was impractical for it to obtain sufficient replacement work. Only where the above exacting requirements can be satisfied will a contractor be entitled to *Eichleay* damages.

In *P.J. Dick, Inc. v. Principi*, 324 F.3d 1364 (Fed. Cir. 2003), the Department of Veterans Affairs (DVA) selected P.J. Dick, Inc. (PJD) to construct the Clinical Addition to the DVA Medical Center in Ann Arbor, Michigan. Under the contract, PJD was due to complete the work by January 12, 1998. During the contract the government issued over 400 orders changing the contract and causing various delays to different aspects of the project. These modifications increased the contract price by over 5 percent and caused the DVA to grant 107 days of additional contract performance time. In accepting the additional days to complete the contract, PJD reserved its right to seek additional impact and suspension costs. PJD completed the contract on September 29, 1998, 260 days after the original contract completion date and 153 days after the revised date. PJD presented to the government contracting officer claims for a time extension for all 260 days of delayed performance and sought field and home office overhead damages for most of these delays.

After the CO denied the claims, the Board concluded that PJD was not entitled to *Eichleay* damages because it failed to prove that it was placed on standby. The Board found that PJD was not placed on standby because "PJD was able to progress other parts of the work during the time periods it alleges it was suspended." The Board determined that PJD's direct billings remained substantial during the several suspension periods and that it accelerated the work on the contract was irrelevant. The Secretary and PJD each appealed different aspects of the Board's decision.

First, PJD argued the Board committed legal error because a contractor was automatically on standby any time there was a government-caused delay of an uncertain duration extending the performance of the contract, at the end of which the contractor can be required to immediately resume work. Second, PJD argued that substantial evidence did not support the Board's findings that its direct billings "showed no appreciable diminution during the alleged suspension periods" and that it accelerated performance of the contract.

The Board applied the correct legal standard for standby. The contractor must show a suspension, whether formal or functional, of all or most of the work on the contract. Consequently, PJD's legal argument failed. The evidence, at worst, showed that in one of these delay periods PJD billed 53 percent less than it had the month before. There was, however, no evidence that PJD's direct billings were less than they would have been absent the suspensions—that was the controlling test. A comparison of pre- and intra-delay billings or intra- and post-delay billings was not the test. Regardless, PJD's direct billings during the delay periods could hardly be characterized as "minor." At worst, PJD's direct billings during one of the periods of suspension were 47 percent of what they were in prior months; thus, PJD continued to perform substantial amounts of work on the contract during the suspension periods. The Board's conclusion that PJD was not on standby was affirmed.

Page 12-39, add after note 152 in second full paragraph:

Another board of contract appeals supplemented the proof requirements for recovery of *Eichleay* damages with "disruption in stream of income" standard and also held that *Eichleay* damages cannot be recovered after substantial completion. In *Young Enterprises of Georgia, Inc. v. General Services Administration*, GSBCA No. 14437, 00-2 B.C.A. (CCH) ¶ 31,148 (2000), the contractor was awarded a contract for the renovation of a historic courthouse. During the course of construction, the contractor alleged its performance was delayed by the need to remediate unanticipated asbestos, various change orders and constructive suspension at the end of the project. The contractor received time extensions and additional compensation for the asbestos delays and change orders, but the government denied the constructive suspension claim. The government's defenses to the constructive suspension claim were that there were concurrent subcontractor delays, and that some of the delays occurred after substantial completion. Before the board of contract appeals, the government also argued that the contractor should not be able to recover *Eichleay* damages where there was no disruption of cash flow from the contract.

The board agreed with the government in these two points. It ruled that the recovery of unabsorbed home office overhead is not appropriate for delays after substantial completion because the contractor did not demonstrate that it was unable to increase its bonding capacity or obtain additional work. Thus, the board recognized that substantial completion is a milestone event, precluding the contractor from recovering *Eichleay* damages as well as precluding the owner from assessing liquidated damages.

"The stream of income" standard was applied to preclude recovery of home office overhead damages. The board found that because the contractor received continuous progress payments, and performed change order work during the alleged delay period, there was no basis to find that the contractor's home office overhead was unabsorbed or that there was disruption in the stream of income emanating from the contract.

Page 12-41, add after second full paragraph:

In *Melka Marine, Inc. v. United States*, 187 F.3d 1370 (Fed. Cir. 1999), the Federal Circuit court determined that a contractor's ability to bid or perform other work during suspension or delay did not by itself indicate replacement for delayed or suspended project. The Navy awarded Melka a contract for dredging work, construction of a breakwater, repairs to an existing boat ramp, quay wall, finger pier, and various other repair work. The contract required the government to obtain a dredging permit prior to the commencement of the dredging and breakwater work, but not for the repair work. The government notified Melka that the permit had not yet been obtained, but was expected at any time. The government noted that construction repairs were not dependent on the permit. Melka mobilized all of its equipment to the job site by November 15, 1994. On November 16, 1994, Melka sent a letter to the government stating that Melka's proposed schedule for completing the work would be negatively impacted by the government's failure to timely obtain a permit and that it would re-sequence the work to avoid inactivity. On November 21, Melka began the re-sequenced work by starting work on the boat ramp. Work was suspended the next day, however, due to the discovery of a six-inch pipe in the ramp and the discovery of a prior boat ramp beneath the existing ramp, which required a revised ramp design. Modification P00003 was a unilateral modification compensating Melka for overhead and field costs associated with performing the work out of sequence and for its dredging and breakwater equipment for 12 days.

As a result of these delays and the delay in obtaining a permit, on November 29 the government issued a formal "Suspension of Work" order. The order suspended the commencement of dredging and breakwater work for an indefinite period. Two days later, the government revised its Suspension of Work order to permit work on the repairs to the quaywall and the finger pier, and also suggested that equipment on site that would not be utilized on the repairs to the boat ramp, quaywall, or finger pier could be demobilized. The government requested that Melka take any other steps necessary to minimize costs due to the suspension of permit related work. By December 27, 1994, Melka had demobilized its dredging and breakwater equipment while it continued the re-sequenced repair work, which was substantially completed by January 4, 1995. Melka later mobilized its dredging and breakwater equipment for use on another project from

January 4–18, 1995, and a second project from January 20, 1995 through March 31, 1995. Melka originally scheduled these jobs to be performed after the government project, but the government-imposed delay allowed Melka to perform these jobs earlier. Melka also bid on other projects during this period that it did not receive, but that it would have been able to perform had they been awarded them.

On February 2, 1995, the government informed Melka that work on the dredging and breakwater, rather than commencing in February 1995, could not begin until October 15, 1995, and requested that it agree to perform this work under the original contract price. Melka agreed and proposed a completion date of February 29, 1996. All work was completed by February 1996.

Melka asserted that it was entitled to recovery of its unabsorbed overhead from November 16, 1994 through March 30, 1995 as calculated by the *Eichleay* formula. The Court of Federal Claims denied Melka's claims and dismissed Melka's complaint. Melka appealed. From November 16 (the beginning of Melka's alleged damages period) to November 29, 1994 (the date the government issued the formal Suspension of Work order), the Court of Federal Claims found that no government-imposed delay in fact occurred and Melka could not have been on "standby." Although the government had failed to obtain a permit for the dredging and breakwater work, the parties had agreed that Melka would re-sequence its work so that work not related to the permit could be done. From November 29, 1994 to January 4, 1995 (the approximate end of Melka's performance of all those contract tasks not dependent upon issuance of the permit), the Court of Federal Claims found that although the government issued a suspension of work order at the beginning of this period, that order directed Melka to perform non-permit-related work and indicated that the dredging permit would not be obtained until 1995. After the Suspension of Work order, Melka continued to work on the contract (with re-sequenced work) and did so on a virtually uninterrupted basis through January 3, 1995. Melka argued that because the dredging and breakwater work accounted for a significant amount of the work in the government contract, once such work was suspended indefinitely, Melka was on standby.

The Federal Circuit Court found that Melka could not establish that it was on standby by dividing the contract into one portion that was able to be performed (the repairs) and another that was on standby (the dredging and breakwater work), because the contract required that both types of work be performed. If work on the contract continued uninterrupted, albeit in a different order than originally planned, the contractor was not on standby. Given these facts, the Court of Federal Claims did not clearly err in finding that Melka was not on standby and not entitled to *Eichleay* damages during this period. From February 2, 1995 (when the government informed Melka that permit-related work could not begin until October 15, 1995) to March 30, 1995 (the last date for which Melka sought recovery), the Court of Federal Claims found that Melka could not have been on standby because it knew then with certainty that it could not be called on to perform the work before October 15. The Federal Circuit found this was not clearly erroneous because "standby" requires an uncertain delay period when the government can require the contractor to resume full-scale work at any time. Melka knew that it would not be called on to work on the contract until October 15 and the government did not require Melka to keep its dredging and breakwater equipment at the site. Thus, *Eichleay* damages could not be awarded for this time period either. The Federal Circuit affirmed those portions of the Court of Federal Claims decision.

January 4, 1995 (when Melka completed those contract tasks not dependent upon receipt of the permit) to February 2, 1995 (the date when the government and Melka agreed that the permit-related work would not commence until October 15, 1995) presented a different issue because the government conceded that Melka was on standby of indefinite duration. The issue became whether the government satisfied its burden of showing that it was not impractical for Melka to obtain replacement work, or that Melka's inability to obtain replacement work was due to circumstances not related to the government-caused delay. In an effort to satisfy its burden, the government argued that the work Melka was able to perform on the two projects constituted replacement work. Those projects, according to the government, could not have been performed if Melka had been performing the dredging and breakwater parts of the government contract because it could only take on one large dredging project at a time. The government also argued that Melka would have taken on more work had its bids been successful and the economy been stronger, both reasons unrelated to the government's delay and suspension of work. Melka argued that the projects it took on were not replacement work because they were relatively small in comparison to the work on the government contract and were merely moved up in its schedule in order to mitigate its losses as a matter of good faith.

The Court of Federal Claims noted that Melka performed several other projects while on standby and bid unsuccessfully on several others, and it held that Melka was not entitled to any *Eichleay* damages for this time period. The Federal Circuit held that the Court of Federal Claims applied the wrong legal test when it focused on Melka's ability to take on any other work during the delay period as a basis to deny any *Eichleay* damages. The appropriate inquiry was whether Melka was able to take on replacement work, not just any additional work. Because the Federal Circuit was unable to discern from the record whether the work performed or bid on by Melka during this time frame was replacement work, producing sufficient support to absorb all the overhead costs that the government contract would have if work had not been suspended, the case was remanded to the Court of Federal Claims for a determination of that issue.

The court stated that a heavy burden falls on the government once a contractor has established its prima facie case of entitlement to *Eichleay* damages because in very few cases will the government be able to demonstrate that it was not impractical for the contractor to take on replacement work. Nevertheless, the government may be able to show that Melka was not injured by the delay. For example, the government argued that part of Melka's inability to obtain work was due to cold winter weather, rather than the government-caused delay and requirement to stand by. Such evidence related directly to the issue of what caused Melka's inability to obtain adequate replacement work; if in fact it was unable to obtain such work, or if bad weather prevented performance of work that was obtained. Bidding on other projects can also support, rather than undercut, a contractor's argument that it was impractical for it to obtain replacement work, particularly if the projects that were bid on could not be commenced and/or completed within the government-caused delay.

Melka argued that it could not obtain replacement work by bidding on other projects because such work involved substantial lead times that would render it impossible to begin the project in a reasonable period of time, much less before the suspension is lifted. For example, three-and-a-half months elapsed from the time the project was advertised until Melka actually commenced work on the delayed contract. Such commercial realities in the market may be highly material when determining whether it was impractical to obtain any replacement work or sufficient replacement work. Finally, the relative sizes of the projects at issue in terms of dollar amount, timing, and duration are relevant when determining whether a particular project was in fact replacement work, as opposed to merely additional work in the ordinary course of business for a contractor that routinely performs multiple contracts. The key was the potential for the new work to contribute a comparable amount to the contractor's overhead, as the government contract would have absent the suspension, assuming that the time durations are also comparable. However, replacement work need not be identical in size, duration, or type as compared to the delayed work. If the same amount of money was not contributed to the overhead costs in the same period by the replacement work, then the contractor should be able to obtain at least some *Eichleay* damages. Partial replacement entitles the contractor to reduce its *Eichleay* claim, not substitute for it.

Another board of contract appeals case indicates that change order work that is subcontracted out may not constitute "replacement work" for the prime contractor for purposes of determining the recovery of *Eichleay* damages. In *Able Products Co. v. Department of the Treasury*, GSBCA No. 15167-TD, 01-2 B.C.A. (CCH) ¶ 31,554 (2001), the contractor, Able, was awarded a contract to renovate the HVAC system, and during the course of performance discovered the presence of asbestos that had to be abated. The asbestos abatement work was performed by a subcontractor under a change order to Able because Able's personnel were not properly qualified to do the task. Able filed a delay claim alleging that its own HVAC renovation work was on standby while the subcontractor performed the asbestos abatement. The delay costs included unabsorbed home office overhead calculated according to the *Eichleay* formula.

The government moved for summary judgment as to the *Eichleay* damages, arguing that the change order work for the abatement was replacement work for the delayed HVAC work. The Board denied summary judgment, reasoning that there was a factual question as to whether the change order for asbestos abatement by the subcontractor was truly replacement work.

Performance of a new project during portion of delayed contract does not indicate that the contractor could obtain substitute work, preventing any recovery under the *Eichleay* formula.

In *Gladwynne Construction Company v. Mayor and City Council of Baltimore*, 2002 Md. App. LEXIS 168 (Ct. Sp. App. Sept. 25, 2002), Gladwynne

Construction Company renovated several laboratories at the Polytechnic High School for the City of Baltimore. The work was to complete within 180 calendar days. Construction was completed 309 days late. The contract was substantially completed on September 11, 1998. Nevertheless, change orders continued until December 1998, and the work was not fully completed until January 31, 1999. Gladwynne was to replace the utility lines located in a crawlspace under a concrete slab in Room 29 of the School. Gladwynne planned to replace the utility lines by drilling two four-inch holes into the concrete crawlspace and then working inside the crawlspace to insert new utility lines. However, there was no crawlspace. Instead, Gladwynne broke the concrete slab in Room 29, dug a trench to hold new utilities, and then installed them in poured concrete. While digging the trenches, it severed existing cable lines, which had to be repaired. The tile flooring was also damaged from the trenching and had to be removed. When the existing tiles were removed, Gladwynne found asbestos, which also had to be removed. That process was not completed until Gladwynne was 170 days into a 180-day contract. As a consequence of the change in design, Gladwynne was unable to install the plastic utility pipes that were stored on site. Instead, Gladwynne ordered a more durable utility pipe, known as Duriron, but had to wait until January 1998 for its arrival. In February 1998, after the trenching work was completed, and the new Duriron lines were installed. Rather than patching the four-inch holes, as originally anticipated, Gladwynne had to smooth the surface of the entire floor of Room 29 to prepare it for the new vinyl floor. When Gladwynne discovered leaks in the laboratory windows, the city hired another contractor to do the repair work. Gladwynne had to cease installation of the casework until the leaks were remedied. More than 50 changes to the construction plans were made by the architect. The unforeseen problems and changes extended the job and caused Gladwynne to incur additional field and home office expenses. After the city failed to make full payment under the contract, Gladwynne filed suit. Gladwynne's claim included extended field costs and home office overhead costs. The trial court awarded Gladwynne $66,083.55 for extended field costs for a period of 195 days, rather than the 309 days claimed. Moreover, the court declined to award any home office overhead expenses because the contractor failed to satisfy the *Eichleay* formula.

On appeal, Gladwynne argued that the court erred in failing to award any damages for extended home office overhead. Gladwynne asserted there was no way that Gladwynne could guess when the city would complete the redesign process, and thus Gladwynne had to remain ready to perform changed and additional work until the final change order was completed in January 1999. The city maintained that Gladwynne failed to meet its burden to satisfy the *Eichleay* criteria because Gladwynne failed to present evidence as to the inability to assume other work during the delay. The city contended that a significant portion of Gladwynne's alleged damage amount occurred after Gladwynne started work on a replacement project in November 1998. Gladwynne could not stop working until the City stopped redesigning. The change order work slackened in Fall 1998, to the point where Gladwynne was able to take on a replacement job in November 1998.

But, prior to that time, Gladwynne was working on a contract that was supposed to have expired in February 1998, and was receiving no additional compensation from the city. Gladwynne also insisted that it was unable to take on additional replacement work due to bonding restrictions, which were not "eased" until November 1998. In November 1998 Gladwynne began a $1 million project for the Corps of Engineers called "28 Units." The project was almost finished in November 1998, when Gladwynne began work on the project for the Corps of Engineers. Nevertheless, Gladwynne continued to work on the project through January 1999. Gladwynne submitted a bid for a $1 million project for the federal government six months to a year before that project began, between November 1997 and May 1998. Even if Gladwynne submitted a bid for another project during its performance of the project, this would not establish that Gladwynne was able to take on other work during the delay period. Gladwynne showed that its bonding restrictions were not eased until November 1998, so that it could not take on a new project prior to that date. Gladwynne presented evidence to satisfy the *Eichleay* formula for at least some portion of the total delay. Nevertheless, the trial court declined to award any delay damages for overhead expenses, on the ground that Gladwynne did not meet all of the elements of *Eichleay*, but it failed to identify which element was not satisfied. Because it was not apparent why the court determined that the *Eichleay* formula was not applicable for any period of time, the appellate court vacated the judgment. On remand, the court should reconsider whether the *Eichleay* formula was satisfied, and, if so, for what period of time and at what daily rate.

Page 12-42, add after first paragraph:

The second prerequisite to use of the *Eichleay* formula requires that the contractor be unable to take on additional work. In *Safeco Credit v. United States*, 44 Fed. Cl. 406 (Ct. Cl. 1999), the Court of Claims found that the contractor failed to prove it was not able to take on additional work. Safeco cited a statement made by its president in response to a government interrogatory that the suspensions prevented it from undertaking any other work because its resources (employees, equipment, and finances) were committed to the completion of this project. The government stated that the fact that Safeco was able to do the change order work demonstrated that the contractor was not prevented from doing additional work as would be required to use the *Eichleay* formula. The court determined that Safeco's proof was not sufficient to show that they could not have taken on any other jobs during the contract period.

The court found the modifications to the contract required Safeco to perform change order work for which it was compensated; the revised completion dates were clearly set forth in the modifications. Furthermore, aside from the unsubstantiated conclusory allegation of Safeco's president, Safeco failed to show an inability to take on other work. Unlike other cases, the Federal Circuit had decided that the element of uncertainty was absent in *Safeco.* Accordingly, the Navy was determined not to be liable for *Eichleay* damages.

Page 12-42, add after note 161 in second full paragraph:

For example, in *Idela Construction Co.*, ASBCA No. 45070, 01-2 B.C.A. (CCH) ¶ 31,437 (2001), the contractor demonstrated that it repeatedly encountered uncharted and unanticipated water lines, and was required to stop work on the installation of a sewer line until the government determined whether the water lines were active. This required the contractor to suspend work for an uncertain duration, and extended the completion of its work beyond the contractually required completion date.

However, if the contractor is able to perform some of the contract work, more than "minor" work items, it may not be in a true standby ngmode. For example, in *Carousel Development, Inc.*, ASBCA No. 50719, 01-1 B.C.A. (CCH) ¶ 31,262 (2001), the contractor was awarded a contract to renovate 24 military family housing units, to be released to the contractor in groups of four units every 21 days, and each group to be completed by the contractor within 75 days of release. The government became dissatisfied with the contractor's work, and did not release additional units to the contractor until it completed work on 12 units that were only about 50 percent done. The contractor filed a claim alleging suspension and requesting *Eichleay* damages. Although the Board found that the government suspended work on the 12 remaining units by refusing to release them, the contractor was performing contract work on first 12 units and never came to a standstill. Thus, the contractor was not entitled to recover *Eichleay* damages.

Page 12-42, add after second paragraph:

An owner may avoid paying the contractor's unabsorbed home office overhead by permitting a remobilization period after the suspension ends. For example, in *Complete General Construction Co. v. Ohio Department of Transportation*, 94 Ohio St. 3d 54, 760 N.E.2d 364 (2002), Complete General Construction filed a complaint to recover the unabsorbed home office overhead costs it incurred for the Ohio Department of Transportation's two extensions of contract. The court of claims awarded the contractor $184,947 on its claim for unabsorbed home office overhead. The owner appealed.

According to the owner, the court of claims erred because the contractor failed to establish that it was on standby during the suspension period. The owner conceded that the length of the suspension period was uncertain and that the contractor could have been called back to work at any time during the suspension period. However, the owner contended that the contractor was not on standby when work on one part of the project was suspended but the contractor continued to work on other parts of the project.

The appellate court stated that whether the contractor was able to perform other work on the contract during the suspension period was not relevant to whether the company incurred unabsorbed overhead during the extension period. The uncontroverted evidence showed that the suspension period was of uncertain duration, that the contractor was subject to being called back to work at any time during the suspension period, and that the contractor was required to work on the contract beyond the original completion date due to the suspension. This was sufficient to establish a prima facie case that the contractor incurred unabsorbed overhead as a result.

The owner further contended that the court of claims erred in utilizing *Eichleay* because the evidence showed that the contractor was able to reallocate whatever unabsorbed overhead it incurred as a result of the delay to replacement work. The contractor had successfully bid on $78 million worth of work during the contract, and the annual value of the contractor's work, as well as the company's bonding capacity, increased by $10 million. The appellate court stated that the evidence did not establish that the contractor had "replaced" the work lost during the delay, thereby offsetting unabsorbed overhead. The rule that work obtained as part of a contractor's normal, ongoing bidding process does not constitute replacement work permits few instances in which the owner can successfully rebut a prima facie showing of entitlement to *Eichleay* damages, because the responsibility for a contractor being on standby rests with the owner. By fixing, at the outset of the suspension period, a future date on which the contractor may return to work, or by allowing the contractor a remobilization period at the end of the suspension period, the owner may avoid keeping the contractor on standby during the suspension and thus avoid liability for unabsorbed overhead expenses in the future. Judgment affirmed in part and reversed in part, and case remanded.

Page 12-43, add after first paragraph:

In *Safeco Credit v. United States*, 44 Fed. Cl. 406 (Ct. Cl. 1999), the Claims Court repeated that change orders that extended the performance period did not suspend the work as required to collect unabsorbed home office overhead damages under the *Eichleay* formula. The dispute arose from a 1985 contract for berthing improvements to a naval dock. Safeco Credit was the surety to Fraley Associates, Inc., and the contractor and succeeder to its responsibilities and interests. The contract was divided into seven construction phases identified as 1A, 1B, 2A, 2B, 3A, 3B, and 3C, and a mobilization phase. The Navy added or changed work on the project by 53 modifications to the contract. Many of the modifications extended the time period for completion of phases 1A, 1B, 2A, and 2B—those phases of the project which (with the exception of phase 1B, which was to begin before the completion of phase 1A) were to run consecutively. The parties disagreed about the effect of these modifications on the contract completion dates and, in turn, on Safeco's possible liability for failure to complete the project on time.

Safeco contended that it was entitled to reimbursement for home office overhead as a result of extensions to the project for $663,707.47. The government asserted that Safeco failed to apply the *Eichleay* formula correctly. The court stated that the first prerequisite to use of the *Eichleay* formula was the occurrence of a government-imposed delay, which placed the contractor on standby. Safeco asserted that *Eichleay* applied to contract extensions, which cause uncertainty as to future performance on a project, as well as suspensions. Safeco also alleged that the government's conduct, evidenced by the number of modifications to the contract, created uncertainty in the length of performance of the contract. The government contended that Safeco performed additional work throughout the extension period and therefore was not on "standby." The court found that Safeco misinterpreted Federal Circuit precedent by asserting as a general rule that "*Eichleay* applies to contract extensions in addition to suspensions."

The contract was extended as a result of modifications issued by the government. These modifications did not suspend work, but rather required Safeco to perform additional and change order work for which it was given additional compensation. The Federal Circuit had determined that *Eichleay* was not applicable to a situation involving additional work due to contract modification as opposed to a suspension without evidence that the contract changes resulted in a delay which required the contractor to stand by idly and suspend its work. The work performed by Safeco during the period of suspension was significant. Furthermore, the modifications added by the government did not make the length of perfor mance uncertain. Rather, the modifications expressly set forth dates by which the contract was lengthened.

Page 12-43, add at end of second full paragraph:

In some instances, change order work after original project completion that extends the project completion date entitles contractor to additional field office overhead for additional performance periods. The case of *Gladwynne Construction Co. v. Mayor and City Council of Baltimore*, 2002 Md. App. LEXIS 168 (Ct. Sp. App. Sept. 25, 2002), also illustrates the point. The facts of the case were that Gladwynne renovated several laboratories at the Polytechnic High School for the City of Baltimore. The work was to complete within 180 calendar days. Construction was completed 309 days late. The contract was substantially completed on September 11, 1998. Nevertheless, change orders continued until December 1998, and the work was not fully completed until January 31, 1999. Gladwynne was to replace the utility lines located in a crawlspace under a concrete slab in Room 29 of the School. Gladwynne planned to replace the utility lines by drilling two four-inch holes into the concrete crawlspace and then working inside the crawlspace to insert new utility lines. However, there was no crawlspace. Instead, Gladwynne broke the concrete slab in Room 29, dug a trench to hold new utilities, and then installed them in poured concrete. While digging the trenches, it severed existing cable lines, which had to be repaired. The tile flooring was also damaged from the trenching and had to be removed. When the existing tiles were removed, Gladwynne found asbestos, which also had to be removed. That process was not completed until Gladwynne was 170 days into a 180-day contract. As a consequence of the change in design, Gladwynne was unable to install the plastic utility pipes that were stored on site. Instead, Gladwynne ordered a more durable utility pipe, known as Duriron, but had to wait until January 1998 for its arrival. In February 1998, after the trenching work was completed, and the new Duriron lines were installed.

Rather than patching the four-inch holes, as originally anticipated, Gladwynne had to smooth the surface of the entire floor of Room 29 to prepare it for the new vinyl floor. When Gladwynne discovered leaks in the laboratory windows, the city hired another contractor to do the repair work. Gladwynne had to cease installation of the casework until the leaks were remedied. More than 50 changes to the construction plans were made by the architect. The unforeseen problems and changes extended the job and caused Gladwynne to incur additional field and home office expenses. After the city failed to make full payment under the contract, Gladwynne filed suit. Gladwynne's claim included extended field costs and home office overhead costs. The trial court awarded Gladwynne $66,083.55 for extended field costs for a period of 195 days, rather than the 309 days claimed. Moreover, the court declined to award any home office overhead expenses because the contractor failed to satisfy the *Eichleay* formula.

On appeal, Gladwynne complained that the court erred in finding that the project was only delayed for 195 days, rather than the 309 days. Gladwynne asserted that it was delayed by the city at least through December 1998 or January 1999, because the city redesigned the project via change orders in an attempt to remedy its defective plans and specifications. The city continued to issue change orders through December 1998, and Gladwynne continued to work on those change orders through January 1999. The schedule was, along with the architect's design, rendered completely meaningless by the eventual recognition that contrary to the architect's drawings, there was no utility crawlspace. During the time that Gladwynne was attempting to surmount the utility problem, the city added 54 change orders for various other reasons. Gladwynne pointed out that because the crawlspace did not exist, it had to order a different type of pipe, which did not arrive until January 19, 1998, resulting in a four-and-a-half-month delay. Moreover, Gladwynne explained that, since entire floors were being trenched, no other work could be started until the trenches were filled on February 18, 1998. Consequently, the work was delayed by another month. As a result of the trenching in Room 29, Gladwynne had to flash patch the entire floor prior to installation of the vinyl. Only after that was completed, in early May 1998, could Gladwynne begin to install cabinetry. It took until the middle of July to finish the cabinet installation due to two unforeseen asbestos abatements as well as having to stop work while the City mishandled various window leaks. Then, and only then, could Gladwynne measure and place the order for the manufacture of the countertops. Installation of the countertops was completed in mid-August 1998, at which time Gladwynne was able to install the faucets, electrical devices, gas tops, and the pipe chases that the City added midway through the project.

The city contended that it was responsible for delays only through May 30, 1998, because by that date Gladwynne had completed more than 90 percent of the project and had received from the City $430,400 of the total contract amount of $497,000." In asserting that the project was 93 percent complete by May 5, 1998, the city did not consider the change orders. Clearly, Gladwynne could not have completed the project while change orders were still forthcoming. The trial court found that Gladwynne was entitled to recover extended field costs due to delays occasioned by the City, but only for 195 days, representing the period from February 28, 1998, to September 11, 1998. The appellate court determined that trial court erred in limiting the period of delay for which field costs were compensable to September 11, 1998. The trial court's judgment was vacated and remanded for further proceedings as to the length of delay for which Gladwynne was entitled to compensation for extended field costs.

Page 12-44, add after eighth sentence in carryover paragraph:

It is unreasonable to expect a contractor who has been delayed to factually prove that, except for a delay on one contract, he was not able to bid for another contract and as a consequence thereof was unable to generate sufficient income to fund his overheads. In reality, except where a contractor carries out only one contract at a time or where the delayed contract is a high proportion of the company turnover, it is almost impossible to obtain the necessary definitive, contemporaneous evidence to prove that a contractor could have obtained profitable work elsewhere that he was unable to do, but for the claimed delay. Such evidence that can be obtained is usually inconclusive. Notwithstanding these problems, it is clear that if a contractor has entered into a contract that generates turnover and contribution to overheads and is prevented from earning these amounts, overheads will be underfunded and he will incur a loss. J. Pettet, *Claims for Head Office Overheads—Alternatives to Formulae,* 65 Arbitration 130, at 139 (No. 2, May 1999).

[C] Adjusting the Formula Calculations

Page 12-45, add after first sentence in first full paragraph:

However, a court may not adjust or perform the *Eichleay* formula calculation if the contractor uses the wrong periods or otherwise provides the raw data in making an incorrect calculation. For example, in *Orlasky, Inc. v. United States*, 68 Fed. Cl. 296 (2005), the contractor was awarded a contract to replace high voltage circuit breakers on an island for the Navy. The contractor's work was delayed by the government in many ways: lack of power to do the work, lack of access to the work, and changes. The contractor was unable to demobilize from the project until eight months after the scheduled completion date. The contractor was on "standby" according to the Court of Federal Claims even under the strict standards of the decision in *P.J. Dick, Inc. v. Principi*, 324 F.3d 1364 (Fed. Cir. 2003), and experienced government-caused delays on an unknown duration, and the contractor had no idea when it would be able to resume work, and was unable to take on any other work, and further wasn't able to leave the island until its work was complete. However, the court denied the contractor's claim because in making the *Eichleay* calculation it used costs and revenues for a period that did not correspond to the contract period: the contract performance was from November 23, 1995, to February 4, 1997, but the contractor submitted cost data for April 1, 1996, to

March 30, 1997. The court declined to make an adjustment or recalculate the damages stating that the court "cannot undertake to prepare evidence, even if plaintiff had provided the raw data."

Page 12-46, add after first paragraph:

The period for calculating extra home office support expenses is not limited to the contract duration plus any allowable extensions when the notice to proceed is delayed by the designer's error. For example, in *Crown Corr, Inc. v. Wil-Freds Construction, Inc.*, 2000 U.S. Dist. LEXIS 17961 (N.D. Ill. Dec. 11, 2000), Bertrand Goldberg Associates, Inc., was hired as the architect and construction coordinator of the Wright Junior College Project by the City of Chicago, acting through its Public Building Commission. Wil-Freds was hired to construct the floors of five buildings. The contractor was delayed and submitted a claim. Wil-Freds's claims included loss of productivity, wage and benefit increases, and extended general conditions. Wil-Freds and the city compromised on Wil-Freds's claims. The settlement agreement included a table that allocated dollar amounts of the settlement into various categories. Wil-Freds sought $701,798 from Goldberg for extra home office support. Goldberg argued that Wil-Freds had used the wrong starting date in calculating its damages, utilizing the contract award date as opposed to the date of notice to proceed. Goldberg also argued that the appropriate time period to be used was the contract period plus any extensions allowed. Part of Wil-Freds's claim against Goldberg for extra home office support was premised on the allegation that certain notices to proceed were unreasonably delayed because of Goldberg's failure to coordinate the Wright Project properly. Based on this allegation, it would be unreasonable to require Wil-Freds to compute damages beginning with its receipt of the notice to proceed. The reviewing court found the starting date used by Wil-Freds to be reasonable. This was not a breach of contract case whereby a contractor was suing an owner for delay damages. This case involved a contractor suing a construction coordinator (designer) for delay damages. The appropriate time period to be utilized to determine extra home office support damages was the equitable period already applied by Wil-Freds.

[D] *Hudson* Formula in English Law Cases

Page 12-48, add after note 185 in first full paragraph:

In *City Axis Limited v. Daniel P. Jackson* 1998 CILL 1382, the plaintiff contracted to carry out renovation work for the defendant. A dispute arose and the plaintiff claimed for an extension of time to the contract period together with loss and expense including overhead costs based on the use of the *Emden* formula. The claim was rejected by the court on the basis of the conditions of contract and the lack of factual evidence. In arriving at this decision the judge made reference to the factual criteria necessary to establish entitlement to such damages as defined in *Ellis Don Ltd. v. The Parking Authority of Toronto,* 29 B.L.R. 28 (1978) and *Alfred McAlpine v. Property and Land Contractors, Ltd.,* 51 B.L.R. 59 (1995). In particular, that the plaintiff failed to prove (i) that other work was available which could have been executed to maintain turnover; (ii) that he was unable to deploy its resources to take on this other work; (iii) there was no change in the market that would have adversely affected its business. The judge stated that the evidence provided was statistical only and the plaintiff had failed to prove any loss or damage. This judgment would appear to contradict the reasoning given in *Amec Building Ltd. v. Cadmus Investments Co. Ltd.,* 51 Con. L.R. 105 (1996), where the factual background was not satisfied; but nonetheless a (partial) formula was accepted. It should be established as a matter of fact that actual loss was suffered. It is suggested this could only be done by analyzing the anticipated gross profitability of the particular contract from the documents calculating the tender price to check if the figure was realistic and comparing this against the actual gross profitability achieved for the three years preceding the claim and the claimed period.

In *Beechwood Development v. Stuart Mitchell,* (2001) 18-CLD-04-04, the court used the actual average overhead and profit from the contractor's cost records to determine the amount of recovery for unabsorbed overhead rather than the bid overhead and profit when applying the *Hudson* formula. In *Beechwood,* the contractor was hired by the developer to build residential housing in a tract development. The contractor hired Stuart Mitchell, a land surveyor to provide a topographical survey of the properties on which the houses were to be built. Before starting work, the contractor noted certain discrepancies in the survey, and requested Stuart Mitchell to resurvey the property. Errors were acknowledged and corrected, and as a result the building footprints had to be moved and adjusted, delaying the building permits and start of construction by approximately 10 weeks. The contractor sued the surveyor for: (1) additional direct site costs, (2) loss of overhead and profit, (3) and increases in the cost of materials. The second item of damages was characterized as a "reduction in turnover, overhead and profit" in the years impacted by the delays. According to the contractor, the delay resulted in a reduction of turnover (sales) that otherwise would have been earned, and that would have contributed to the recovery of head office overheads and to the generation of profit in the affected years. Those resources would otherwise have been able to generate turnover for other construction work. The contractor alleged that there was no practicable means of assessing this loss other than by a formula using the percentage of the contractor's turnover represented by overheads and profit as follows:

Overhead & Profit % × Contract Sum/Contract Period × period of delay.

The amount of this claim was £75,042.20. An approach requested by the contractor other than the formula would be an evaluation of the loss of gross profits for the "lost" period based upon the average sales (turnover) of the preceding two financial years. The court noted that this claim was essentially for reimbursement of contributions to overheads and net profits that would have been made but for the delays, and was calculated by the *Hudson Formula*, well recognized and accepted in the case of *J.F. Finnegan Ltd. v. Sheffield City Council*, (1988) 23 B.L.R. Rather than use the amount of the contract, the court found that the average of the two prior years turnover gives a "more reliable basis for valuing 'lost' revenue than does the contract price . . . The latter is, indeed, the work which, but for the defender's breach of duty, would have been in train during the relevant period. That particular work was in the event postponed rather than wholly lost but the postponement has a knock-on effect in relation to the later work." The court also used the average of the two prior years' overhead and profit in the calculation of recovery rather than the estimated contract overhead according to the strict application of the *Hudson Formula.*

Page 12-50, add to note 196:

Manshul Constr. Corp. v. Dormitory Auth., 79 A.D.2d 383, 436 N.Y.S.2d 724 (1981) (court used formula similar to the *Hudson* formula to calculate home office overhead after project completion).

[E] *Eichleay* and Change Order Overhead

Page 12-50, add after note 197 in first full paragraph:

Contractors and owners have debated what is an acceptable markup rate for change order work and whether the change order overhead is adequate to compensate the contractor for the overhead costs associated with delays due to change orders. A sampling of change order overheads on projects in the United States indicates rates as low as 10 percent and as high as 35 percent on change orders. Whether the change order overhead should cover delay-related overhead costs may be governed by the contract in the general conditions or in the supplementary general conditions or special conditions. It is advisable for owners in preparing theses contract conditions to define the major and minor items of direct and indirect costs, and to explicitly address how and whether they apply differently to change orders that impact the critical path of the schedule or otherwise cause delay.

In general, the direct costs of change orders includes labor, materials, equipment, and subcontractors. The indirect costs of change orders includes taxes, profit, risk, contingencies, field general conditions, and overhead. The general conditions may include salaries of the various levels of supervision (project director, construction manager, project manager, general superintendent, trade superintendents), salaries of the engineering staff (chief engineer, office engineer, schedule engineer, material engineers, draftsmen, field engineers, survey party chief, instrument persons), salaries or wages of other support staff (quality assurance engineers, safety engineers, mechanics, plant operators, first aid workers, secretaries and clerical workers, computer operators, warehouse staff, security guards, janitors, and runners), automotive costs (company vehicles, ambulances, tractor trailers, special purpose vehicles, fuel trucks), project facilities (project office, warehouses, carpenter shops, pipe fabrication shops, weld testing facilities, electrical shop, first aid station, tool sheds, storage sheds, machine shop, equipment shop, concrete batch plant, quarry, training building), temporary construction (access roads, construction bridges, drainage structures, laydown areas, fencing, gates, parking areas, environmental protection), support systems (water supply, compressed air, electrical service, site communications, gas, oxygen), and other general expenses (office furniture, engineering supplies, engineering equipment, printing reproduction, computers, CPM scheduling, telephone, utilities, portable toilets, signs, safety equipment, permits and licenses, advertising, job travel expenses, testing and laboratory, legal fees, audit fees, medical supplies, progress photos, building rental, building and grounds expenses, maintenance costs, exterior lighting, drinking water and ice, taxes, bonds, insurance, payroll burden, back-charges, consultants, weather protection, general office supplies). Items, such as the above, are generally classified as indirect costs and may need to be clarified and defined for change order pricing. *See* McDowell, *Defining Mark-up Rates and Their Cost Impact* (Masters Paper, University of Southern California, M.B.A. Program 2003).

It is not unusual for contractors to charge project overhead for many of the items noted above on a percentage basis when a change order does not cause delay, and charge project overhead costs on a daily basis when the change order causes delay. This may be related to claims for home office or regional office costs. Project office or regional office overhead costs may increase as the volume of work increases as the result of change orders, and recovering for project overhead as a percentage markup may allow a contractor to recoup these costs if the volume of change orders is high. If the dollar amount of change orders is low but the change order causes delay to the work, the percentage markup on change orders may not allow the contractor to adequately recoup its project overhead. But according to a case before the Armed Services Board of Contract Appeals, a federal government contractor cannot have it both ways if it involves field office overhead, but it may be able to use both a per diem rate for home office overhead if it meets the tests for the application of the *Eichleay* formula. In *Caddell Construction Co., Inc.*, ASBCA No. 49333, 00-1 B.C.A. (CCH) ¶30,702 (1999), *aff'd on recon.,* 00-1 B.C.A. (CCH) ¶30,859 (2000), Caddell was awarded a $17.1 million contract by the Army Corps of Engineer (COE) for the construction of a wastewater treatment plant. At a pre-construction conference, the government requested that the contractor express its overhead both as a percentage of direct costs and as a daily rate, and the contractor stated that it had been its practice for many years to calculate field office overhead was as a per calendar day cost and sometimes as a percentage of cost. When change orders were being negotiated, the government took the position that field overhead costs are not allowed when contract time is not extended because the government had previously paid field office overhead on a per diem basis. The contractor reserved its rights to claim field overhead in various change order modifications, and later filed a claim with the contracting officer for field office overhead. The contracting officer responded to the claim stating the field overhead costs as a percentage of direct costs would not be allowed, and denied the claim. The claim was appealed to the Armed Services Board of Contract Appeals, and the government moved for summary judgment. The government based its arguments on the prior Board decision in *M.A. Mortensen Co.*, ASBCA Nos. 40750, 40751, and 40752, 98-1 B.C.A. (CCH) ¶ 29,658 (1998) *aff'd on recon.,* 97-1 B.C.A. (CCH) ¶ 28,623 (1997), 30,702 (1999), where the Board held that the Federal Acquisition Regulations (FAR) required the use of a single distribution base, not switching between a time distribution base (per diem rate) and a direct cost distribution base (percentage markup) for allocating job site overhead costs to changes. The contractor argued that the Board in past cases provided that field office overhead recovered as a percentage markup had to be deducted from the field overhead pool to calculate per diem rates, the Board chose to abide by its decision in *Mortensen.* The use of more than one method was required even if the government had directed the contractor to use the per diem rate.

Upon reconsideration, the Board examined the contractor's argument that there were cases from the United States Court of Appeals for the Federal Circuit that allowed home office overhead to be recovered under the *Eichleay* formula which is a per diem method not withstanding the contractor's recovery for home office overhead on a percentage basis during the course of contract performance. The Board noted that home office overhead recovery under the *Eichleay* formula was an essential aspect of government contract law that it requires more than a regulatory action (such as FAR) to overrule or nullify it as was the case with the FAR restrictions on the recovery of field office overheads. Further, the Board noted that the recovery of field office overheads are not subject to the judicially imposed restrictive conditions for the recovery of home office overhead under the *Eichleay* formula. The Board was not persuaded by the contractor's arguments, perhaps because *Eichleay* recovery is an "extraordinary remedy."

Page 12-51, add after second sentence of first paragraph:

For example, in *Babcock Energy Limited v. Lodge Sturtevant Limited*, 1994 CILL 981, the contractor's claim contained a request for head office overheads arising from a breach of contract on a power station. Babcock provided detailed evidence of head office staff who had been engaged sorting out the problem and had been diverted from other work that would have generated income. It also proved that at the time the head office had plenty of work and it had also been necessary for some work dealing with the breach of contract to be farmed out to agency staff. In arriving at his judgment, the Official Referee observed that Babcock had kept detailed and "uncontraverted" records of all staff and managerial time spent dealing with the matters arising from the breach of contract and that such particularization fully met the required criteria for recovery of such costs.

Page 12-51, add after first paragraph:

The award of unabsorbed home office overhead according to the *Eichleay* formula must credit the portion of change order markup attributed to home office overhead. In *Complete General Construction Co. v. Ohio Department of Transportation*, 94 Ohio St. 3d 54, 760 N.E.2d 364 (2002), the contractor recovered its unabsorbed home office overhead costs, incurred for two time extensions, in the Ohio Court of Claims. The public owner appealed, arguing that the court of claims should have deducted a portion of the more than $2 million worth of change order work performed on the contract from the contractor's unabsorbed overhead award. Given the court of claims finding that change orders for additional work included a 5 percent markup for overhead, the court was required to deduct 5 percent of the change order work from the award. The appellate court agreed. The appellate court stated that the trial court's exclusion of a 60-day extension period from the general contractor's compensable period showed that the court of claims should have deducted a portion of the change order overhead on the contract from the contractor's unabsorbed overhead damage award. Given the trial court's finding that home office overhead was built into change order payments for additional work, to the extent that the change orders also granted the contractor compensation for additional work, the change orders granted compensation for additional overhead. The court of claims' failure to deduct 5 percent of the value of any of the change orders from the amount of the contractor's unabsorbed overhead damage award resulted in excessive compensation for unabsorbed overhead. Nonetheless, deducting the entire 5 percent would result in an excessive reduction of the damage award, as the total value of all contract change orders included payments to subcontractors. On remand, the court of claims had to determine the amount by which change orders compensated the contractor for extra work, and deduct 5 percent of that amount from the *Eichleay* damage award. Judgment affirmed in part and reversed in part, and case remanded.

[F] Alternative Calculation Methods

Page 12-52, add at end of first full paragraph:

In situations in which contract performance has not yet begun, but the government terminates the contract for convenience after a period of delay, the contractor may recover unabsorbed overhead costs without using the *Eichleay* formula. In *Nicon, Inc. v. United States,* 2003 U.S. App. LEXIS 11498 (Fed. Cir. June 10, 2003), the Department of the Army, Mobile District Corps of Engineers, awarded Nicon a $1.4 million contract to repair Dorm 371 at MacDill Air Force Base. A disappointed bidder filed a bid protest, which suspended action on the contract before any of the repair work was commenced. The government notified Nicon of the protest and instructed it to "take no further actions as to the preparation and forwarding of submittals to the Resident Office." Nicon responded immediately, informing the government that it would "cease any further mobilization efforts associated with the . . . project." The bid protest was dismissed, but the government never directed Nicon to proceed with the contract. Nicon wrote a letter to the government to complain about the delay in issuing the notice to proceed and received no response. Nicon wrote again to request permission to proceed, but the government still failed to issue a notice to proceed. After this second letter, Nicon met with government representatives to discuss revised pricing for the contract. The government then terminated the contract for convenience. This termination occurred before the government ever issued a notice to proceed and before Nicon secured payment and performance bonds for the contract. A period of 107 days elapsed between the March 30, 1998, award of the contract to Nicon and dismissal of the bid protest, and another 181 days (for a total of 288 days) elapsed thereafter before the government's termination of Nicon's contract for convenience.

After the termination, Nicon sought direct costs and associated overhead and profit, as well as unabsorbed home office overhead for the time period between award of the contract and termination. Nicon used a modified version of the *Eichleay* formula to calculate its unabsorbed home office overhead damages. After negotiation, Nicon was awarded in direct costs, related overhead, and profit. However, the contracting officer denied Nicon's claim for unabsorbed home office overhead for the 288-day alleged delay period.

Nicon then filed suit in the Court of Federal Claims to recover the unabsorbed overhead damages. The Court of Federal Claims granted summary judgment in favor of the government, concluding that Nicon was seeking to apply an alternative to the accepted *Eichleay* formula and that it could not modify the formula as Nicon requested. Nicon appealed the Court of Federal Claims' decision.

The government terminated the contract after a lengthy delay, and the contractor was never permitted to perform any aspect of the contract. This is an unusual situation that did not fit neatly into the *Eichleay* formula. Nicon admitted that it had no actual "contract billings" or "days of performance" to use in the formula, but proposed to substitute constructive figures instead. It asserted that had Nicon been permitted to commence the repair work on schedule it would have completed the project within the time required by the contract and it would have received the full amount of the contractual billings.

The Court of Federal Claims refused to substitute the original contract price for "contract billings" in the formula and refused to substitute the anticipated period of performance for "days of performance." According to Nicon, this effectively added the requirement that performance had begun as an additional prerequisite to the formula.

Constructive figures might not be substituted into the *Eichleay* formula. The formula was not intended to simply compensate for any government-caused delay; rather, *Eichleay* damages were only available when the delay caused contract performance to require more time than originally anticipated. The Court of Federal Claims was therefore correct in concluding that the *Eichleay* formula was only applicable in situations in which contract performance had begun. However, that did not mean that a contractor, who was required to remain on standby because of a government-caused delay but was never allowed to begin performance, might not receive some of its unabsorbed home office overhead as part of its termination for convenience settlement by some other method of allocation. There was no bar to the award of home office overhead in a termination for convenience settlement, provided a reasonable method of allocation was available on the particular facts of the case.

Although unabsorbed overhead was not specifically listed in the FAR's "Termination for Convenience of the Government (Fixed-Price)" clause as one of the costs that would be paid under a settlement, it was also not excluded anywhere and could be asserted under the category of "costs incurred in the performance of the work terminated, including initial cost and preparatory expense allocable thereto." 48 C.F.R. § 52.249-2(g)(2)(i) (2002). Furthermore, although unabsorbed overhead was not mentioned in 48 C.F.R. § 31.205-42 as a cost peculiar to a termination situation, the costs listed there "are to be used in conjunction with the other cost principles in subpart 31.2," none of which appeared to prevent the award of unabsorbed overhead damages in a termination for convenience settlement if they were properly allowable and allocable. One other possible method a contractor in Nicon's situation may use to obtain unabsorbed overhead damages as part of its termination for convenience settlement was an equitable adjustment under the contract's suspension of work clause.

It would be inappropriate in the termination for convenience setting to rigidly apply a formula developed in different factual circumstances and thereby deny the contractor fair compensation for unabsorbed home office overhead.

On remand in this case, the Court of Federal Claims was to determine if a bid protest extending from the period of April 3, 1998, through July 15, 1998 was solely the fault of the government, and therefore whether Nicon might recover damages for this period. Nicon took on other contracts during the period of delay, and it must be determined on remand if these contracts qualified as replacement work that prevented Nicon from being on standby. The Court of Federal Claims also was to determine whether, on the facts of this case, there was a reasonable method of allocating some portion of Nicon's unabsorbed overhead for the delay period to this contract.

Page 12-52, add after note 208 in third full paragraph:

A different measure of damages for home office overhead may be applied for delays that occur after the scheduled completion date. In *Manshul Construction Corp. v. Dormitory Authority*, 436 N.Y.S.2d 724, 79 A.D.2d 383 (1981), the court compensated the contractor for the home office overhead incurred during delays subsequent to the project completion by applying a formula involving a four step process similar to the *Hudson* formula. (See **§ 12.06 [D]**.) In the first step, the court estimated the direct cost of the work performed after the scheduled completion date. This assumed that the contractor used a 15 percent overhead and profit factor.

So 100/115 or 87 percent of the billings during the post-completion period were used as the basis for calculating the direct cost of the work. In the second step, the court applied an overhead factor. Assuming that the contractor's 15 percent factor was for both overhead and profit, the court determined that overhead was 7.25 percent and profit was 7.25 percent. In the third step, the court multiplied the direct cost of the work times the 7.25 percent overhead factor to calculate the additional overhead that the contractor incurred for post-completion delays. In the fourth step, the court determined that the contractor was responsible for 5 percent of the delays, so it multiplied the post-completion overhead by 95 percent to calculate the additional overhead that the contractor was entitled to recover for the delays caused by the owner that forced the contractor to continue work after the scheduled completion date. This complex calculation was used because of concerns with the application of the *Eichleay* formula based upon the New York Court of Appeals case in *Berley Industries, Inc. v. City of New York*, 45 N.Y.2d 688, 385 N.E.2d 281, 412 N.Y.S.2d 589 (1978).

Page 12-53, add after first full paragraph:

In *Complete General Construction Co. v. Ohio Department of Transportation*, 94 Ohio St. 3d 54, 760 N.E.2d 364 (2002), the Ohio Supreme Court accepted the *Eichleay* formula to calculate unabsorbed home office overhead damages in public construction delay cases. The Ohio DOT argued that the *Eichleay* formula allowed contractors to recover for breach of contract without establishing causation. To the contrary, before the formula could be applied, a contractor must prove an extraordinary set of circumstances that demonstrated causation and damages. A contractor may recover only if there is an owner-caused construction delay. Moreover, the "standby" character of the delay must also be caused by the owner, and must prevent the contractor from finding replacement projects to cover the overhead. Thus, *Eichleay* did no violence to contract law. A court in Ohio could also utilize the direct cost formula. The direct cost method compared the direct costs actually attributed to a project as a portion of all of the direct costs incurred by the business over a particular period. The result was a ratio by which the percentage of indirect costs could be calculated, including home office overhead applicable to a particular project.

Page 12-53, add at end of subsection:

A contractor generally is not allowed to use its estimated monthly home office overhead as the basis for recovery of extended home office overhead. For example, in *MacQuesten General Contracting, Inc. v. HCE,* 296 F. Supp. 2d 437 (S.D.N.Y. 2003), MacQuesten was awarded a general contract for the construction of multi-family housing in Bronx, New York. MacQuesten in turn subcontracted the foundation work to HCE. During performance of the foundation work, HCE encountered unanticipated subsurface rock that necessitated revisions in the drawings that delayed HCE's work for two months. When HCE was unable to obtain an adjustment to its subcontract amount, HCE sued MacQuesten for extended home office overhead. At trial, HCE calculated its extended home office overhead by taking its estimated overhead for the planned five-month performance period, computing a monthly overhead rate and multiplying the monthly rate by the two months of compensable delay that it experienced. In setting aside the jury's award the court held that HCE's method was legally insufficient because there was no evidence that the delay actually increased HCE's total overhead cost, and there was no evidence that HCE's actual overhead costs were unavailable. The matter was remanded to the trial court on the issue of damages.

[G] Unallowable Overhead Costs by Contract

Page 12-55, add after second paragraph:

A contractor will inevitably incur overhead costs that have no relevance to the particular contract on which the claim is being made. Such items will therefore have to be excluded from the aggregate balance of overheads. Such non-applicable exclusions may be:

- Administration of non-contracting activities such as manufacturing, main tenance work, repair services, domestic work, property development, management services, housing developments, and other similar activities;
- Management resources dealing with previous claims, delays, and abnormal problems on earlier and concurrent contracts;
- Allowance for grants;
- Redundancy payments;
- Interest and finance charges;
- Losses on other contracts;
- Goodwill costs.

J. Pettet, at 139.

[1] By Regulation

Page 12-56, add after second full paragraph:

In *Complete General Construction Co. v. Ohio Department of Transportation*, 94 Ohio St. 3d 54, 760 N.E.2d 364 (2002), the Ohio Supreme Court concluded that the costs prohibited by Federal Acquisition Regulations (FAR) should not be recovered in claims that involve Ohio public contracts. Complete General

Construction Company constructed part of I-670 for the Ohio Department of Transportation (ODOT). Early in the project, design errors attributable to ODOT caused a 14-month delay. Complete General sued ODOT to recover unabsorbed home office overhead, among other costs. The Court of Claims awarded Complete General $184,947 for unabsorbed home office overhead. ODOT appealed the part of Complete General's award based upon the *Eichleay* formula. According to the Ohio Supreme Court, differences between federal and Ohio public contracting law allowed contractors to recover inappropriate costs when the *Eichleay* formula was applied in Ohio. The federal government adopted the FAR to determine which costs were allowable and which were not. FAR prohibitions included interest on borrowings, entertainment expenses, contributions and/or donations, bid and proposal costs, and bad debts. The idea that the government should fund a contractor's parties, sports tickets, political contributions, or other expenses that brought nothing tangible to the government's project was unreasonable. Such costs should not be recoverable in an Ohio case applying the *Eichleay* formula. The trial court erred in applying the *Eichleay* formula without allowing ODOT to dispute items of overhead.

§ 12.07 IDLE EQUIPMENT COSTS

Page 12-59, add after fourth full paragraph:

The contractor must connect the cause of idled equipment during winter shutdown with the acts of the owner. In *Complete General Construction Co. v. Ohio Department of Transportation*, 94 Ohio St. 3d 54, 760 N.E.2d 364 (2000), Complete General Construction filed a complaint to recover its idle and extended equipment costs incurred during the Ohio Department of Transportation's two extensions of contract. The court of claims awarded the contractor $62,622.50 on its claim for idle equipment costs. However, the court of claims found against the contractor on its claim for extended equipment costs. The owner appealed.

At trial, the contractor claimed three types of equipment costs as a result of the delay: (1) the costs incurred as a result of having equipment idle on the job site during the suspension period; (2) the costs incurred as a result of having equipment idle on the job site during the extension period; and (3) the costs of owning equipment used during the extension period. The owner challenged the court of claims' damage award for idle equipment costs incurred during the suspension period. The contractor had presented no evidence to establish a causal link between the delay and the equipment that was idled during the suspension period. The appellate court stated that the actual delay had been fully resolved by the end of December 1991. The fact that the owner did not require the contractor to resume work on the project until after April 29, 1992, or that the owner granted the contractor a 12-month extension for what was actually a 7-month suspension, was not sufficient to make the owner responsible as a matter of law for equipment being idle during the period after the actual delay ended. Although it is possible that the equipment was idle until April 29, 1992, as a result of some lingering problem related to the delay, the record was devoid of any evidence establishing such a causal connection. No evidence was presented to show that the equipment was on the job site due to the delay, as opposed to some other cause such as a scheduling choice by the contractor. Judgment affirmed in part and reversed in part, and case remanded.

[C] Challenging Equipment Rates

Page 12-64, add after carryover paragraph:

Under a cost reimbursement contract, a contractor is entitled to recover rental costs paid for equipment it rents. When a contractor uses its own fully depreciated property, the rental value of that property ordinarily is not an allowable cost. However, the government may agree to pay the contractor a reasonable amount for the use of the contractor's fully depreciated property. In *Union Boiler Works, Inc. v. Caldera*, 156 F.3d 1374 (Fed. Cir. 1998), the Court of Appeals for the Federal Circuit affirmed the Court of Federal Claim's denial of a contractor's claim for the rental costs of fully depreciated property when the contractor had not entered into an agreement with the government regarding payment of those costs.

In 1989, the Army contracted with Union Boiler Works, Inc. to repair three boilers at Aberdeen Proving Ground, Maryland. After work began, the contracting officer directed Union Boiler to replace insulation on two of the three boilers. As part of the contract modification, the government granted Union Boiler a 90-day extension to accommodate the changed work.

Because it was running behind schedule, in October 1989, Union Boiler installed a temporary boiler in the affected building. Union Boiler owned the temporary boiler, which was fully depreciated on Union Boiler's books. Union Boiler completed the repairs on the boilers affected by the contract change 70 days after the 90-day extension period had passed.

The ASBCA sustained its earlier decision that the government's direction to replace the boiler insulation caused Union Boiler to install a temporary boiler 90 days earlier than the contract required. The decision left a quantum of compensation up to the parties to resolve. In July 1995, the contracting officer denied Union Boiler's claim for cost to hook up and start up the temporary boiler, as well as the rental value of the temporary boiler. The government argued that because Union Boiler did not complete the added work within the allocated 90-day period, it would have incurred hookup and start-up costs even without the change. The board agreed and granted the government's motion for summary judgment.

With respect to rental costs, Union Boiler contended that it was entitled to recover the rental value of its boiler, $4,500 a month, as lost opportunity costs for rents it could have collected had the equipment not been needed by the government. To assess the allowability of the rental costs, the board first looked to the recovery of costs for construction equipment under FAR § 31.105(d)(2). This section of the FAR permitted the government to specify a schedule of predetermined rates to be used when actual cost data was not available. The board found that Union Boiler had no evidence of actual costs, and the contract did not prescribe a predetermined rate schedule. Accordingly, the board concluded that the allowability of the costs claimed would be determined under FAR Subpart 31.2.

§ 12.08 LOST PRODUCTIVITY

[B] Measured Productivity Method

Page 12-77, add at end of section:

A measured mile analysis compares work performed in one period, not impacted by events causing a loss of productivity with the same or comparable work performed in another period, that was impacted by productivity affecting events. *Clark Construction Group, Inc.*, No. VABCA-5674, 2000 VA B.C.A. LEXIS 4 (Apr. 5, 2000). The measured productivity method has also been called the "measured mile analysis." Although there are different ways to calculate the measured productivity, the goal of the measured productivity method is to compare productivity during undisrupted periods (or locations) with productivity during disrupted periods (or locations). Ideally, the comparison will be made on the same project, but some practitioners will make a comparison between what they consider to be similar projects. The comparison may be made between similar work on the same project. In *P.J. Dick, Inc.*, VABCA No. 5597, 01-2 B.C.A. (CCH) ¶ 31,647 (2001), P.J. Dick, the general contractor, was awarded a contract by the Department of Veterans Affairs (DVA) to construct an addition to a medical center. Its electrical subcontractor, Kent, alleged that the electrical design contained numerous deficiencies, latent errors, and other problems that necessitated numerous Requests for Information (RFIs) and extensive revisions to Kent's electrical work. P.J. Dick requested a time extension to deal with the electrical design problems, but DVA refused and threatened liquidated damages if P.J. Dick did not complete in a timely manner. P.J. Dick then ordered Kent to accelerate its work. Kent claimed that the acceleration and the design problems disrupted and otherwise impacted its work, especially the installation of the branch circuits. P.J. Dick settled the claim with Kent and then brought a claim against the DVA that was taken to the Department of Veterans Affairs Board of Contract Appeals. Before the Board, the claimant's expert used a "measured mile" analysis to demonstrate the impact of the various disruptions. The expert opined that, although the impacts disrupted the installation of the branch circuit work, there was no time period when the branch circuit installations were unaffected. The expert testified that the electrical feeder installation was similar to the branch circuit installation because the feeder work was also done by union crews at the same time under the same weather conditions. Therefore, the expert used the installation of the electrical feeder work as the basis of its "measure mile" productivity analysis and developed an efficiency factor by dividing the number of labor hours actually spent on the feeder installation (adjusted for change order work) by the budgeted labor hours. The budgeted hours for branch circuit installation was multiplied by the efficiency factor to obtain a revised budget for the branch circuit work. The revised budgeted hours for the branch circuit work were then subtracted from the actual labor hours expended on the branch circuit work. The difference was the productivity loss due to the actions of the DVA. The DVA countered that the claimant's expert used different kinds of work in developing the measured mile and, therefore, the expert's measured mile was inappropriately used in this matter. For example, the feeder circuits required larger conduit and lengthier conduit runs than the branch circuits. This would distort any comparison between the feeder work and the branch circuit work. However, the board noted that a measured mile analysis does not require the comparison of exactly the same work, but the work should be reasonably similar. Measured productivity obviates the need for the claimant to prove the reasonableness of its estimated productivity because estimated productivity is not used as the basis for comparison.

Courts have accepted the measured productivity method as a way to establish disruption and to quantify the productivity loss. For example, the court decisions in *General Insurance Co. v. Hercules Construction, Natkin & Co. v. George A. Fuller Co.*, and *Groves-Black* (all discussed in the main text) all accepted lost productivity calculated by the measured productivity method. In addition, in *U.S. Industries, Inc. v. Blake Construction, Inc.*, 671 F.2d 539 (D.C. Cir. 1982), the plaintiff, U.S. Industries, sued its joint venture partner, Blake Construction, alleging disruption, among other things, from the failure to schedule and coordinate the work. U.S. Industries submitted financial and statistical evidence along the lines of a measured mile to prove that its labor costs increased dramatically when it was disrupted by the defendant.

Experts are often engaged to perform and present measured productivity analyses. In *Danac, Inc.*, ASBCA No. 33,394, 97-2 B.C.A. (CCH) ¶ 29,184 (1997), *aff'd*, 98-1 B.C.A. (CCH) ¶ 29,454 (1998), the claimant's expert prepared an analysis purportedly using the measured mile approach. The expert's analysis calculated the ratio of the number of labor hours expended to the number of units completed in various time periods. According to this analysis, the claimant experienced a 55 percent productivity loss in one period and a 25 percent productivity loss in another period when compared to one of the more efficient time periods that was used as the baseline for measuring productivity.

The use of the measured productivity approach has its limitations, according to some courts. For example, the comparison to work in different time periods or locations should be similar in terms of complexity or difficulty of the work activities, skill level of the workers and supervisors, and work environment. If the work tasks are not comparable, courts may reject or adjust the measured mile analysis. In *United States ex rel. Shields v. Citizens & Southern National Bank*, 367 F.2d 473 (4th Cir. 1966), it was determined that the work in the "disrupted" areas was by nature more time-consuming than the work in the measured mile areas, and the analysis was neither a reasonable comparison nor a reasonable basis for quantifying disruption damages. Industry commentators have also identified other limitations, adjustments, or problems with the use of the measured productivity:

- The lack of an unimpacted time period or location to develop a measured mile baseline, because disruption occurs throughout the project.
- "Learning curve" or worker orientation and familiarization should be taken into account.
- Time and price requirements, such as accelerated work, change order work, and time and material compensation, may require special consideration.
- Environmental factors, such as temperature, wind, and other weather conditions, may affect the comparison.
- The measured mile period or location may not represent a reasonably sustainable level of productivity.

Thomas & Sanvido, *Quantification of Losses Caused by Labor Inefficiencies: Where Is the Elusive Measured Mile?* 1 Constr. L. & Bus. 3 (Summer 2000); Patton & Gatlin, *Claims for Lost Labor Productivity*, Constr. Law. 21 (Apr. 2000); Loulakis & Santiago, *Getting the Most Out of Your Measured Mile Approach*, Civil Eng'g 69 (Nov. 1999); Ginsburg & Dickson, *Loss of Efficiency/Edition II*, Constr. Briefings No. 99-11 (Oct. 1999).

Some of these factors might be adjusted by using a baseline approach rather than a strict measured productivity analysis, according to the studies and recommendations of Penn State University professors Thomas and Sanvido. Thomas & Sanvido, *Quantification of Losses Caused by Labor Inefficiencies: Where Is the Elusive Measured Mile?* 1 Constr. L. & Bus. 3 (Summer 2000). The baseline approach uses a comparable period of work time when the contractor achieves good productivity even though it may be hampered by its own inefficiencies. Nor does the baseline period have to be free of impacts caused by the other party, as is required when using a measured mile analysis. Further, the baseline time frame need not be comprised of consecutive work periods—the baseline may include sporadic workdays, weeks, or months. To ensure that the baseline is a realistic representation of the contractor's potential productivity, the professors recommend that it comprise at least 10 percent of the time when the measured work activities were being performed.

The baseline approach described by Thomas and Sanvido may raise concerns because it is an analysis based upon the contractor's "best productivity." This may exclude conditions reasonably encountered on a project, such as learning curves, normal corrective work, and reasonably anticipated mobilization requirements. Therefore, a baseline productivity analysis may be bolstered by a demonstration that it represents a reasonably attainable and sustainable level of productivity, taking into consideration the field conditions anticipated by the contract documents and other conditions inherent in the work.

§12.10 INTEREST

Page 12-82, add to end of section:

States may also provide for interest in the event payment is delayed or ignored. For example, Missouri has statutes for refusing to pay in bad faith and without reasonable cause (Missouri's Public Works Prompt Payment Act (MPWPPA)) and when no interest is provided under a contract for delayed payment (Missouri's Private Prompt Payment Act (PPPA)). Under Missouri's PPPA, a subcontractor is entitled to 9 percent interest on the amount of a contractor's wrongfully withheld retainage, minus the contractor's offset. For example, in *Jerry Bennett Masonry, Inc. v. Crossland Constr. Co.*, 2005 Mo. App. LEXIS 1104 (2005), Jerry Bennett Masonry subcontracted with Crossland Construction to perform masonry work for the construction of a school building. The subcontractor completed its work and presented an invoice to the contractor. The contractor withheld 10 percent retainage from its payment to the subcontractor. The contractor asserted that the subcontractor maintained an insufficient number of personnel at the project site, which resulted in delays and necessitated extra work by the contractor to prevent its being assessed liquidated damages under its contract with the school district. The subcontractor countered that any delays were caused by weather, soil conditions, other subcontractors, and the contractor's errors. The subcontractor sought judgment under the Missouri Public Works Prompt Payment Act (MPWPPA) and the Private Prompt Payment Act (PPPA) for the withheld retainage together with attorney's fees and interest. The trial court found that both parties were at fault due to poor decisions regarding staffing and inadequate planning. The trial court awarded damages in the amount of the retainage to the subcontractor, but determined that the contractor was entitled to a set-off per its counterclaim. The trial court did not assess interest on the retainage. The subcontractor appealed.

The subcontractor maintained that the trial court erred in failing to assess 18 percent interest under the MPWPPA or, alternatively, 9 percent interest under the PPPA. The subcontractor argued that the retainage amount constituted a liquidated sum to which interest applied. The contractor asserted it did not withhold the retainage amount in bad faith and without reasonable cause, which was a prerequisite to an entitlement of 18 percent interest under the statute. The trial court made no express determination that the contractor acted in bad faith without reasonable cause in withholding payments from the subcontractor. Lack of sufficient manpower by the subcontractor contributed to the contractor's delay in completing the project in a timely fashion. The trial court did not err in not awarding 18 percent prejudgment interest pursuant to the MPWPPA. However, the trial court erred by not awarding the subcontractor 9 percent interest pursuant to the PPPA. The withheld retainage was readily ascertainable comporting with the requirements of the PPPA. The contractor's offset did not preclude the recovery of interest. Because the subcontract provided for no interest payments after completion of the project, the subcontractor was entitled to 9 percent interest on the principal amount, which was the retainage minus the contractor's offset, from the date that the contractor was paid its own retainage by the school district.

[C] Prejudgment Interest for Denied Use of Funds

Page 12-87, add after second paragraph:

A flow-through agreement in a subcontract does not include an interest clause in the general contract when the subcontract work is limited to the work provisions of the general contract. For example, in *Thor Electric, Inc. v. Oberle & Assocs.*, 741 N.E.2d 373 (Ind. Ct. App. 2000), the Richmond Hotel Limited retained Oberle as the general contractor on a renovation project. Oberle retained Thor as a subcontractor to perform specified work on the project, including plumbing, heating, ventilation, and air conditioning. Under the contract, Oberle would pay Thor 95 percent of the invoiced amounts five days after Oberle was paid by the owner. Thor would receive the 5 percent retainage upon satisfactory completion of the work. Should Oberle fail to pay Thor, the subcontract provided that Thor could stop work until payment of the amount owed, and the contract price would be adjusted for the additional costs of shutdown, delay, and startup. In its cross-claim, Thor sought payment of its retainage from Oberle. Thor was entitled to $82,633.30 in retainage under the subcontract. Additionally, Thor sought prejudgment interest on the retainage. The trial court rendered its judgment, awarding Thor attorneys' fees, prejudgment interest, and damages against Oberle.

Thor argued that its subcontract with Oberle contained a flow-through provision that incorporated the provisions of the contract between Oberle and the owner. Based on the flow-through provision, Thor contended that the proper interest rate was the local prime rate plus 2 percent. The language of the flow-through provision clearly provided, however, that incorporation was limited to the work provisions of the general contract as applicable to Thor. The work provisions were limited to the specific performance of the work—materials used, labor, and timing—and as such did not include the payment provisions of the Oberle/owner contract. Consequently, the flow-through provision did not incorporate the interest-rate provision of the Oberle/owner contract into the subcontract. Furthermore, whereas the general contract provided for interest in the event of late payment, the subcontract provided for stoppage of work in the event of late payment and was silent on the issue of interest. In addition, the court recognized a public policy interest in allowing subcontractors to negotiate a price without being constrained by the contractor and owner's negotiated price and method of payment. A subcontractor would look to the agreement between the owner and contractor for the project specifications and materials requirements in submitting a bid to the contractor. However, the subcontractor is free to negotiate its own financial arrangements. This subcontract did not provide the interest rate to be applied in calculating prejudgment interest.

[D] Prejudgment Interest Authorized by Statute

Page 12-90, add after note 335 in third full paragraph:

In *Dugan & Meyers Construction Co. v. State of Ohio Department of Administrative Services*, 2005 Ohio 3810, 2005 Ohio App. LEXIS 3503 (July 28, 2005), the State of Ohio Department of Administrative Services (ODAS) engaged Dugan & Meyers Construction Co., Inc. (D&M) (prime contractor) for construction of three buildings at the Fisher College of Business at The Ohio State University (OSU). The contract included a provision that time was of the essence and another provision for the assessment of liquidated damages in the amount of $3,000 per day. The work was delayed significantly, eventually completing some six months after the contract completion date. OSU assessed liquidated damages against D&M for part of the delay in the amount of $325,000. D&M filed a claim to recover its contract balance, reversal of the liquidated damages, and extra amounts attributed to delay damages. After OSU rejected D&M's claim, D&M instituted an action for breach of contract. OADS asserted counterclaims seeking liquidated damages. D&M recovered $602,745 as the balance of its contract; a complete reversal of the liquidated damages; and $730,760 in "cumulative impact" damages, plus interest on those damages.

ODAS argued that the trial court erred in ordering them to pay prejudgment interest at the rate of 10 percent per annum when the contract specified a particular method of calculating interest. Additionally, ODAS pointed out, the contract provided that the owner may deduct any liquidated damages to which it was entitled from any application for payment. ODAS argued that because OSU deducted from D&M's pay applications the liquidated damages it felt it was owed, when the trial court ordered these sums returned, interest on these sums should have been calculated in accordance with the contract. They contended that it was error for the trial court to impose an interest rate of 10 percent per annum, which was the rate applicable only where the damages flow from the breach of a contract that itself did not specify an interest rate. There was no authority that dealt with the precise issue whether deductions from contractor payments, which were taken to satisfy a liquidated damages assessment later judicially determined to be unwarranted, were subject to the contract's "late payment" interest provision, or whether they were subject to the statutory prejudgment interest amount. Despite legitimacy in both positions, the better choice was to apply the statutory interest rate because a payment made following entry of a judgment awarding return of wrongfully assessed liquidated damages, rendered sometimes years after the fact, strained the definition of "late payment," even when the liquidated damages were originally assessed by deducting the amount from a contractor payment. Accordingly, there was no error in the trial court's award of prejudgment interest at the statutory rate. The trial court's judgment was affirmed in part, reversed in part, and remanded with instructions.

[F] Computing Prejudgment Interest

Page 12-95, add after third sentence in second full paragraph:

For example, in *Thor Electric, Inc. v. Oberle & Assocs.*, 741 N.E.2d 373 (Ind. Ct. App. 2000), the court held that retainage due on substantial completion is readily ascertainable for calculation of prejudgment interest. Richmond Hotel Limited retained Oberle as the general contractor on a renovation project. Oberle retained Thor as a subcontractor to perform plumbing, heating, ventilation, and air conditioning. Under the contract, Thor was required to submit monthly invoices to Oberle. Oberle would then pay Thor 95 percent of the invoiced amounts five days after Oberle was paid by the Owner. Thor would receive the 5 percent retainage upon satisfactory completion of the work. Should Oberle fail to pay Thor, the contract provided that Thor could stop work until payment of the amount owed, and the contract price would be adjusted for the additional costs of shutdown, delay, and start-up. The contract did not expressly provide for the recovery of attorneys' fees or other damages. Thor sought payment of its retainage from Oberle. Thor was entitled to $82,633.30 in retainage under the Oberle/Thor contract. The trial court awarded Thor prejudgment interest. Oberle appealed.

Oberle argued that the award of prejudgment interest was improper because the damages and the due date were not readily ascertainable. Oberle contended that the damages were not readily ascertainable because Thor had not completed approximately $10,000 of the work. Further, Oberle argued that the due date was not ascertainable because it was uncertain when Thor satisfactorily completed the work. Here, the damages were readily ascertainable. Oberle retained 5 percent of the contract amount until Thor satisfactorily completed the work. Thus, the amount of damages could be easily determined by a simple mathematical computation. The due date was also readily ascertainable: Thor was entitled to payment upon completion of the work. A project manager for Oberle created a punch list of things to be completed, which Thor fulfilled. Thor notified Oberle of the completion, and Oberle acknowledged receipt of operation manuals and drawings. Another building inspector testified that Thor had not completed approximately $10,000 of the work. However, $10,000 represented approximately one-sixth of 1 percent of the contract price, and therefore, Thor had substantially completed the work. Consequently, Thor was entitled to its retainage on March 26, 1987. As the damages were readily ascertainable, there was no error in the trial court's conclusion that Thor was entitled to prejudgment interest.

§ 12.11 PUNITIVE DAMAGES

Page 12-97, add after first full paragraph:

Punitive damages are appropriate for a subcontractor's reprehensible conduct in interfering with a contract between the contractor and owner. In *Envtl. Energy*

Ptnrs., Inc. v. Siemens Bldg. Techs., Inc., 2005 Mo. App. LEXIS 1568 (Oct. 25, 2005), Environmental Energy Partners (contractor) contracted to perform work on property owned by St. John's Regional Medical Center (owner). Siemens Building Technologies (subcontractor) served as a subcontractor on the project. The subcontractor failed to complete its work on schedule. The subcontractor refused to cooperate or communicate with the contractor during the course of the project. The subcontractor refused to meet with the contractor and other subcontractors to assess the status of the project's work. The subcontractor failed to provide information that would have permitted the contractor to receive payments from the owner to which it was entitled. The subcontractor imposed a lien on the owner's property and erroneously represented that its work on the project was complete when it was not. Although the contract provided that the subcontractor was not to be paid until the contractor was paid, the subcontractor surreptitiously caused the owner to divert the payment intended for the contractor to the subcontractor. The subcontractor brought suit against the contractor. Seeking actual and punitive damages, the contractor filed a counterclaim against the subcontractor for interference with contract and other claims. The trial court returned a verdict in favor of the subcontractor on its claim. Verdicts were returned in favor of the contractor on all its claims against the subcontractor. The jury awarded actual and punitive damages to the contractor. The subcontractor appealed and the contractor cross-appealed.

The subcontractor argued that the award of punitive damages was a denial of its right to due process of law, but the subcontractor's actions were reprehensible and without justification. Although the subcontractor was 617 days past its completion date and was liable to the contractor for liquidated damages, the subcontractor negotiated an agreement with the owner that resulted in payment to it of funds that were contractually owed to the contractor. Upon securing payment of the sum that the owner was obligated by contract to pay the contractor, the subcontractor entered into an agreement with the owner under which the payment would not be made known to the contractor. Punitive damages were appropriate to deter such conduct in the future. The ratio of punitive damages to total compensatory damages was approximately 4:1. Given the subcontractor's reprehensible conduct in interfering with the contract, that ratio did not constitute a violation of due process of law. The trial court's judgment was affirmed in part, reversed in part, and remanded with directions.

§12.12 LOST PROFITS

Page 12-99, add after first full paragraph:

Unlike specific job costs, overhead and profit often are not susceptible to definite and particular documentation and calculation. For example, in *Aniero Concrete Co. v. New York City Constr. Auth.*, 2003 U.S. Dist. LEXIS 23126 (S.D.N.Y. Dec. 23, 2003), the New York City School Construction Authority

(SCA) contracted with the P.J. Carlin Construction Company (Carlin) for modernization and renovation of the Morris High School. Aetna had issued a performance bond obligating Aetna to complete the Project at its expense if Carlin failed to do so. Carlin was terminated from the Project after little more than a year. Aetna retained Aniero Concrete to act as replacement general contractor. Aetna and Aniero purported to enter into a written Completion Agreement. Aniero ceased working on the uncompleted project on and, complaining of Aetna's delays in making payments under the purported contract, commenced an action against Aetna. Aniero claimed that Aetna owed Aniero $4,007,620. The court held in an earlier opinion that the Completion Agreement was invalid because Aetna failed to obtain the SCA's written consent of Aetna's assignment of the Carlin contract to Aniero. Aniero's claim against Aetna sounds in quantum meruit. Aetna denied Aniero any recovery in quantum meruit. Aniero included in its quantum meruit claim a claim for overhead and profit. It calculated overhead and profit on the basis of 15 percent of its incurred actual job costs. Aniero made no effort to break down that 15 percent factor between overhead and profit. Aetna contended that Aniero's proof was insufficient to sustain a claim for either overhead or profits.

That their recovery should be calculated according to a percentage of job expenses was only reasonable. Aniero gave considerable trial testimony in support of its claim for overhead, including an extensive and plausible list of costs constituting Aniero's standing overhead during the period the company worked on the project; that the project made use of those standing facilities, including construction vehicles, Aniero's construction yard, and the clerical, managerial, and secretarial services of Aniero employees. Aniero did not undertake to allocate the 15 percent figure between overhead and profit. Aetna argued that the failure demonstrated that the claimed overhead and profit were both speculative. Aniero paid its bills and whatever was left constituted profit. This was a perfectly reasonable approach, specifically approved by the New York construction quantum meruit cases.

In his testimony Aetna's consultant referred to an annual construction industry publication, the *Means Construction Handbook*, which he described as "the bible of the construction industry estimating for costs." From that publication Guerrero derived a 3½ percent figure for profit as being at the high end. The difficulty with this testimony is that, as demonstrated during cross-examination, Guerrero testified at his pre-trial deposition that the "appropriate markup" for profit of between 2½ and 4 percent he derived from Means referred to the services of construction managers and not general contractors (like Aniero); and it is readily perceivable that on a project of this magnitude, a general contractor on a fixed price contract runs a greater economic risk of loss than does a construction manager, and consequently could expect a greater reward if the job goes well.

Aniero Concrete Company, Inc. was entitled to a judgment against Aetna Casualty and Surety Company in the total amount of Aniero's actual job costs incurred on the project and 15 percent of the amount as an allowance for overhead and profit.

That a claim of 15 percent of costs covers both overhead and profits combined is a matter of law. Recoveries of a combined overhead and profit at

15 percent have been approved by numerous cases decided by courts in the Second Circuit and by New York courts. *Aniero Concrete Co. v. New York City Construction Authority*, 2003 U.S. Dist. LEXIS 23126 (S.D.N.Y. 2003), citing *Manshul Const. Corp. v. Dormitory Authority of New York*, 79 A.D.2d 383, 436 N.Y.S.2d 724, 730 (1981) (recognizing reasonableness of a 15 percent addition to claim to account for overhead and profit); *U.S. for Use of Tower Masonry, Inc. v. J. Kokolakis Contracting, Inc.*, No. 93 Civ. 6369, 1995 WL 539742, at *5 (S.D.N.Y. Sept. 8, 1995) ("These costs include a 10% add-on for overhead and 10% for profit, percentages which the Court finds to be normal and reasonable."); *Westcott v. State of New York*, 264 A.D. 463, 36 N.Y.S.2d 23 (3d Dep't 1942) (15 percent of costs to cover profit and overhead deemed reasonable despite lack of specific accounting); *Fehlhaber Corp. v. State of New York*, 69 A.D.2d 362, 419 N.Y.S.2d 773, 778 (3d Dep't 1979) (11.22 percent of costs to cover profits and overhead was found to be reasonable); *D'Angelo v. State*, 41 A.D.2d 77, 341 N.Y.S.2d 84, 88 (N.Y. App. Div. 3d Dep't 1973) (recognizing a 10 percent addition to a quantum meruit claim to account for "overhead and profit"). *Cf. Clifford R. Gray, Inc. v. New York*, 251 A.D.2d 728, 674 N.Y.S.2d 440, 442 (3d Dep't 1998) (20–27 percent addition for profit and overhead was not reasonable).

Page 12-100, add after carryover paragraph:

A contractor seeking to recover lost profits in an action for tortious interference must present sufficient evidence from which lost net profits can be calculated. In *Thille & Son Constr., Inc. v. Thille*, 2005 Neb. App. LEXIS 207 (Sept. 6, 2005), Russell and Brenda Thille joined Howard and Cynthia Helm to form Thille & Son Construction, Inc. (subcontractor one) for the business of laying sanitary and storm sewer pipe in housing and commercial developments. Russell and Brenda Thille then formed a new company, R & B Excavating (subcontractor two), to perform the same type of services. On April 25, 2001, Russell Thille submitted bids for subcontracting work laying pipe on the Quail Run project. The bid document made no specific reference to indicate on whose behalf the bid was being submitted. On April 26, the Thilles resigned from subcontractor one and signed a settlement agreement with the Helms on April 27. Subcontractor two executed a subcontract agreement on the Quail Run project on May 17, 2001. Subcontractor one contacted the primary contractor, Vrana & Son Construction Co. (contractor), on the Quail Run project, claiming that Russell Thille's bid was submitted on behalf of subcontractor one, who should therefore have been awarded the contract. The contractor then voided the subcontract and subsequently awarded it to Cedar Construction Co. (Cedar). Subcontractor one filed an action alleging subcontractor two had tortiously interfered with subcontractor one's contract opportunity on the Quail Run project. Based on a jury verdict for subcontractor one, the trial court awarded damages. Subcontractor two appealed.

Subcontractor two argued that subcontractor one failed to present sufficient evidence to support an award of damages and that it failed to present evidence from which the jury could calculate lost net profits, as opposed to lost gross profits.

Subcontractor one presented no reliable financial data to support testimony that it suffered lost profits from not being awarded the Quail Run project. A vice-president for Cedar testified that he was not sure if Cedar made any profit on the project. The owner of subcontractor one testified that he disagreed with Cedar's estimates of costs, that subcontractor one could have obtained materials at lower costs, and that subcontractor one would have done the project differently and used some different materials. There was no financial data presented to support any of subcontractor one's testimony concerning the cost of materials or its estimates of daily labor costs. The district court erred in instructing the jury to calculate subcontractor one's lost profits in a fashion that considered only lost gross profits. The proper method of calculating damages for lost profits was based on lost net profits and not lost gross profits. Where subcontractor one presented evidence of only gross profits and failed to provide evidence of expenses and overhead costs from which net profits could have been calculated, it failed to present sufficient evidence of lost profits.

§ 12.14 RECOVERY OF ATTORNEYS' FEES IN STATE COURT ACTIONS

Page 12-108, add to note 434:

Thor Elec., Inc. v. Oberle & Assocs., 741 N.E.2d 373 (Ind. Ct. App. 2000) (subcontractor may recover attorneys' fees for contractor's failure to pay retainage only if permitted by agreement, statute, or rule; because there was no contract provision or applicable statute authorizing an award of attorney fees, trial court erred in awarding attorneys' fees to subcontractor).

Page 12-111, add after carryover paragraph:

For the purposes of Tennessee's Prompt Pay Act, a contractor is responsible for attorneys' fees when it acts in bad faith by discharging a subcontractor long before a contract deadline. *Southeast Drilling & Blasting Servs., Inc. v. Hu-Mac Contractors., LLC*, 2003 Tenn. App. LEXIS 636 (2003), Hu-Mac Contractors, LLC, was the general contractor of a development project. Southeast Drilling and Blasting Services, Inc. submitted a proposal to the contractor. On June 8, 1998, the work began. Southeast Drilling had a drilling rig on site by June 22, 1998. The movement of the drilling rigs resulted in mechanical breakdowns. On June 23, 1998, both of subcontractor 1's drilling rigs were on location, but neither rig was operable. All week, Southeast Drilling had problems with workers and equipment. On June 29, 1998, Southeast Drilling moved additional workers and had both drilling machines working. The contractor notified Southeast Drilling it was being terminated. On June 28, 1998, the contractor contacted Explosive Engineering to make a proposal to drill and blast the roadway. The contractor signed subcontractor 2's proposal on June 30, 1998. Southeast Drilling believed it was not behind schedule in spite of a slow start. Subcontractor 2 worked for seven weeks to complete the job. Neither of the agreements between the contractor and Southeast Drilling or subcontractor 2 had a completion date or work schedule, nor was any verbally communicated to either. No complaint was made regarding the quality of work that subcontractor 1 performed. Southeast Drilling brought suit against the contractor for breach of contract. The Chancellor entered a judgment against the contractor. The Chancellor determined that Southeast Drilling was entitled to its lost profits, prejudgment interest, and attorneys' fees pursuant to the Prompt Pay Act of 1991. The contractor appealed.

The contractor argued there was no showing that it acted in bad faith. There was no evidence the quality of the work performed by Southeast Drilling was below industry standard. Rather than give Southeast Drilling an opportunity to avoid additional defective performance, the contractor chose to meet with and hire subcontractor 2. Once the contract was made with subcontractor 2, Southeast Drilling was summarily dismissed. The contractor failed to deal in good faith, common equity, and fairness with Southeast Drilling by discharging it seven days after they moved on the premises and long before the contract deadline would have to be met. The court determined there was bad faith exercised on behalf of the contractor. The award of attorneys' fee to Southeast Drilling was sustained.

§12.19 OWNER'S DAMAGES

Page 12-122, add after second full paragraph:

If a contractor's defective work causes the performance of the contract to be more expensive or burdensome, the owner is entitled to damages including compensation for stopping business while corrected or delayed work is completed. For example, consider *Pelshaw Co. v. City Motors*, 2003 Iowa App. LEXIS 266 (2003). R.L. Pelshaw Co. agreed to build a new 175-foot metal building addition for City Motors. Pelshaw proposed to receive $30,000 for construction of the addition, plus $39,000 for installation of the concrete foundation and flooring, including a floor drainage system. The floor drainage system failed to work and the parties disputed the best method to correct the system. City Motors concluded the building was not completed in a workmanlike manner and declined to pay Pelshaw. Pelshaw filed a claim for $35,280 based on breach of contract. City Motors filed a counterclaim for damages resulting from Pelshaw's poor workmanship and delays caused by the installation of a new drainage system. The district court awarded Pelshaw $34,280, but also awarded City Motors $24,000 as replacement costs for the floor and $15,000 as reasonable compensation for closing the business while the floor was repaired. Pelshaw has appealed, and City Motors has cross-appealed.

Pelshaw contended the district court erred in awarding City Motors $15,000 as compensation for down time resulting from any business closure due to efforts to correct the faulty drainage system. Pelshaw contended that while there was no dispute the floor's installation was defective, City Motors was not entitled to extra compensation for "down time" since it was denied the opportunity to correct its mistakes. City Motors initially determined such a shutdown would cost it over $129,527 over a 15-day period. The district court concluded that $24,000 would put City Motors in a position the same as if the floor been correctly installed and that $15,000 was reasonable compensation for a five-day shutdown. Pelshaw's poor workmanship caused considerable delays in the building's construction, and the failure to properly install the drainage system caused City Motors to incur additional expenses in correcting it. Affirmed.

Page 12-123, add after second full paragraph:

On the other hand, the simplicity of owner damages does not excuse an obligation to calculate and prove them. For example, in *Cut Bank School District No. 15 v. Rummel*, 2002 Montana 248 (S. Ct. 2002), Cut Bank School District No. 15 solicited bids from area contractors to repair and patch the steps and sidewalk in front of the Cut Bank High School, resurface the steps and sidewalk with a concrete texture finish, and apply an epoxy sealant on all the surfaces. Rummel's bid proposing to complete the work at a cost of $7,950 was accepted. The District desired that the work be completed on or before August 29, 1999, because students and staff would be returning to the school on the following day. There was no written documentation that this timeliness element was a requirement of the contract between the parties, but the District contended that Rummel guaranteed the work would be finished by August 29, 1999. In contrast, Rummel contended he made no guarantee, but told the District he would make his best effort to complete the project by that date. On August 15, 1999, the District accepted another bid from Rummel for $1,500 to repair and patch a second sidewalk on the north side of the high school. Again, the District contended Rummel guaranteed completion of the project by August 29, 1999. On August 30, 1999, the District terminated the contracts with Rummel because the work had not been completed. The District subsequently filed a complaint asserting that Rummel's failure to complete the projects by August 29, 1999, was a breach of the contracts and requesting monetary damages. The district court entered an order granting judgment in favor of Rummel based in its determination that, even assuming Rummel's failure to complete the projects was a breach of the contracts, the District failed to present evidence establishing that damages resulted. The District appealed.

The District argued that it proved damages resulting from the breach of the first contract via Rummel's testimony. On cross-examination, Rummel was asked to give a rough estimate of what it would cost in materials and labor to complete the first contract project from the point at which he left it on August 29, 1999. In response, Rummel testified "I don't know off the top of my head . . . between 3 and 4,000, total, maybe." The District presented no evidence of the actual cost to complete the project originally contemplated in the first contract because the District contracted in the summer of 2000 with another construction company to repair the front steps and sidewalk using an entirely different method and materials. Thus, the only evidence relating to the District's damages for breach of the first contract was Rummel's purely speculative statement. Such speculation was insufficient to prove damages to a reasonable degree of certainty and, consequently, the District Court's finding of fact in that regard was not clearly erroneous.

[A] Elements in Owner's Damages

Page 12-128, add at end of section:

In New York, an owner can seek damages on behalf of its contractor based on a liquidation agreement while at the same time asserting independent claims. In *N. Moore St. Developers, LLC v. Meltzer/Mandl Architects, P.C.*, 2005 N.Y. App. Div. LEXIS 8316 (Aug. 4, 2005), North Moore Street Developers (owner) retained Meltzer/Mandl Architects (architect) to provide architectural and related design services for the renovation and rehabilitation of an eight-story building. The owner then hired R.C. Dolner, Inc. (contractor) to perform the general contracting services for the project. The owner filed a lawsuit that in the first two counts alleged damages in the amount of $4.1 million due to the architect's breach of contract and professional malpractice. The third and fourth causes of action asserted similar claims on behalf of the contractor in the amount of $2.45 million. The owner had previously entered into a Liquidating Agreement with the contractor, which provided that the owner admitted liability to the contractor, that the amount of such liability was liquidated to the amount of any recovery by the owner from the architect, and that the owner could prosecute the contractor's claims on its behalf.

The architect filed a motion for summary judgment to dismiss the third and fourth causes of action on the grounds that the contractor had no recognizable claim against the architect and the owner could not bring the claims on the contractor's behalf. The trial court granted the motion, finding the liquidation agreement was unenforceable because the owner had asserted its own claims against the architect and not just claims on behalf of the contractor. The owner appealed.

The owner argued that the lack of privity between the architect and the contractor did not bar it from asserting the contractor's claims since the very purpose of a liquidating agreement was to bridge the privity gap. The owner further argued it had accepted responsibility for the contractor's damages caused by the architect, and, even absent the liquidating agreement, the owner could have asserted the fourth cause of action for professional malpractice on the contractor's behalf given the relationship, sufficiently equivalent to privity, that existed between the architect and the contractor. The court of appeals found that the owner/general contractor agreement clearly stated that the owner was responsible to the contractor for delays caused by the owner in excess of 40 days, or 70 days in the aggregate. This included delays caused by the owner and those under its agency and control, such as its architect. In the complaint, the owner alleged that the architect's breach of the architect/owner agreement and its negligence during the project caused a delay in the final completion of the project for a period of approximately 12 months, a period clearly in excess of 70 days. The complaint set forth additional claims against the architect in the third and fourth causes of action, which alleged damages to the contractor as a result of the architect's negligence and breach of contract. In the complaint, the owner admitted liability to the contractor, providing a legitimate basis for the contractor's pass-through claims. The court found the liquidation agreement had all the necessary elements and that the trial court was mistaken in its notion that the owner's assertion of independent claims rendered the pass-through claims unsustainable.

[B] Limitations to Owner's Delay Damages

Page 12-128, add after note 534 in carryover paragraph:

General damages are those that are so usual an accompaniment of the kind of breach or wrongdoing that the mere allegation of the wrong gives sufficient notice. 2 McDonald, Texas Civil Practice 124, § 6.17.1 (1970). If an item of damage is not peculiar to only a particular plaintiff, the damage is not consequential, but direct. Consequential damages may be recovered only if proved to be "the natural, probable, and foreseeable consequence" of the defendant's breach. *CMC Steel Fabricators, Inc. v. Harrop Constr. Co.*, 131 F. Supp. 2d 882, *citing Mead v. Johnson Group Inc.*, 615 S.W.2d at 687. For example, in *Roanoke Hospital Assoc. v. Doyle and Russell, Inc.*, 215 Va. 796, 214 S.E.2d 155 (1975), the owner claimed three types of interest costs: (1) added interest costs (including expenditures on borrowed funds and interest revenue lost on invested funds) during the construction period arising from the longer term of borrowing necessitated by the contractor's unexcused delay (extended financing costs); (2) added interest costs during the construction period attributable to higher interest rates during the extended term (incremental construction interest costs); and (3) added interest costs for the permanent loan attributable to higher interest rates (incremental permanent interest costs). The court concluded the extended financing costs were direct damages because customarily construction contracts, particularly large contracts, required third-party financing and a delay in completion ordinarily required an extension of the term of construction financing. The interest costs incurred and the interest revenue lost during such an extended term were predictable results of the delay and were, therefore, compensable direct damages. Only the incremental construction interest costs were consequential damages. *See also Hemenway Co. v. Bartex, Inc.*, 373 So. 2d 1356 (La. Ct. App.), *cert. denied*, 376 So. 2d 1272 (La. 1979). Similarly, in *Chandler v. Madsen*, 197 Mont. 234, 642 P.2d 1028 (1982), the supreme court explained that the necessity of moving from a residence undergoing massive repairs and of renting a temporary residence was likely to result from design errors that caused major damage to a structure. The court found that costs for moving, storage, and temporary rental were improperly deleted from the damage award. In *Miracle Mile Shopping Center v. National Union Indemnity Co.*, 299 F.2d 780 (7th Cir. Ind. 1962), the district court allowed

$8,500 for the increased cost of mechanical work that resulted from the contractor's abandonment of the project. The appellate court distinguished the additional cost of completion from special or consequential damages. *Miracle Mile Shopping Center, citing Johnson-Johnson Inc. v. Farah*, 123 Ind. App. 87, 108 N.E.2d. 638 (1952). *See also Walker Builders, Inc. v. Lykens*, 628 So. 2d 923 (Ala. Ct. App. 1993) (homeowner's mental anguish damages were actual, not special damages); *Grossman v. Sea Air Towers, Ltd.*, 513 So. 2d 686 (Fla. Dist. Ct. App. 3d Dist. 1987) (architect was held liable for lost rents when tenants vacated owner's building due to noise and dust of construction resulting from collapse of an improperly designed parking garage without mentioning consequential damages); *Apfel v. D.A. Bennett, Inc.*, 109 A.D.2d 1039, 487 N.Y.S.2d 159 (1985) (costs awarded for the owner's excess heating fuel consumption due to the contractor's defective insulation installation without mentioning consequential damages); *Lenz Constr. Co. v. Cameron*, 207 Mont. 506, 674 P.2d 1101 (1984) (temporary rental costs for comparable equipment awarded as compensatory damages in negligence action, where the owner's equipment was damaged by the contractor).

"The potential for delay damages should always be within the contemplation of parties entering into a construction contract." *Construction Law* § 6.10[5] (S. Stein, ed. 2001). Many industry commentators agree and include those damages that should be expected from acts that delay a construction program or project. A treatise on Texas construction law suggested those damages may be:

> If there is no liquidated damage clause in the contract, the owner is entitled to recover actual damages suffered because of the delays. The measure of damages is normally the value of loss of use of the building for the duration of the delay, *or expenses caused by the delay.* J. Canterbury, Texas Construction Law Manual § 10.20 (2d ed. 1992) (emphasis added) citing *Ryan v. Thurmond*, 481 S.W.2d 199, 1972 Tex. App. LEXIS 2458 (1972) (A contractor who has not performed the contract in due time is generally liable in damages for the delay. The measure of damages for failure to complete the work within the time fixed by the contract is generally the value of the use of the building).

Professor Justin Sweet in his work, *Legal Aspects of Architecture, Engineering and the Construction Process*, stated that:

> Direct losses can occur in any construction contract. They are usually measured by conventional objective formulas, such as lost use for unexcused contractor completion delay, cost of correction or diminished value for other contractor defaults, interest for owner-delayed payment, the balance of the purchase price for nonpayment, and the additional costs caused by owner disruption or delay.
>
> Sometimes a claimant claims consequential damages, those that go beyond *direct* losses. Consequential damages are abnormal, related to the particular transaction, and more subjective. . . .

Sweet, *Legal Aspects of Architecture, Engineering and the Construction Process* 608 (5th ed. 1994).

The American Institute of Architect's (AIA) Standard Form of General Conditions of the Contract for Construction, AIA Document A201 (1997 ed.), includes at Paragraph 4.3.10 a mutual waiver of consequential damages that defined consequential damages to include, "... damages incurred by the Owner for rental expenses, for losses of use, income, profit, financing, business and reputation, and for loss of management or employee productivity or of the services of such persons." Similarly, the Associated General Contractor's of America's (AGC) Standard Form of Agreement and General Conditions Between Owner and Contractor, AGC Document No. 200 (2000 ed.), contract also contains a mutual waiver of consequential damages at Paragraph 10.2. The AGC defines consequential damages,

> ... including but not limited to the Owner's loss of use of the Project, any rental expenses incurred, loss of income, profit or financing related to the Project, as well as the loss of business, loss of financing, principal office overhead and expenses, loss of profits not related to this Project, or loss of reputation.

Assuming these definitions are a good bright line for the industry generally, the AIA and AGC standard excludes most traditional delay damages in their definition of consequential damages: cost overruns occasioned by the delay, payments to contractor's for delay claims, escalated costs, and all the others.

In *H. Shurter v. M. Butler*, 43 Tex. Civ. App. 353, 94 S.W. 1084 (1906), the appellate court stated that delay damages that included a payment to a third party, additional labor caused by the delay, and security costs were *not* too remote to be "direct" damages. H. Shurter contracted with the city of Houston to build sewers in 120 days, or in default to pay $10 per day liquidated damages. M. Butler subcontracted to furnish the brick and knew the terms of the contract. By the terms of the subcontract, the bricks were to be furnished at such time and in such quantities as they were called for. Instead of completing the work in 120 days it was not done for nearly a year, the delay caused by the failure to furnish brick as required for the work. The contractor claimed delay damages, including the $10 per day paid to the city for the delay in completing the sewer; the expense of a watchman required to watch the excavation made for the sewer, for the time the work was delayed; a large sum expended for extra labor rendered necessary by the delay; and the amount expended for re-excavation occasioned thereby; as well as the difference between the amount paid for brick, which the contractor was compelled to buy elsewhere. The trial court refused proof as too remote of any of the delay damages claimed. According to the reviewing court, the delay damages were not too remote to be recovered as matter of law. The subcontractor knew that the work had to be completed in 120 days and that the completion of the work depended upon furnishing the brick promptly as needed in the prosecution of the work. The subcontractor was liable for such damages as were the proximate result of the breach,"that is, for all such damages as resulted directly from such breach and which might be reasonably supposed to have been in contemplation of the parties, at the time of making the contract, as likely to result therefrom." *Id.* at 1085. Reversed and remanded.

In *U.S. for the Use and Benefit of CMC Steel Fabricators, Inc. v. Harrop Constr. Co.*, 131 F. Supp. 2d 882, 891 (S.D. Tex. 2000), the Federal District Court stated that "the direct damages that are available . . . are virtually the same under any theory of liability and causation" and stated a wide variety of incidental damages were also recoverable. *Id., citing Kish v. Van Note*, 692 S.W.2d 463, 466–67 (Tex. 1985). The court identified as direct damages "offsets" in the form of direct damages it took to correct subcontractor's errors.

In *Hess Die Mold, Inc. v. American Plasti-Plate Corp.*, 653 S.W.2d 927 (1983), the court allowed an owner the cost to replace a plastic mold that the contractor failed to supply. American Plastic was in the business of injection molding of plastics. Hess manufactures die molds. One of American's molds no longer performed satisfactorily, and it solicited bids for a new one. Hess submitted a quote to American to design and build a suitable mold at a price of $23,445. Hess never completed a mold that would perform or operate properly in American's business. As a result, American was forced to contract with another manufacturer for the production of a suitable mold at a contract price of $47,500. The cost to American for this mold, over and above its contract price with Hess, was $24,065. American sued Hess. The court rendered judgment for $24,065 and Hess appealed. Appellant's points of error all attacked the award of $24,065 on the ground that such losses were "special damages," as opposed to "general damages." The trial court stated the damages represented the benefit of the bargain for American. The Texas 12th District Court of Appeals found the damage award of $24,065 was properly characterized as an item of general damages. Hess contracted to replace an existing mold, which American was actively using in its manufacturing process. It should have been clearly foreseeable to Hess that if it failed to supply a suitable new mold to American within the time specified, American would be forced to obtain a mold from another manufacturer in order to meet its commitments to its customers. Since the cost was an item of general damages, it followed that allegations and/or findings of contemplation of said loss were not necessary to support the judgment. Appellant's points one through three were overruled.

Costs for re-performance are recoverable even with a limitation of liability clause that limits a party's liability to re-performance. In *CBINA-CON, Inc. v. UOP Inc.*, 961 S.W.2d 336 (1997), Fina contracted with UOP for the designs for a fluidized catalytic cracking unit (FCCU) at Fina's Port Arthur refinery. UOP's liability to Fina was limited to re-performing at UOP's expense. UOP's contract with Fina also waived any special, consequential or indirect damages, including loss of use, loss of profits, plant or unit downtime, or suits by third parties. Fina also contracted with CBI to construct a catalyst cooler, which was part of the FCCU. UOP did not have a contract with CBI. Fina also entered into contracts with other parties, to modify the design, inspect, install, fabricate and manufacture certain components of the catalyst-cooling unit. The catalyst-cracking unit stopped operating properly because the catalyst cooler failed. After the unit failed, Fina filed suit against CBI and the other companies, but not UOP, to recover damages in replacing the catalyst cooler and for $17,450,000 in consequential damages until replacement of the catalyst cooler. Fina made claims against CBI for negligence, breach of contract, breach of warranty, and strict products liability. Fina alleged CBI negligently breached warranties in connection with the construction of the catalyst-cooling unit. CBI filed a third-party claim against UOP for contribution. UOP moved for summary judgment. The trial court granted UOP's motion for summary judgment. CBI appealed.

The Fina/UOP contract expressly limited UOP's liability to Fina for any breach, loss or damage to re-performance. The only direct damages recoverable by Fina against UOP for negligence was re-performance *or the cost of re-performance.* If Fina had pursued UOP for negligence in designing the catalytic cooler, UOP would be obligated to re-perform at UOP's expense. If UOP had not re-performed the work, then Fina would have been limited to a breach of contract claim for UOP's refusal to re-perform the work. CBI's claim was also so limited. Even if CBI could bring a tort claim, based on the derivative nature of CBI's claim, CBI's recovery was limited to that of Fina. Affirmed.

In *Anderson Development Corporation v. Coastal States Crude Gathering Co.*, 543 S.W.2d 402 (1976), Anderson Development agreed to construct two steel pipelines for Coastal States Crude Gathering Company. Anderson sued to recover damages for breach of contract, payment of retainage withheld by Coastal, and payment on an invoice for work performed. Anderson alleged Coastal's failure to supply proper right-of-way as required by the contract caused work to be done in the inclement fall months instead of the more desirable summer months; forced Anderson to construct in a piecemeal fashion; prevented completion of the job until 76 days after the date required in the contract; and caused Anderson $267,525.33 in excess of the amount Coastal was obligated to pay under the contract. Anderson's petition alleged that Anderson spent $267,525.33 over the $544,799.47 Coastal owed Anderson under the contract's Unit Price Schedule, that these sums represented the reasonable value of work done for Coastal, and that Coastal was obligated to pay the excess because its failure to supply proper right-of-way caused the expenditures. Trial to the court for recovery of the retainage and for the amount of the invoice resulted in a $69,943.59 judgment for Anderson, plus interest and costs. Anderson appealed.

Coastal asserted that Anderson claimed special damages, which were not specifically stated. According to the Texas Court of Appeals, because Coastal contracted to furnish proper right-of-way, which was a prerequisite to Anderson's doing any work, it was foreseeable that its failure to supply right-of-way would keep Anderson from progressing with construction and would cause Anderson additional expense in completing work. These were general damages. Anderson's points of error were sustained.

The date of the contract (and if amended, the date of the last modification) may also be important to distinguish direct from consequential damages. When the breach includes an unexcused delay in completion, if the completion date has been altered by consensual amendment, the contemplation of the parties is to be determined as of the date of amendment. *Roanoke Hospital Assoc. v. Doyle and Russell, Inc.*, 215 Va. 796, 214 S.E.2d 155 (1975). Knowledge that is not available at the time of contracting cannot be used later to prove that a contractor could foresee the damages. *Wills Elec. Co., Inc. v. Mirsaidi*, 2001 Tenn. App. LEXIS 915 (2001) (contractor held back $19,013 because the subcontractor had not fully paid all its suppliers or obtained lien releases. The subcontractor borrowed $25,000. The trial court ruled that the contractor was liable for consequential damages for the amount of interest associated with the subcontractor's loan. The appellate court found no evidence that at the time of contracting the contractor knew or suspected that failure to make complete payment would force the subcontractor to borrow money. After contracting, the contractor became aware that the subcontractor faced some financial difficulties).

All courts recognize that the difference in value of the building caused by the contractor or designer's breach is a direct damage. Whenever an item of owner damage strays from that formula the owner risks a court concluding the damage is consequential. For example, in *Wright Schuchart, Inc. v. Cooper Industries, Inc.*, (not published) 1994 U.S. App. LEXIS 31520 (9th Cir. 1994), an unpublished decision that also involved a waiver of consequential damages, the Ninth Circuit Court of Appeals rejected the contractor's claims for "direct damages" against an owner as calculated by its claims consultant as consequential despite the contractor's argument that otherwise it could not be made whole (because consequential damages are necessary to make any plaintiff whole); the direct damages were foreseeable (because foreseeability was immaterial); and if it could not collect its direct damages the contractor's exclusive remedy failed (because the contractor's lack of a remedy was the consequence of excluding consequential damages in the contract). The contractor sought:

> project supervisory personnel kept on payroll an extra 15 months; labor necessary to service and maintain equipment supporting the overall project; additional employee benefit costs (i.e., insurance, travel, rest breaks); provision of room and board to island personnel (46 man months); escalation in room and board costs that would not have occurred if project had been complete according to SWS schedule; additional electricity, telephone use, facsimile charges; opportunity losses from equipment tied up on the overall project; decline in value of assets due to harsh weather conditions on the island; costs incurred in demobilizing and mobilizing the island while Cooper [owner] redesigned the exciter control panels; loss of labor productivity and efficiency during delays and disruptions; loss of learning on the part of Wright employees due to disruption; additional site cleanup costs; additional subcontractor costs incurred due to change of schedule from the disruption; costs of preparing the claim against Cooper; and home office general and administrative costs.

According to the court, these damages were undeniably typical examples of consequential damages as that term was commonly understood. The court agreed with the contractor, however, that any costs it incurred by directly contributing to the repair of defective owner-furnished equipment were recoverable as direct damages. *Wright, citing Canal Elec. Co. v. Westinghouse Elec. Corp.*, 756 F. Supp. 620, 624, 627 (D. Mass. 1990). The contractor alleged direct repair damages under the label "back charges." The court did not agree that the contractor could recover amounts it expended facilitating the repair. The contractor could not also recover costs such as those associated with the demobilization and mobilization, for example.

However, restoration and replacement costs as means of determining damages may be allowed if diminution in the fair market value measure was unreasonable. In *Massachusetts Port Authority v. Sciaba Constr. Corp.*, 437 Mass. 1104, 766 N.E.2d 118 (2002), Massachusetts Port Authority was legislatively mandated to build a public park. The plan was to use an existing pier upon which to construct recreational improvements. The authority hired Sciaba Construction to perform the platform renovations. While the contractor was working with torches on the platform during a heavy windstorm, hot metal fell onto the creosote-treated wood members comprising the platform and its underlying pilings and supports, igniting them and causing severe damage. The authority filed suit, claiming that the contractor was negligent in performing "hot work" on a windy day without proper permission and without adequate safety precautions. The authority contended that roughly 61 percent of the platform, other than the pilings, needed to be replaced. The authority identified three options for completing the project. The first option was to completely remove the existing damaged structure and replace it with a new, entirely concrete, structure. The second option was to remove only the timber structure and replace it with concrete, but to leave the wooden pilings. The final option was to replace the timber members of the structure with new timber, while also leaving the existing pilings in place. A consultant arrived at net cost estimates for each option after deducting the amounts that would be saved by not proceeding under the pre-fire schedule of work. The contractor objected to evidence of the cost of repairs or replacement because the only relevant evidence on the question of damages was the difference in fair market value caused by the fire. The judge instructed the jury that ordinarily the measure of damages was the diminution of market value resulting from a wrongful act, but explained that an alternative method could be used. The jury awarded $1,049,130, which was identical to the net cost of replacing the pier as estimated under option 1. The contractor moved for a judgment notwithstanding the verdict and for a new trial. The judge denied the motions. The contractor appealed.

According to the appellate court, diminution in fair market value would not have been a fair and reasonable measure of damages. Even assuming that comparable sales of similar piers were available, or that expert testimony could have established the property's approximate fair market value before and after the fire, the judge had discretion to determine that the resulting figure would not compensate the authority for the special value the property held as a place to build a park, and would not have fairly represented the actual injury suffered by the authority. Replacing or restoring the pier was necessary because the fire had caused substantial damage. The authority was not in a position to sell the property and to find a new location on which to construct a park. Not only was the property part of a larger plan taking into account the surrounding area, but the authority was acting under legislative direction to build a park on the specific location. Prior to the fire, the pier was reasonably well adapted to the special purposes for which it was intended. The pier possessed a value to its owner that could not be properly weighed by using the ordinary methods of damage calculations. It was reasonable to allow the jury to hear evidence of restoration and replacement costs and to instruct them on the availability of those methods. Affirmed.

By presenting an owner's increased costs as part of the benefit of the bargain formula, a court fit owner delay costs into an accepted theory and avoid the risk of the court finding an owner's damages are consequential. The increased costs may be added to the square foot costs the building would cost, claiming the difference between the increased square foot costs to complete the building and the square foot costs estimated to complete the buildings. The added expenses also represent the *decreased* appreciation of the value of the space. Just because the space is worth more than now should not deprive the owner of the reduced appreciation costs. The court in *Noble v. Tweedy*, 90 Cal. App. 2d 738, 203 P.2d 778 (1949) (landlord refused to complete bowling alley for tenant. The court recognized the difference in value between the property as built and the value that it would have had if it had been built in accordance with contract plans and specifications as the exclusive measure of damage), allowed something similar when it affirmed an owner award for the difference in value of a bowling alley that included the additional costs to complete the project. Although no substantive difference from presenting its completion costs as "direct" costs as in *Wright*, the translation into a square foot costs permitted the appellate court to endorse the classic formula for calculating owner damages and eliminated the need for legal support for failure of remedy and other similar arguments.

Page 12-130, add at end of section:

Despite contractor delay, owner is not entitled to installation of more expensive fixtures that are installed quicker. For example, in *Harley Paws, Inc., d/b/a Three Dog Bakery v. Mohns, Inc.*, 249 Wis. 2d 488, 639 N.W.2d 203 (Wis. Ct. App. 2001), Mohns, Inc. and Harley Paws, Inc., contracted for the remodeling of a retail store. Mohns told Harley Paws that the construction should be completed in approximately five weeks, and estimated the cost to be $52,000. Based on that information, Harley Paws scheduled its grand opening for September 26, 1998. Construction commenced in early August 1998 and continued until September 11, when disputes developed primarily involving: (1) the modification, possible delay, and additional costs involving the construction of the store's cabinetry; and (2) the rescheduling of the faux finisher, who was scheduled to do the finishing painting on September 12, but could not because Mohns had failed to complete the preliminary painting. When Harley Paws discovered defects in the drywall and learned from Mohns' president, that the store would not be completed by the grand opening date, Harley Paws fired Mohns and arranged for other contractors to complete the remodeling. Harley Paws claimed that it cost $42,175.68 to complete Mohns' contract and sought damages of $28,958.82, including more expensive countertops necessary to complete the project on time. The replacement cabinetmaker could only guarantee the timely installation of Surrell countertops, as opposed to the less expensive Laminate countertops, for which Harley Paws had contracted with Mohns. Harley Paws subsequently sued Mohns for breach of contract. The court concluded "that the termination on the 11th of the contract and the hiring of replacement contractors was reasonable."

The trial court found that the subsequent cabinetmaker's use of the more expensive countertops required a $2,000 reduction in Harley Paws damages. The trial court's careful considerations were not erroneous, arbitrary, or excessive.

An owner's damages for cost to complete a project for a contractor may also have to be reasonable. The actual cost to complete may not be sufficient proof. Consider *Driver Pipeline Co. v. Mustang Pipeline Co.*, 2001 Tex. App. LEXIS 8403 (Dec. 20, 2001) (opinion withdrawn). Mustang wanted to build a 200-mile gas pipeline. Mustang hired Driver to dig the trenches and lay the pipe for the north 100 miles of pipeline. Because Mustang wanted to finish the pipeline by a specific deadline, it prepared a contract stating that the prices were based on 11 hours per day and 7 days a week to meet a 14-week schedule (98 construction days), and specifying that the final completion of the pipeline within that time was of "great importance to the Company." And then the rains came. At the end of 58 days, Driver had finished only 15 miles of the pipeline and had laid off some of its crews and suspended operations. Driver proposed that Mustang give it additional time and money to finish the project. Driver took the position that rain delayed its work and that all parties understood that his bid was based on no weather delays. The owner of Driver testified that his bid took into account a 15 percent contingency for rain days. At trial, however, Driver testified that this was a "best-time" contract with no rain days included and that he relied on a clause allowing (but not requiring) Mustang to give them extra time in the event of weather delays. There was evidence that the builder of the south portion of the line sought and received a 30-day extension on its construction.

Mustang sued Driver for breach of contract, seeking costs to complete the project. Driver countersued for wrongful termination, seeking damages for termination, lost profits, and attorney's fees. Mustang sought damages of $4,065,286 because it cost that much more than the original contract price with Driver to complete the pipeline. At the conclusion of an eight-day trial, the jury found: Driver failed to comply with the contract; Mustang was not justified in terminating its contract with Driver; $2,104,601 would compensate Mustang for damages caused by Driver's failure to comply with the contract; and $2,515,958 would compensate Driver for damages caused by Mustang's unjustified failure to comply with the contract. The trial court found that the verdict was for Driver and that there was no evidence of probative force to sustain the jury's finding with respect to the damages question for Mustang. The court then rendered judgment in favor of Driver and Seaboard Surety against Mustang. The court awarded Driver actual damages of $3,146,728.33.

Mustang contended that the trial court erred by disregarding the jury's answer to its damage issue. There was substantial evidence of the amount that it cost Mustang to complete the pipeline. The jury charge contained language limiting any recovery to the reasonable and necessary costs incurred to complete the pipeline less the amount agreed to between Mustang and Driver to construct the pipeline. The position taken by Driver and accepted by the trial court was that this evidence was legally insufficient because there was no evidence before the jury to show that the amount that was admittedly paid was "reasonable and necessary" to complete the work. Although there was substantial evidence about the extent of damages, there was no evidence from any source to suggest that the actual expenditures were reasonable or necessary—merely that those were the amounts expended in completing the pipeline. Mustang argued correctly that it was not necessary for the witnesses to repeat the magic words "reasonable and necessary" in order to provide evidence that expenses are in fact "reasonable and necessary." As long as the evidence sufficiently showed that the expenses were reasonable and necessary, it was not required that particular "magic words" be used by the witnesses. However, a plaintiff cannot recover damages just by proving how much it spent. It must also provide evidence that the amount spent was reasonable and necessary. Mustang presented no testimony that showed that the amount paid was reasonable or necessary. Evidence of the failure to complete the contract and the completion by another contractor did not suffice to show that the cost of the work was reasonable. The trial court correctly granted a JNOV.

Chapter 13

ESTABLISHING THE BASIS OF LIABILITY

§ 13.02 BREACH OF EXPRESS PROMISES IN CONTRACT

Page 13-6, add after note 32 in second full paragraph:

A project owner does not breach a site development agreement by taking over the developer's work where concern about timely completion is reasonable, even without a project schedule. In *Novi Promenade Assocs. Ltd. P'ship v. Target Corp.*, 2005 U.S. Dist. LEXIS 22068 (E.D. Mich. Sept. 30, 2005), Landon Development (developer) agreed to perform site work for property on which a Target (owner) store would be built. The parties signed a site development agreement (SDA), which provided for the owner's takeover of the site work if the developer did not meet dates set forth in the master schedule. Preliminary work on the owner's pad was performed by the developer's contractor. The operating engineers union began picketing the site because the contractor was non-union. Union problems, weather, and difficulty with the bidding process led to delays. The owner notified the developer that it intended to take over the work site. Target took over the remaining site work. The parties cross-claimed against each other for breach of contract.

Both parties argued that the other party breached the SDA. The owner asserted that the developer breached the SDA on the basis of its untimely site development work. The developer claimed that the owner improperly took over the site development in bad faith. The developer complained that the owner breached its duties by contacting potential subcontractors, thus preventing the developer's completion of the bidding process. The developer also maintained that the owner agreed to pay excessive sums to contractors in order to meet its schedule. The owner's concern about the developer's commitment and ability to complete its work prior to the opening date was reasonable. Due to the delays and problems, the owner was justified in believing that the store's opening date was in jeopardy. The owner's contacts with subcontractors impaired but did not preclude the developer's ability to successfully bid the work. The owner's contacts with subcontractors were reasonable considering its concerns about the developer's ability to complete its work on time. The owner did not breach the SDA by its takeover. The developer could not be held liable for breaching the SDA due to its untimely work because there was no master schedule. Both parties, however, breached the SDA by refusing to make payments for development costs. The owner failed to pay some invoices for the developer's site improvement fee. The developer failed to pay the owner's costs. In general, the owner did not spend an unreasonable amount, as it anticipated a significant increase of expenses formerly allocated to the developer when it took over the work. The owner's takeover necessitated certain change orders and modifications, but the owner overpaid for grossly overcharged storm work and for overtime. The owner was awarded the total site work costs, minus unreasonable overpayments to a contractor. The judgment for the developer for the unpaid invoices was accounted for as an offsetting obligation. Judgment was entered for both parties on their breach of contract claims.

[D] Prime Contractor as Third-Party Beneficiary

Page 13-12, add after first carryover paragraph:

However, an electrical prime contractor was not a third-party beneficiary to the general prime contractor's contract with the owner as a result of clauses that made prime contractors liable to other primes for delay in *Kinback Corp.v. Quaker Constr. Man., Inc.*, 2001 U.S. Dist. LEXIS 16845 (M.D. Pa. 2001). Quaker was awarded the general trades contract and Kinback the electrical contract. The contract's general and supplementary conditions indicated that, "Costs caused by delays or by improperly timed activities or defective construction shall be borne by the party responsible therefor." The electrical contractor could perform its electrical work only after other trades had completed certain portions of their work. The construction project experienced delays as a result of the contractor being unable to meet any of its construction deadlines. The project began to operate out of sequence. Electrical work could not be installed in a productive manner until the building structure was reasonably complete. Because of the delays caused by the contractor, the electrical contractor had to accelerate staffing levels and work overtime. As a result, the electrical contractor incurred substantial additional costs by not being able to complete the project by its final completion date. The electrical contractor instituted an action against the contractor. The complaint alleged that the contractor breached its contract with the owner by not meeting its construction deadlines. As a result of the breach, the electrical contractor, an alleged third-party beneficiary to the deadlines in the contractor-owner contract, suffered injury.

The contractor filed a motion to dismiss because there was no privity between the electrical contractor and the contractor and the electrical contractor was not a third-party beneficiary of the contractor's contract with the owner. The electrical contractor contended that it was an intended third-party beneficiary to the time limits set forth in the contract between the contractor and the owner. The language did not explicitly confer third-party beneficiary status upon the electrical contractor. The benefit for the construction to be completed in a timely manner ran to the owner, not to the electrical contractor. There was no privity of contract between the electrical contractor and the contractor and the complaint's contract claim was dismissed.

[E] Other Third Parties as Beneficiaries

Page 13-14, add after carryover paragraph:

In *Jenne v. Church & Tower, Inc.*, 814 So. 2d 522 (Fla. Dist. Ct. App. 4th Dist. 2002), the court concluded that a course of dealing could not create third-party beneficiary status by. Church & Tower, Inc. contracted with Broward County to design and construct a detention center. The contract provided that the project was to be substantially completed within 548 days of the "project initiation" date. The anticipated completion date was October 4, 1997. Although the county began to occupy the facility in February 1998, a final Certificate of Occupancy was not issued until October 28, 1998. Sheriff Ken Jenne sued the contractor for breach of contract, alleging that the delay in completing the facility caused him to incur labor costs and expenses associated with transporting inmates to other counties, resulting in over $13 million in damages. The sheriff contended that he was the intended third-party beneficiary of the contract between the county and the contractor. The contractor moved to dismiss the complaint, arguing that the sheriff was not an intended third-party beneficiary and that damages were barred by the liquidated damages clause of the contract. The trial court dismissed the complaint with prejudice. The sheriff appealed.

The sheriff alleged that the parties intended that he primarily and directly benefitted from the contract. There was a federal court decree that ordered the sheriff to develop reorganization and jail population management plans for the jail system. The county requested construction proposals, which stated, "It is the intent of the County and the Broward Sheriff's Office that the program and schematic plans comply with the foregoing applicable laws." There was a proposal submitted by the contractor that the sheriff claimed "acknowledges that the contractor needed the sheriff's approval for performance under the contract and acknowledged that it would be working together with the sheriff throughout the project." The sheriff alleged that he purchased furniture for the jail, appointed his office personnel to the selection committee, and made decisions on site location and management. The factors did not confer third-party beneficiary status on the sheriff. Negotiations and dealings between the parties cannot modify a written contract to create the parties' "intent," when the lack of such intent was evident from the contract. The parties to the contract were the county and the contractor, and the contract involved the construction of a public facility funded by the county. The contract was intended to directly and primarily benefit the citizens of the county, not the sheriff. The sheriff received only incidental or consequential benefit from the contract's enforcement, so he was precluded from suing for a breach of it. An article concerning "Defective Work," gave the county, not the sheriff, the right to notify the contractor to make corrections. The liquidated damages clause allowing the county to recover delay damages indicated that the parties to the contract did not intend that the sheriff enforce a separate remedy for the same delay. Affirmed.

A subcontractor that is protected by owner's insurance coverage is not a third-party beneficiary of the contractor's contract with the owner. In *Donald B. Murphy*

Contractors, Inc. v. King County, 112 Wash. App. 192, 49 P.3d 912 (2002), Frank Coluccio Construction Company contracted with King County to construct a waterway project. The contractor subcontracted with Donald B. Murphy Contractors for the construction of a concrete access shaft. The contract assigned the contractor the responsibility of presenting subcontractor claims to the county for coverage. Any insurance proceeds were to be paid to the contractor, who would then have the obligation of paying to each subcontractor "a just share." The subcontractor encountered difficulties on the site, and completing the shaft cost more than anticipated. The subcontractor notified the contractor of the extra expenditures. The subcontractor submitted a claim to the county, through the contractor, for coverage under the county's builder's risk insurance policy. The county notified the contractor that it rejected the subcontractor's claim, stating that no coverage was available because the information supplied by the subcontractor did not describe any direct physical loss or damage to covered property. The subcontractor filed an action directly against the county. The basis of the action was a contractual provision between the county and the contractor, in which the county promised to procure and maintain "All Risk" builder's risk insurance on the project, including the interests of subcontractors. The county moved for summary judgment based on another contractual provision that expressly disclaimed any intent that subcontractors would be third-party beneficiaries to the contract. The trial court granted the county's motion for summary judgment. The subcontractor appealed.

On appeal, the subcontractor contended that the county's agreement to procure and maintain builder's risk insurance, including the interests of subcontractors, created an enforceable contract interest in favor of it under third-party beneficiary principles. The Washington appellate court explained that the plain language of the contract disclaimed any intent to have subcontractors be third-party beneficiaries. The contract provided instead for all claims by subcontractors to be handled through the general contractor. The procedure envisioned by the contract was clear, and the subcontractor initially followed it. The subcontractor corresponded with the contractor about its claim for a number of months before filing suit directly against the county. The provision for insurance including the interests of subcontractors showed the county's desire to benefit the subcontractor, but a desire to benefit a third party was not the same as an intent to assume a direct obligation to that third party. Because the contracting parties did not intend it, the subcontractor was not a third-party beneficiary to the contract. The subcontractor alternatively contended that the county breached an implied covenant of good faith and fair dealing. Because the county did not owe a contractual duty to the subcontractor, the claim was properly dismissed.

[F] Good Shepherds

Page 13-15, add after first full paragraph:

In breach of contract actions against the federal government, prime contractors have long been permitted to present subcontractors' claims on a pass-through basis against the government, even though the no-privity rule has barred subcontractors from recovering directly against the government. *Interstate Contracting Corp. v. City of Dallas, Texas*, 2003 U.S. App. LEXIS 2083 (5th Cir. 2003), *citing* Henry R. Kates, *Facilitating Subcontractors' Claims Against the Government Through the Prime Contractor as the Real Party in Interest*, 52 Geo. Wash. L. Rev. 146, 150 (1983); *see also Severin v. United States*, 99 Ct. Cl. 435 (1943), *cert. denied*, 322 U.S. 733, 88 L. Ed. 1567, 64 S. Ct. 1045 (1944) (in a breach of contract claim against the federal government, a prime contractor may recover damages on behalf of its subcontractor only if the prime contractor suffered actual damages); *J.L. Simmons Co. v. United States*, 158 Ct. Cl. 393, 304 F.2d 886, 888 (Ct. Cl. 1962) (a prime contractor suffers actual damages if the prime contractor (1) has reimbursed its subcontractor for the subcontractor's damages; or (2) if the prime contractor remains liable for such reimbursement in the future); *Folk Constr. Co. v. United States*, 2 Cl. Ct. 681, 685 (1983) (limiting the application of the *Severin* doctrine and holding that "a prime contractor is precluded from maintaining a suit on behalf of its subcontractor only when a contract clause or release completely exonerates the prime contractor from liability to its subcontractor").

Similarly, a number of states have permitted pass-through claims in cases involving state government entities. *Interstate Contracting Corp. v. City of Dallas, Texas*, 2003 U.S. App. LEXIS 2083 (5th Cir. 2003), *citing Clark-Fitzpatrick, Inc./Franki Found. Co. v. Gill*, 652 A.2d 440, 449 (R.I. 1994); *Frank Coluccio Constr. Co. v. City of Springfield*, 779 S.W.2d 550, 551–52 (Mo. 1989) (en banc); *St. Paul Dredging Co. v. State*, 259 Minn. 398, 107 N.W.2d 717, 724 (Minn. 1961); *Roof-Techs Int'l, Inc. v. Kansas*, 57 P.3d 538, 550–53 (Kan. Ct. App. 2002); *Metric Constructors, Inc. v. Hawker Siddeley Power Eng'g, Inc.*, 121 N.C. App. 530, 468 S.E.2d 435, 438–39 (N.C. Ct. App. 1996); *Schiavone Constr. Co., Inc. v. Triborough Bridge & Tunnel Auth.*, 209 A.D.2d 598, 619 N.Y.S.2d 117, 118 (1994); *Board of County Comm'rs v. Cam Constr. Co., Inc.*, 300 Md. 643, 480 A.2d 795, 795 (1984); *Kensington Corp. v. Michigan*, 74 Mich. App. 417, 253 N.W.2d 781, 783 (1977); *Buckley & Co., Inc. v. New Jersey*, 140 N.J. Super. 289, 356 A.2d 56, 73–74 (N.J. Super. Ct. Law Div. 1975); *D.A. Parrish & Sons v. County Sanitation Dist. No. 4*, 174 Cal. App. 2d 406, 344 P.2d 883, 888 (1959).

At least five states have not allowed a pass-through claim. *Interstate Contracting Corp. v. City of Dallas, Texas*, 2003 U.S. App. LEXIS 2083 (5th Cir. 2003), *citing Board of Governors for Higher Educ. v. Infinity Constr. Servs., Inc.*, 795 A.2d 1127, 1129 (R.I. 2002); *Farrell Constr. Co. v. Jefferson Parish*, 693 F. Supp. 490, 498 (E.D. La. 1988); *Dep't of Transp. v. Claussen Paving Co.*, 246 Ga. 807, 273 S.E.2d 161, 164 (Ga. 1980); *Barry, Bette & Led Duke, Inc. v. New York*, 240 A.D.2d 54, 669 N.Y.S.2d 741, 743 (1998); *APAC-Carolina, Inc. v. Greensboro-High Point Airport Auth.*, 110 N.C. App. 664, 431 S.E.2d 508, 511–12 (1993).

Page 13-17, add after note 79 in carryover paragraph:

If the general contractor has entered into a liquidating agreement with the subcontractor, but fails to submit the subcontractor's claim, it may be liable to the subcontractor for the amount of the claim. *Shomar Construction Services, Inc. v. Lawman*, 693 N.Y.S.2d 784, 262 A.D.2d 956 (1999).

Some states permit pass-through claims only when there is a liquidating agreement in place that meets certain requirements, *Interstate Contracting Corp. v. City of Dallas, Texas*, 2003 U.S. App. LEXIS 2083 (5th Cir. 2003), *citing Bovis Lend Lease LMB Inc. v. GCT Venture, Inc.*, 285 A.D.2d 68, 728 N.Y.S.2d 25, 27 (2001) (a liquidation agreement has three basic elements: (1) the imposition of liability upon the general contractor for the subcontractor's increased costs, thereby providing the general contractor with a basis for legal action against the owner; (2) a liquidation of liability in the amount of the general contractor's recovery against the owner; and (3) a provision that provides for the "pass-through" of that recovery to the subcontractor); *Barry, Bette & Led Duke, Inc.*, 669 N.Y.S.2d at 743 (holding that a showing of actual contractual liability on the part of the contractor is necessary in order for a pass-through claim to be permissible), while other states permit pass-through claims when the prime contractor pleads the suit on behalf of the subcontractor and has an obligation to render the recovery to the subcontractor. *Interstate Contracting Corp. v. City of Dallas, Texas*, 2003 U.S. App. LEXIS 2083 (5th Cir. 2003), *citing Frank Coluccio Constr. Co.*, 779 S.W.2d at 551–52.

Page 13-17, add at end of first full paragraph:

A no damage for delay clause does not preclude a contractor from entering into a liquidation agreement with a subcontractor for the subcontractor's delay claim. In *Bovis Lend Lease LMB Inc. v. GCT Venture, Inc.*, 285 A.D.2d 68, 728 N.Y.S.2d 25 (1st Dep't 2001), GCT Venture, a developer retained Bovis Lend Lease LMB for the restoration and renovation of New York's Grand Central Terminal. The contractor was responsible to the developer for the performance of its subcontractors, although no contractual relationship was created between the developer and the subcontractors. The contractor had to submit a list of subcontractors for pre-approval to the developer for retail work. The subcontractors' contracts provided that the subcontractors' only recourse with respect to delays was to seek an extension of time for performance. The contractor alleged that the developer materially breached the general contract, which delayed and impeded the subcontractors' work, causing them to incur additional costs. The contractor alleged that it had entered into a Liquidation Agreement in which it had acknowledged its liability to the subcontractors for such claims, liquidated the amount of liability, and obligated itself to recover such amounts on their behalf. In its motion to dismiss, the developer asserted that the contractor had no contractual liability to its subcontractors due to the no damage for delay clauses in the subcontracts. The court granted the developer's motion. The contractor appealed.

The clause that the contractor would not be liable for any additional costs incurred as a result of delays caused by the developer was consistent with the general rule that, absent a contractual commitment to the contrary, a prime contractor was not responsible for delays that its subcontractors incurred unless the delays were caused by some agency or circumstances under the prime contractor's direct control. The contractor assumed liability in an agreement subsequent to the original general contract and subcontracts. While the developer and owner had a right of pre-approval as to any specific subcontractor who would perform retail work, there was no generic right to pre-approve contracts between the contractor and its subcontractors, or a specific right to pre-approve any liquidation agreement. There was no authority cited or found for requiring the contractor to have obtained permission from the developer as a condition precedent to entering into a liquidation agreement. The no damage for delay clause did not prevent the contractor from subsequently entering into the liquidation agreement. Since the elements for a liquidation agreement were met, the developer's motion should have been denied.

Page 13-19, add at end of section:

A pass-through agreement between an owner and contractors to allow the contractors to pursue the owner's claim against a designer may also be valid. For example, in *Roof-Techs International, Inc. v. State of Kansas*, 2002 Kan. App. LEXIS 959 (Nov. 1, 2002), Brent Bowman and Associates Architects, P.A. (BBA) entered into a contract with the State of Kansas to provide architectural design and construction administration services for the Farrell Library Expansion and Renovation Project. The Law Company entered into a contract with the State for construction of the project. Thereafter, Law subcontracted mechanical, electrical, and roofing work to Central Mechanical Construction Co., Inc., D.L. Smith Electrical Construction, Inc., and Roof-Techs International, Inc. Central filed suit against the State, BBA, and Law to recover damages incurred as a result of delays and problems in the project. Law answered the suit brought by Central and Smith, and filed a cross-claim against the State. Law's claims against the State were breach of contract, breach of implied warranty, and misrepresentation. The State asserted cross-claims against both Law and BBA for breach of contract, breach of warranty, indemnity/contribution, and negligence. Roof-Techs filed a separate lawsuit that mirrored that of Central and Smith and sought to recover damages sustained as a result of defects in the plans and delays on the project. The petition set forth a breach of contract claim against Law, a professional negligence claim against BBA, and a negligent misrepresentation claim against Law, the State, and BBA, collectively.

Roof-Techs' initial petition contained a professional negligence claim against BBA and a negligent representation claim against Law, the State, and BBA, collectively. The professional negligence claim was based on the allegation BBA breached its duty of care by failing to provide adequate and accurate drawings and information upon which Roof-Techs had relied in formulating its bid. Additionally, Roof-Techs alleged BBA had failed to timely and adequately respond to requests for information and proposed change orders. The negligent representation claim was based on the allegations that BBA falsely represented that the drawings were adequate and sufficient to allow Roof-Techs to formulate an accurate bid on the project and BBA falsely represented that asbestos would be removed by others prior to the start of work. BBA, the State, and Law all filed motions for summary judgment against Roof-Techs on the tort claims.

The State, Law, Central, and Smith entered into a settlement and liquidation agreement. As part of the agreement, the State paid $750,000 to settle all claims with Law, Central, and Smith that were unrelated to BBA's conduct. The agreement assigned to Law, Central, and Smith all of the claims the State may have had against BBA. In addition, under the terms of the agreement, the breach of contract claims of Central, Smith, and Law asserting defects in the architectural plans were purportedly passed through the State to be asserted against BBA. In other words, the agreement intended to allow the contractors to step into the shoes of the State to present their claims directly against BBA. The agreement authorized the contractors to pursue those claims either in their own names or in the State's name. In the present case, however, the subcontractors and general contractor passed their claims through the owner against the architect. As an added twist, the agreement in the instant case purported to assign the State's claims against BBA to Law, Central, and Smith. Nothing in Kansas law specifically prohibited the assignment of the State's claims for breach of contract against BBA to Law, Central, and Smith. Although a contract for personal services cannot be assigned, a breach of contract claim, where the underlying contract was for non-legal personal services, can be assigned.

In Paragraph 2.1 of the agreement, the State acknowledged its liability to Law, Central, and Smith for claims they asserted against the State based on the acts or omissions of BBA, but only to the extent that Law, Central, and Smith recover from BBA. Further, the State's liability to Law, Central, and Smith, and Law's liability to Central and Smith were liquidated to the amounts recovered from BBA and its insurers. Finally, the agreement contained a pass-through provision whereby the State agreed that Law, Central, and Smith were entitled to any amount recovered on the claims against BBA. The concept of the instant agreement seemed to follow the general guidelines for pass-through claims and liquidation agreements. There is nothing in the record that indicated any of the parties to the agreement signed any sort of waiver that implicated a *Severin* problem, i.e., an agreement which released the State from liability to the contractors. The wording in the agreement appeared to be sufficient to retain liability on behalf of the State for the amount of recovery against BBA. The district court's dismissal of the State's claims was reversed, and Law was allowed to proceed by means of a pass-through process.

In some cases, a subcontractor may be able to recover for its delay costs caused by the government even though the prime contractor was not delayed. For example, in *E.R. Mitchell v. Danzig*, 175 F.3d 1369 (Fed. Cir. 1999), a subcontractor was allowed to recover against the owner for its unabsorbed home office overhead costs. The subcontractor incurred delay to the subcontractor's performance schedule due to the owner's correction of defective drawings, even though the owner's actions did not delay the performance of the prime contractor or the overall project completion schedule. The ruling may have no application in the case where the subcontractor's performance time is indicated on the overall project schedule but not specifically approved by the owner.

The burden of proof also varies among jurisdictions. *Interstate Contracting Corp. v. City of Dallas, Texas*, 2003 U.S. App. LEXIS 2083 (5th Cir. 2003), *citing Frank Coluccio Constr. Co.*, 779 S.W.2d at 552 (holding that the government/owner has the burden of proving that the prime contractor has been relieved of responsibility to pay any recovery to the subcontractor); *Cam Constr. Co., Inc.*, 480 A.2d at 799 (same); *Kensington Corp.*, 253 N.W.2d at 784 (same) with *Claussen Paving Co.*, 273 S.E.2d at 164 (holding that prime contractor must prove that it is liable to the subcontractor before it can assert a pass-through claim); *Wexler Constr. Co. v. Housing Auth.*, 149 Conn. 602, 183 A.2d 262, 264–65 (Conn. 1962) (same).

In *Interstate Contracting Corp. v. City of Dallas, Texas*, 2003 U.S. App. LEXIS 2083 (5th Cir. 2003), City of Dallas, Texas and Interstate Contracting Corporation (ICC) entered into a fixed sum contract for the construction of levees around a City water treatment plant; the excavation of two areas to create storm water detention lakes; and some miscellaneous work including trash removal, surveying, and linear depth checking. In turn, ICC subcontracted with Mine Services, Inc. (MSI) for the levee construction and the excavation of the storm water detention lake. The subcontract between MSI and ICC provided, in part:

> In the event SUBCONTRACTOR has a claim for which the Owner may be responsible, the CONTRACTOR, in its sole discretion, may initiate with the Owner, at the SUBCONTRACTOR'S expense and which shall include attorney's fees, any dispute or claim procedures provided for in the Contract Documents for the use and benefit of SUBCONTRACTOR: otherwise SUBCONTRACTOR shall have full responsibility for the preparation of its claims and shall bear all expenses thereof, including attorney's fees . . . CONTRACTOR shall be liable to SUBCONTRACTOR only to the extent of the amount, if any, actually awarded as a result of the disputes process: SUBCONTRACTOR shall be entitled only to the amount, if any, actually awarded as a result of the disputes process: and such amount when received by CONTRACTOR from the Owner shall satisfy and discharge CONTRACTOR from any and all liability to SUBCONTRACTOR for or on account of the acts or omissions of the Owner or its Architect or Engineer.

Hence, ICC was given the sole discretion to bring a claim against the City on behalf of MSI at MSI's expense. If such a suit was brought, MSI agreed to release ICC from further liability in exchange for whatever ICC recovered from the City.

The material excavated for the lakes was to be used, to the extent it met specifications, to construct levees. The Interior Borrow Lake (IBL) was one of the "borrow" sites the City designated as a source of fill material. Shortly after work began, MSI discovered that the materials in the IBL differed from what it expected. Due to a lack of suitable material, MSI was forced to manufacture material by mixing sand with the limited quantities of clay. The contract was silent on the issue of manufacturing fill material. Manufacturing material substantially decreased MSI's productivity and increased its costs. The parties discussed using fill from other sites that MSI believed were not designated as borrow sites under the contract. But using these sources was more expensive than the manufacturing process and, therefore, MSI continued to manufacture fill.

ICC informed the City of MSI's increased work, but the City indicated it would deny any claim contending that manufacturing fill was beyond the scope of the contract. ICC subsequently sought MSI's direct costs from the City and re-notified the City of its protest. The City also denied claims for costs regarding trash removal, linear depth checking, surveying, and "extended performance." All of the claims presented, except the trash removal, were injuries to MSI.

ICC filed its suit on behalf of MSI against the City for breach of contract, quantum meruit, breach of implied warranty, and fraudulent inducement. Prior to the commencement of the action, ICC and MSI entered into a detailed "Claims Presentation and Prosecution Agreement" concerning MSI's claims for significant project costs overruns due to the City's failure to disclose anticipated difficulties. The Agreement provided:

> II. RECITALS
> . . . E. MSI represents that its Claim is based on the conduct of the City and that it has no other claims against Interstate except MSI's claim that is the subject of this Agreement. . . .
> G. Interstate and MSI agree that it is in their mutual best interests for Interstate and MSI to pursue a claim against the City of Dallas in the name of Interstate ("the Claim"). The Claim shall consist of MSI's Claim in the estimated amount of $4,062,584.00 plus a 15% markup for profit and overhead for Interstate and Interstate's own costs claim in the amount of $222,798.18 for jobsite costs and jobsite general and administrative costs associated with extended performance caused by the City's interference with performance of the Contract ("Interstate Claim"). The parties desire, therefore, to agree upon a procedure through which they will coordinate the preparation, presentation and prosecution of the Claim against the City of Dallas.
> III. TERMS
> . . . 1. MSI may pursue the Claim against the City in Interstate's name. Interstate shall cooperate fully with MSI including, but not limited to, passing on the Claim to the City and executing such documents that may be required to further the Claim to the City and executing such documents that may be required to further the Claim; and MSI shall cooperate fully with Interstate. MSI shall have the responsibility for the preparation of any claim, the presentation and prosecution of any such claim, and the conduct of any litigation. . . .
> 3. MSI shall diligently pursue the Claim. The right to abandon, settle, compromise or dismiss the Claim shall be shared by MSI and Interstate. Interstate and MSI shall each not settle the Claim without the others['] prior written approval. . . .
> 4. All costs, fees and other expenses (including expert and attorney fees) incurred by MSI in connection with the preparation, prosecution and litigation of the Claim shall be paid by MSI. MSI shall have no responsibility for any attorney fees or expenses that Interstate may elect to incur. . . .

> 6. From any amount paid in settlement of the Claim, Interstate and MSI shall be paid first for their respective markup (including profit and overhead). If the percentage of markup actually paid as part of any settlement or judgment is specified, the markup payable shall be calculated according to the June 21, 1995 Letter of Understanding agreement. If the markup is not specified then the settlement amount will be presumed to include a total markup of 15% and each party shall receive one-half of the markup or 7.5%. . . . The remainder of the settlement funds shall be prorated and paid between Interstate and MSI in the same percentage that each party's portion of the Claim (exclusive of markup) is to the total Claim (exclusive of markup) submitted to the City. . . .

The district court allowed ICC to bring its claims on behalf of MSI. Following an 11-day jury trial, the jury found that the City breached its contract with ICC. The jury also found the City breached an implied warranty to provide accurate and suitable plans and specifications in light of subsoil conditions at the project.

On appeal, the City argued that the district court erred in concluding that ICC could seek damages on behalf of its subcontractor MSI because there was a lack of privity of contract between the City and MSI. There was no privity of contract between MSI and the City. Under Texas law, privity of contract was an essential element necessary to any recovery in an action based on contract. However, ICC argued that even though there was a lack of privity between MSI and the City, the district court was correct in permitting ICC, under Texas law, to present MSI's claim on a pass-through basis. No controlling precedent from the Supreme Court of Texas or the Texas Courts of Appeals existed to resolve the issues created by the conflicting contentions of the parties. Accordingly, in light of the absence of Texas authority on these issues and the varied interpretations of these issues by other state courts, the court of appeals certified questions to the Supreme Court of Texas regarding whether Texas law recognized pass-through claims, and if so, what were the requirements that needed to be satisfied for a contractor to assert a claim on behalf of its subcontractor.

§13.07 NEGLIGENCE

Page 13-31, add after note 152 in first full paragraph:

Negligence or contributory negligence by a contractor may also be established by expert testimony but must address the specific elements of negligence such as duty, failure to conform to the required standard, causal connection to the injury, and resulting actual damage. Failure to adequately address any of these elements may cause the negligence theory to fail. For example, in *MCI Constructors v. Hazen & Sawyer, P.C.*, 2005 U.S. Dist. LEXIS 30537 (M.D.N.C. Nov. 28, 2005), the court held that testimony by a contractor's expert on damages that some fault was attributable to the contractor for cost overruns does not establish the contractor's contributory negligence.

MCI Constructors (contractor) contracted with the City of Greensboro for the upgrade and expansion of a wastewater treatment plant. Hazen & Sawyer (engineering firm) was to act as the project supervisor. The engineering firm had designed the upgrade and expansion and had participated in compiling information for potential bidders. Weather and unexpected site conditions caused the contractor to spend more than planned and caused the project to fall behind schedule. When the city determined that the contractor could not comply with a revised schedule, the city terminated the contract. Alleging negligence and other claims, the contractor filed suit against the engineering firm. The engineering firm filed for summary judgment.

The engineering firm argued that it was entitled to summary judgment on the claim of negligence because the contractor was contributorily negligent. The engineering firm based its motion on testimony by the contractor's expert witness on damages. The expert testified that the contractor bore some of the fault for the actual costs exceeding the budgeted costs in each category. The engineering firm argued that if the contractor's negligence caused any part of the excess cost in a particular category, the firm was relieved of any liability for negligence that led to additional costs in that category. According to the engineering firm, the contractor's negligence did not need to cause all of the contractor's injury to serve as a bar to recovery, as long as it caused some of its injury. The expert's conclusion that some fault was attributable to the contractor did not equal the conclusion that the contractor was negligent. It was possible that the contractor used all necessary care and yet made a decision that resulted in additional costs. Although the fault would lie with the contractor in that situation, the contractor would not have been negligent. To establish contributory negligence, the engineering firm needed to demonstrate that all additional costs that fell into a single category constituted the same injury. The expert's testimony was not sufficiently specific to determine whether there was one injury or two different injuries. Without additional information on how the contractor's alleged negligence contributed to the injuries caused by the engineering firm's negligence, the court could not determine that contributory negligence barred recovery by the contractor, and the engineering firm's motion for summary judgment was denied.

[B] Economic Loss Limitations

Page 13-35, add after second full paragraph:

A construction manager's tort claim for a subcontractor's breach of duty may also be barred by the Economic Loss Rule where it is purely for economic loss, whereas its contract claim for indemnification is allowed. For example, in *Town of Clinton v. Geologic Servs.*, 2005 Mass. Super. LEXIS 505 (Oct. 7, 2005), the Town of Clinton (Town) entered various contracts with Geologic Services Corp. (construction manager) to design and construct a water filtration system. The construction manager retained Garrett Engineering, P.C. (engineer) to design the project and Methuen Construction Co. (contractor) as the general contractor. The contractor hired New Stress International, Inc (subcontractor) to fabricate prestressed concrete floor planks. The construction manager was not a party to the subcontract. After the subcontractor fabricated the planks, the construction manager objected to their installation because the planks did not meet the required specifications. The subcontractor installed the planks over the construction manager's objections. The construction manager informed the town that the installed planks were defective. The resulting dispute caused 14 months of delays, at which time the construction manager terminated its contract with the Town. The delays cost the Town $1.2 million. The Town filed a suit against the construction manager and engineer for breach of contract, negligence, fraud, and deceptive practices. The construction manager filed a cross-claim against the subcontractor, who filed a motion for summary judgment.

The construction manager argued that the subcontractor owed a duty of care under its subcontract with the contractor that extended to the construction manager. It further argued that the subcontract included express language whereby the subcontractor agreed to indemnify the construction manager from any claim for damages caused by the subcontractor. The Economic Loss Rule stated that a plaintiff asserting a tort claim could not recover for purely economic losses in the absence of harm to the plaintiff's person or property. The construction manager maintained that the subcontractor caused a delay in construction, not harm to a person or property. Because of the delay, the Town lost the $1.2 million for which the Town sought to recover from the construction manager. Despite the substantial loss, a delay did not constitute personal or physical harm. Therefore, because the construction manager asserted purely economic damages which were not recoverable under the Economic Loss Rule, the subcontractor was entitled to summary judgment on the tort claim for breach of duty. The Economic Loss Rule, however, did not apply to a contract claim. Because the construction manager's indemnification claim was based on the subcontract between the contractor and subcontractor, the Economic Loss Rule did not apply and thus did not bar that claim. Summary judgment was granted as to breach of duty claim and denied as to indemnification claim.

§13.08 FRAUD AND MISREPRESENTATION

Page 13-40, add at end of section:

In *Frontier-Kemper Constructors, Inc. v. American Rock Salt Co.*, 2002 U.S. Dist. LEXIS 16894 (Sept. 9, 2002), American Rock Salt Co. LLC (ARSCo) is the owner of the Hampton Corners Salt Mine Project. Plaintiff Frontier-Kemper Constructors, Inc. and Flatiron Constructors LLC Joint Venture (Frontier-Kemper) contracted to build the mine. In the contract, defendant agreed to pay plaintiff $70,649,000, provided that plaintiff completed the project within 25 months. However, the contract required payment up to $3 million in liquidated damages if certain milestones were not met, in amounts ranging from $15,000 to $21,000 for every day the project was late. The contract prohibited defendant from assessing liquidated damages against plaintiff for delays caused by defendant's change orders: "Due account shall be taken of any adjustment of the CONTRACT time for completion of the WORK as provided by written CHANGE ORDER." Disputes arose, and defendant eventually withheld the full $3 million from plaintiff's progress payments. Plaintiff sought rescission of the contract, compensation in quantum meruit in excess of $27 million, and punitive damages of at least $10 million. Plaintiff contended that rescission of the contract was warranted on fraud in the inducement, fraud during performance of the contract, mutual and/or unilateral mistakes of material fact, and material breaches of the contract by defendant. Plaintiff also sought a declaratory judgment that it was entitled to abandon any additional performance of the contract. Defendant filed a motion to dismiss the complaint. Defendant contended that the various fraud claims must be dismissed since they were nothing more than claims for breach of contract.

Whether because they violate a legal duty separate from the duty to perform under the contract, or because they are collateral to the contract, or both, courts generally hold that misstatements of present fact which induce a contract are actionable as fraud. A plaintiff may pursue a claimed fraudulent inducement of a contract, based upon a misrepresentation of present fact, or upon a false promise to do something besides what is required by the contract. Additionally, a plaintiff may pursue a claim for fraud if the defendant failed to disclose information, when it had a duty to disclose, which would also constitute the breach of another legal duty separate and apart from the duty to perform under the contract.

Plaintiff alleged that defendant withheld important information concerning problems at the site, which it knew at the time would prevent plaintiff from performing the contract within the time allotted under the contract. Plaintiff contended that prior to entering the contract, defendant knew of, but failed to disclose, factors which were likely to delay plaintiff's performance, making it impossible for plaintiff to complete the project in a timely manner, and thereby entitling defendant to liquidated damages. For example, plaintiff alleged that defendant misrepresented the amount of time that would be needed to investigate whether or not the mine was being built upon a Native American burial ground, which resulted in significant delays and additional work, as plaintiff and its subcontractors attempted to "work around" the archaeological sites. Plaintiff maintained that defendant also withheld information regarding the amount of time and work that would be needed to construct an access road to the mine, known as New York State Highway 63. Plaintiff further contended that defendant and its engineering firm delayed in obtaining necessary approvals for the design of Highway 63 from the New York State Department of Transportation (NYDOT), thereby preventing plaintiff from completing its work deadline.

The complaint did not adequately plead that defendant possessed superior knowledge that delays were likely to occur, which it failed to disclose to plaintiff. First, the Amended Complaint did not suggest that plaintiff had any factual basis for alleging that defendant knew, prior to executing the contract, that the archaeological examination would be delayed. Instead, plaintiff merely alleged, in conclusory fashion, and based only upon information and belief, that defendant was aware "of the distinct possibility that Native American burial remains would be found."

The Amended Complaint did not explain why defendant should have known, prior to executing the contract, that a new archaeological investigation would result in significant delays. Plaintiff did not allege that the site actually was a Native American burial site, and in fact, the Amended Complaint did not explain, or allege, the exact reason why the archaeological testing took longer than anticipated. Plaintiff failed to state a claim for fraud. Instead, plaintiff's allegations were for breach of contract, since, according to the amended complaint, "the contract schedule . . . allowed 10 days for 'Archaeological Site Investigation.' " If, as plaintiff maintained, defendant exceeded that time limit, then plaintiff's remedy was for breach of contract. Similarly, plaintiff failed to plead fraudulent inducement with regard to the Highway 63 delays.

As for its contentions of mutual mistake, plaintiff began by alleging that unforeseen problems developed with respect to rock in the salt bed, and differing soil and water conditions, both on the surface and below ground at the mine's underground process gallery. More specifically, plaintiff indicated that it encountered rock where it had been anticipated that there was none, resulting in extra cost and time needed for construction of the mine. Plaintiff stated that, although defendant subsequently agreed to increase the contract price somewhat, it refused to extend the time for performance. Moreover, plaintiff alleged that defendant subsequently assessed liquidated damages against plaintiff due to delays caused by the excavation of the rock. Similarly, plaintiff alleged that the conditions of the site on which it was to build the salt pad were not as originally expected, and that during the excavation of the salt pad, plaintiff discovered an artesian well and flowing sand, which created a quicksand-like condition in the area. Plaintiff indicated that, again, defendant paid a portion of the increased cost but refused to extend the time for performance of the project overall.

The mere fact that a contractor intentionally submits a bid below its costs is not a violation of the federal False Claim Act. In *U.S. ex rel. Bettis v. Odebrecht Contractor of California, Inc.*, 393 F.3d 1321 (D.D.C. 2005), a "whistleblower" filed suit against Odebrecht alleging that it submitted false claims to the government in connection with a public works construction contract. In March 1993, the U.S. Army Corps of Engineers (COE) solicited bids for work on the Seven Oaks Dam in San Bernardino County, California. The COE had an estimate of $203 million for the cost of the work (without profit) and when the bids were opened in July 1993, Odebrecht was the lowest bidder at $167.7 million, some $29 million below the second lowest bidder, Tudor-Saliba. Tudor-Saliba commenced a series of bid protests but in March 1994, a contract was awarded to Odebrecht after the bid protests were resolved. During the course of construction, Odebrecht requested and received several "equitable adjustments" for changes and differing site conditions, but none were based upon its alleged underestimate. By August 2003, the government paid Odebrecht nearly $268 million, $100 million more than the bid price, and Odebrecht maintained that it sustained a loss of over $30 million on the contract performance despite the contract increases. The COE was satisfied with Odebrecht's work on the project.

An employee of the COE's scheduling consultant filed suit under the False Claims Act, arguing that Odebrecht knew or should have known that its bid was inadequate to cover its costs of performance. One count alleged that Odebrecht fraudulently induced the COE to award the contract intentionally submitting its below cost bid with the further intention of seeking price adjustments after award. Another count alleged that Odebrecht submitted an intentionally undervalued bid with the further intention of seeking false price modifications and then submitting false modifications. The U.S. District Court for the District of Columbia granted summary judgment for Odebrecht on all counts. On appeal, the plaintiff argued that Odebrecht fraudulently induced the COE in the contract in three ways: (1) by submitting a bid that did not conform with industry standards for accuracy, (2) by falsely reaffirming its bid despite awareness of rising costs, and (3) by falsely claiming to use cost-savings measures during construction and that fraudulent intent could be inferred by Odebrecht's subsequent failure to use these measures.

Concerning the plaintiff's first argument, the Court of Appeals for the District of Columbia, noted that Odebrecht's bid was merely an offer to enter into a contract, representing that it would perform the required work at the specified bid unit prices in strict accordance with the contract. Odebrecht did not make any representations concerning its costs or method of calculating its bid. There was no proof that Odebrecht did not intend to be bound by the contract unit prices.

When Odebrecht reaffirmed its bid after the bid protests were resolved, it did not represent that its costs had not increased and it did not promise that it would not seek equitable adjustments under the contract. By later submitting an equitable adjustment one cannot infer that Odebrecht had no intention of being bound by its bid unit prices. The fact that it Odebrecht was mistaken concerning its ability to recover increased costs through a change order does not demonstrate fraud.

At the bid opening in July 1993, Odebrecht stated its intention to use several cost-savings measures such as the purchase of new equipment and operating it 20 hours a day, using "rolling construction teams," each working four consecutive 10-hour days, allowing it to work 20 hours a day without paying overtime, and using a "glory hole" to drop excavated rock onto a conveyor belt. The fact that Odebrecht did not actually use these methods throughout the entire construction period did not necessarily mean that Odebrecht's representations were false. The inference was not supported by the trial record because Odebrecht attempted to implement the measures.

The appellate court determined that there were no material facts in dispute and the undisputed facts did not support the conclusion that Odebrecht fraudulently induced the COE to award the contract to Odebrecht, so the judgment of the District Court dismissing the action was affirmed.

§13.10 VIOLATION OF A STATUTE

Page 13-41, add at end of section:

In one case, statutory liability for a contractor's delay damages was based upon a statute that waived immunity for publicly regulated owners. In *J.S. Haren Co. v. City of Cleveland,* 2003 Tenn. App. LEXIS 406 (May 30, 2003), J.S. Haren Company contracted with the State of Tennessee, Department of Transportation to construct improvements on U.S. 11 (SR 2). The improvements primarily consisted of grading, underground drainage facilities, and paving. The effective date of the contract was November 2, 1998, and the completion date was scheduled to be June 15, 2000. The total contract duration of the proposed work was 619 calendar days. This contract duration was established assuming that the Cleveland Utilities Board would establish the location and, if necessary, relocate existing utilities conflicting with the construction of the road improvements prior to commencement of work. The contractor commenced work on the road improvements in or around December 9, 1998. Soon thereafter, the contractor encountered conflicts with existing utility services within the project area that had not yet been relocated and/or were relocated incorrectly. The contractor contended that the board improperly and unlawfully established the existing locations of utilities and improperly and unlawfully relocated conflicting utilities so as to unduly interfere with the contractor's performance of its obligations on the roadway project. As a result of the board's conduct the contractor was forced to delay construction or to perform its work on the project less efficiently and less effectively, thereby causing substantial financial loss. The amended complaint asserted negligence per se based upon the provisions of a Tennessee statute that waived liability of publicly regulated owners. The board filed a motion to dismiss, and the trial court granted the motion. The contractor appealed.

The contractor alleged that it had a "strict liability" claim that fell outside the ambit of the Government Tort Liability Act (GTLA). The court concluded that the contractor had stated a cause of action under the statute. However, the court rejected the contractor's characterization of the claim as one for "strict liability." The court found it more appropriate to refer to the remedy created by the statute as "statutory liability." The pertinent language of the subject statute created a remedy which inured to the benefit of the contractor. The complaint filed by the contractor sought to recover damages under an "independent statute," and not under the GTLA. In creating the remedy, the General Assembly waived the immunity of "owners" such as the board from such suits. The statute created a remedy for entities, such as the contractor, who were damaged by an owner of utilities, such as the board, for failing to do that which was required by the statute. The court vacated and remanded the trial court's order dismissing the contractor's claim to the extent that pertained to the statutory liability claim asserted by the contractor.

§ 13.11 SELECTION OF REMEDIES

[A] Termination and Rescission

Page 13-43, add after note 211 in second full paragraph:

In *Harley Paws, Inc., d/b/a Three Dog Bakery v. Mohns, Inc.*, 249 Wis. 2d 488, 639 N.W.2d 223 (2001), for example, Mohns, Inc. and Harley Paws, Inc., contracted for the remodeling of a retail store. Mohns told Harley Paws that the construction should be completed in approximately five weeks, and estimated the cost to be $52,000. Based on that information, Harley Paws scheduled its grand opening for September 26, 1998. Construction commenced in early August 1998 and continued until September 11, when disputes developed primarily involving: (1) the modification, possible delay, and additional costs involving the construction of the store's cabinetry; and (2) the rescheduling of the faux finisher, who was scheduled to do the finishing painting on September 12, but could not because Mohns had failed to complete the preliminary painting. When Harley Paws discovered defects in the drywall and learned from Mohns' President, that the store would not be completed by the grand opening date, Harley Paws fired Mohns and arranged for other contractors to complete the remodeling. Harley Paws subsequently sued Mohns for breach of contract. The court concluded that Mohns had breached the contract. In its findings of fact, the court explained that, given the need to be ready for the September 26 opening, the parties clearly understood that "time was of the essence." The court concluded "that the termination on the 11th of the contract and the hiring of replacement contractors was reasonable."

Mohns first argued that the trial court erred in finding that it was in breach of contract at the time Harley Paws terminated the contract. Mohns argued that anticipatory breach could not support the trial court's conclusion. Whether the trial court based its decision on an unarticulated theory of anticipatory breach or on a theory of actual breach was not necessarily determinative of the outcome. The facts supported the trial court's conclusion that Harley Paws breached the contract, that the breach could be viewed as an anticipatory one. The work had not progressed as scheduled and by September 11, it had become evident that the work would not be completed within the contracted time. Mohns had substantially underestimated the cost of the cabinetry and, as of September 11, did not know who would construct the cabinets or when they would be installed. The contractors who replaced Mohns testified that Mohns' workmanship had been shoddy and delayed the scheduled subcontractors' work, further indicating that the job would not be completed on schedule. The new contractors worked 12 to 13 hours a day to complete the project by September 23. The court implicitly found that Mohns' words and actions constituted an anticipatory breach.

[B] Abandonment

Page 13-50, add at end of subsection:

This theory may not work for public construction contracts. For example, in *Amelco Electric v. City of Thousand Oaks*, 95 Cal. App. 4th 448, 115 Cal. Rptr. 2d 527 (S. Ct. 2002), the contractor alleged the public owner had abandoned the project in order to support its quantum meruit claim. The Supreme Court reversed, concluding that a public works contract in California could not be abandoned. The City of Thousand Oaks awarded Amelco Electric the prime contract for electrical work at the Civic Arts Plaza. Lehrer McGovern Bovis, Inc. (LMB) managed the project. Amelco expressed concern that the electrical drawings did not identify all revisions, or contain all prior revisions. Amelco requested a change order and $203,759 in additional funds to update the drawings. LMB refused. Amelco accepted this decision and signed a change order for zero dollars and zero additional time, because LMB verbally promised that, "things are going to get better." Things did not get better and Amelco performed a huge number of change orders in order to complete the work. Amelco submitted a $1.7 million total cost claim for costs allegedly resulting from the noncaptured costs of the change orders. While Amelco maintained daily records of its work activities, it was unable to produce documentation of instances in which its performance of change order was delayed or interfered with by LMB's actions, and for which it was not compensated. The person responsible for actually recording the information in the contractor's daily log testified he would not record changes. Amelco's vice-president asserted the sheer number of changes made it "impossible" to keep track of the impact any one change had on the project or on Amelco. Amelco conceded it was inefficient in performing the work, but assigned responsibility for virtually all of that inefficiency to LMB. The jury found the City had both breached and abandoned the contract, and awarded Amelco $2,134,586. The court of appeal concluded that as a matter of law a public works contract could be abandoned.

Under the abandonment doctrine, once the parties ceased to follow the contract's change order process, and the final project had become materially different from the project contracted for, the entire contract—including its notice, documentation, changes, and cost provisions—was deemed inapplicable or abandoned, and the plaintiff may recover the reasonable value for all its work. If such a theory applied in the public works context, the notion of competitive bidding would become meaningless. Even assuming there was substantial evidence of abandonment, the parties could not abandon a public works contract, with its strict statutory requirements. If the City's numerous changes could result in the public contract being set aside in its entirety, Amelco would find itself in no different situation, and should receive no different treatment, than a contractor who has performed under a void contract.

Under the abandonment doctrine, the plaintiff need not demonstrate at what point the contract was abandoned. Amelco's project manager did not even determine Amelco had a claim on the project until after the work was completed.

Under Amelco's approach, a certain number of changes may be permissible, but at some indeterminate point, the next requested change makes the number "excessive," the competitively bid contract was set aside, and the contractor recovered on a quantum meruit basis from the beginning of the project onward. Such a vague definition failed to provide any meaningful guidance as to when that line has been crossed. Public entities would not receive timely notice of claims that would allow them to make project management, budget, or procedural adjustments during the course of construction, thus creating intolerable uncertainty in the budgeting and financing of construction projects.

In addition, allowing contractors to recover in quantum meruit for the actual as opposed to the bid cost of a project encouraged contractors to bid unrealistically low with the hope of prevailing on an abandonment claim based on the numerous changes inherent in any large public works project.

California Public Contract Code § 7105 provided that a public works contract required to be awarded by competitive bid, may be terminated, amended, or modified only as provided in the contract or authorized under another provision of law. The compensation for amendments and modifications were determined by the contract. Amelco contended Public Contract Code § 7105 applied only to compensation for termination, amendment, or modification of a contract, not to damages for its abandonment. Given a public works project must generally be governed by a valid contract, abandonment could not be one of the changes "authorized under provision of law" contemplated by § 7105. Reversed and remanded for retrial on damages.

[C] Quantum Meruit

Page 13-51, add after note 245 in first carryover paragraph:

Generally, if a valid express written agreement exists, a quantum meruit claim is dismissed in favor of a breach of contract claim. However, if a contract has been terminated prior to completion, quantum meruit may be the appropriate theory of recovery. For example, in *MCK Building Associates v. St. Lawrence University,* 301 A.D.2d 726, 754 N.Y.S.2d 397 (2003), MCK was the masonry, millwork, and rough carpentry subcontractor on a project to renovate two buildings on a University campus. MCK substantially completed its work, but according to the general contractor MCK would not complete the remaining work. The general contractor resorted to the default termination process but failed to comply with the provision requiring 10-days' written notice of termination. MCK sued the general contractor and the trial court finding that there was a procedural irregularity in the notice, held that MCK was wrongfully terminated and allowed quantum meruit recovery for MCK's damages. On appeal, the appellate division agreed that quantum meruit was the appropriate measure of damages because the subcontract was terminated prior to completion.

Page 13-52, add after carryover paragraph:

The general rule is that no quasi-contractual claim can arise when a contract exists between the parties concerning the same subject matter on which the quasi-contractual claim rests. When parties enter into a contract they assume certain risks with an expectation of a return. Sometimes their expectations are not realized, but they discover that under the contract they have assumed the risk of those expectations. As a result, they have no remedy under the contract for restoring their expectations. In desperation, they turn to quasi-contract for recovery. The law may not allow this. Generally, courts are hesitant to allow unjust enrichment claims unless there is evidence of fraud or bad faith, there has been a breach of contract or a mutual recission of the contract, when recission is warranted, or when the express contract does not fully address a subject matter.

Two cases illustrate the rule and the exception. In *Commissioners of Caroline County v. J. Roland Dashiell & Sons, Inc.*, 358 Md. 83, 747 A.2d 600 (2000), there was an express contract between the County and the contractor for the renovation of the Caroline County Detention Center. The recovery of money for work performed on the detention center was covered by several valid and enforceable provisions of the written contract between the parties. The contract clearly addressed the process for submitting claims, liquidated damages, and for the waiver of rights and duties as well as covering the basic guidelines for construction of the building addition. The court found the contractor's claim that the County had been unjustly enriched because of its use of the detention center was without merit. Even if the County was enriched, such enrichment was not unjust because it was in strict compliance with the terms of their contract. According to the court, both the County and contractor got what they bargained for by the terms of the contract.

In contrast, in *Abington Constructors, Inc. v. Madison Paper Industries*, 2000 U.S. App. LEXIS 4454 (1st Cir. Mar. 21, 2000), the owner requested bids on a project to reconstruct the Anson Dam waste gate and repair a power station. In order to conduct the repairs, both an "upstream" and a "downstream" cofferdam were necessary because much of the work would be done below the water level of the river. The conditions of the contract stated that a bidder could rely on "the general accuracy of the technical data" in the materials, but authorized only "limited reliance" on drawings. The bidding instructions placed responsibility on the bidder to be familiar with the site terrain. When the contractor visited the site to conduct an inspection, the river flows were too high for it to survey the river. When the owner's project manager later invited the contractor to revisit the site, the contractor declined. Later the contractor agreed to do the work for $808,500.

In its bid form, the contractor represented that it had carefully studied the subsurface conditions at the site. Although the conditions of the contract had been included in the project manual, the parties agreed to a purchase order to signal its acceptance of the contractor's bid, creating a contract. The purchase order referenced the project manual as well as the contractor's bid form in describing the work to be performed. The district court found that Abington committed a breach, although not in bad faith, by failing to conduct its own investigation.

Abington was forced to use 300 bulk bags filled with sand and 5,000 smaller bags to construct the downstream cofferdam, rather than the 180 bulk bags and 250 smaller bags it had initially allocated, due to water levels and river bottom configurations that were different from specifications in the project manual. The route of the cofferdams also had to be altered due to the river bottom configuration. Abington conveyed to Madison its fears that surmounting these obstacles would create additional costs. It later memorialized its concern in a letter to Madison stating that it was compiling the costs associated with overcoming the unexpected conditions and requested that a change order be issued. Madison, however, did not respond. Over the next several months, a series of floods caused extensive delay on the project. The conditions provided that when high water flows caused delay, Abington would be granted an extension of time equal to the length of the delay. The conditions placed responsibility for any overtopping of the cofferdams due to the high water flows on the contractor. Nevertheless, prior to awarding Abington the project, Madison issued an addendum (Addendum I), stating that "in the unlikely event of a major flood, [Madison] will pay the direct replacement costs of cofferdams provided cofferdams have been completed and approved by [Madison] prior to flooding. Madison indicated that it would cover some, if not all, of the additional costs caused by the faulty specifications and high water flows. Abington relied on this statement in continuing to work on the project during subsequent high water flows even though it was entitled to delay performance under the terms of the contract. By mid-October it was working seven days a week in hopes of completing the project on schedule."

The contractor also sent a written memo entitled "Confirm Verbal Agreements," seeking a change order and reiterating Abington's position that the costs associated with repairs and dewatering of the site due to the previous flooding would be covered by Madison pursuant to Addendum I of the contract. Abington subsequently submitted documentation that it had incurred $180,000 in additional costs. The owner's project engineer then cautioned the contractor's project manager that he was not authorized to approve the amount and that new management might not be willing to pay all the additional costs. Moreover, at several subsequent construction meetings, Madison indicated either that it was reviewing the cost overruns or that they would be dealt with at the end of the job, never stating that it refused to pay. Madison's internal documents from this period show that it planned to set aside $100,000 for anticipated extra costs.

The contractor incurred $404,738 in disputed costs above budget. After assuming responsibility for over $121,000 in costs, Abington claimed $283,680 in costs from Madison and an additional 17.8 percent allowance, or $50,495, for overhead and profit, for a total of $334,175. Abington's complaint suggested a number of theories of liability including quantum meruit, unjust enrichment, mutual mistake, impracticability, and breach of contract. The district court found that although the parties had a contractual agreement, Abington was entitled to quantum meruit recovery for the additional work and costs resulting from the defects in the data in the project manual combined with high water flows which were the subject of post-contractual communications between the parties.

It granted Abington an award of $334,175, the reasonable value of its services, but did not reach Abington's remaining theories of liability.

The owner appealed, contending that quantum meruit recovery was not avail able under Maine law when the parties have negotiated a contractual agreement and, even if it were available, the predicate elements for quantum meruit recovery are not found in this case. Where the parties have made an express contract, the court should not find a different one by "implication" concerning the same subject matter if the evidence does not justify an inference that they intended to make one. The fact that an express contract has been made does not prevent the parties from making another one tacitly, concerning the same subject matter or a different one. Further, the fact that a contract has been put into express words does not prevent the meaning and legal operation of those words from being affected by the process of "implication" from the conduct of the parties and from surrounding circumstances. The appellate court believes that Maine law permitted the contractor's claim for quantum meruit relief. If good faith efforts in the field to overcome adverse conditions and get on with the job could always be trumped by the original written contract, the arteries of commerce, particularly in construction, would harden. The contractor was entitled to pursue recovery on the basis of quantum meruit and that the evidence supported the district court's conclusion that the elements of quantum meruit existed.

In California, quantum meruit recovery is unavailable where an unrescinded express agreement exists, even though a subcontractor has allegedly performed services outside the agreement's scope. In *Bohm v. Horsley Co. (In re Groggel)*, 2005 Bankr. LEXIS 2191 (Nov. 10, 2005), the Horsley Company (contractor) subcontracted with Mark Groggel (subcontractor) for the installation of a shelving system as part of a warehouse construction project. The subcontractor did not visit the site or meet with the contractor prior to bidding, so the subcontractor was not aware that the warehouse floor was not level. Contending that he incurred significant costs to remedy the uneven flooring, the subcontractor did not complete its performance until after the completion date. The subcontractor filed a Chapter 7 bankruptcy petition. Seeking damages, the Chapter 7 trustee brought suit against the contractor for quantum meruit.

The trustee asserted that the subcontractor performed services completely outside the scope of the agreement, thus entitling the subcontractor to a recovery under quantum meruit. California law did not provide for quantum meruit recovery in the face of an unrescinded express contract simply because the subcontractor provided services outside the contract's scope. Even if the quantum meruit cause of action could have been brought, the trustee failed to prove that the subcontractor performed any services that could have been considered to be wholly outside the scope of the agreement. The trustee did not prove that the contractor radically redesigned the project or imposed significant changes to the agreement that would have changed the scope of work contemplated by the parties. The trustee did not prove that the parties intended to abandon the agreement. At most, the trustee alleged that there were some changes to the manner in which the installation of shelving was completed and that the subcontractor performed some installation work that exceeded the agreement's scope. The contractor did not impose changes to the agreement of such magnitude that would have altered the scope of the work. Furthermore, the cause of action was time barred by the two-year statute of limitations in Cal. Code Section 339(1). The trustee could not recover from the contractor under a theory of quantum meruit. The court granted judgment in favor of the contractor.

Page 13-52, add after note 254:

Quantum meruit may also be the means for recovery even where there is no agreement between the parties. For example, where a school district refuses to pay for a facility it is using for public purposes, the contractor may be entitled to recover at least the cost of labor and materials. In *Brown v. Penland Constr. Co.*, 2005 Ga. App. LEXIS 1301 (Nov. 23, 2005), a high school baseball coach approached Penland Construction (contractor) about constructing an indoor baseball practice facility. The coach informed the contractor that the facility would be financed through the boosters club. The coach also assured the contractor that the boosters club would do some of the work, such as site preparation, to keep costs down. The Walker County Board of Education (board) approved the facility, but there was no written contract for the project. The contractor had to perform added work, including site preparation. After the contractor completed its work, it submitted numerous invoices to the board, but the board refused to pay for the facility. The contractor filed suit against the board, coach, and school district. A jury found that although no oral contract existed between the parties, the contractor was entitled to recover in quantum meruit in the amount of $150,000. The board, school district, and coach filed a joint appeal.

The board, school district, and coach contended that the trial court should have directed a verdict in their favor because no written contract existed and a recovery in quantum meruit was prohibited by sovereign immunity. They also contended that no evidence was presented regarding the value of the facility. The court did not need to decide whether the contractor was limited in its recovery to restitution because the school district's conduct was a taking for constitutional purposes. The contractor had a valid property interest in the facility it constructed. The contractor paid for the materials and expended the labor to erect the facility. The school district used the facility in the course of providing public education. Since the school district refused to pay for a facility it was using for public purposes, its conduct was a taking and sovereign immunity was not a bar to the contractor's action. The contractor was entitled to enforce its constitutional right to just compensation. The contractor provided uncontroverted testimony that the value of the facility was at least $138,000, based on the value of the materials and labor. The facility generated intangible benefits in recruiting and training athletes for the high school's baseball program. The coach also benefited economically by operating baseball camps in the facility. The jury's verdict of $150,000 was not excessive and was affirmed.

[E] Contractual Remedies

Page 13-56, add to end of first full paragraph:

Interpreting the Differing Site Condition clause that may permit recovery and other clauses in the contract that relate to below-ground interferences that may restrict recovery. Delays in relocating underground utilities are not unknown subsurface or latent physical conditions under a differing site condition clause. *Indiana Department of Transportation v. Shelly & Sands, Inc.*, 726 N.E.2d 1063 (Ind. Ct. App. 2001), the Indiana Department of Transportation agreed to administer federal funds to pay for road reconstruction for the city of Carmel. In preparation, the owner entered into a series of utility agreements to relocate utility facilities. The bid package contained estimates on when the various utilities would relocate their facilities. The owner awarded the contract to Shelly & Sands, Inc. Before the contract was awarded, the owner held a utility coordination meeting to discuss the timetables for utility relocation. Numerous conflicts in the timetable for utility relocation delayed the work. The owner extended the completion date. The contractor later submitted a claim for additional costs caused by the delay. The owner denied the claim. The contractor filed suit alleging the owner failed to timely relocate the utilities. The owner moved for summary judgment, which the trial court denied. The owner appealed.

The owner alleged that the contractor's delay claim was barred by clauses that limited the remedy for utility relocation delays to an extension of time. "Cooperation with Utilities" indicated that no additional compensation would be allowed for delays due to interference from moving utilities, but the time for completion was to be extended. "Contractor's Responsibility for Utility Property and Services" stated that the contractor assumed all risk for any delay or expense caused by public utilities, but also required the contract to be extended. The contractor contended that the language in the "Cooperation with Utilities" clause was internally inconsistent because it established a duty while at the same time relieving the owner from liability if it failed in the task. The clause did not relieve the owner of its duty under the contract. It merely shifted the risk of loss to the contractor and limited the remedy if delays occurred. Limiting the remedy for relocation delays did not render the owner's contractual obligations meaningless. The contractor also asserted that the clause concerning utility relocation was ambiguous when coupled with a differing site conditions clause because delays in utility relocation should be interpreted to be subsurface or latent physical conditions or unknown physical conditions of an unusual nature. The contractor also asserted that the contract was ambiguous because the contract allowed for contract cost adjustments based on unknown physical conditions limiting the remedy for utility relocation delays to time extensions. The utility fixtures, pipelines, and other appurtenances that needed to be relocated for the roadwork to proceed were known physical conditions. The utility fixtures were contemplated in the plans for the project and in the contract. The contract language was not ambiguous.

§ 13.12 CARDINAL CHANGES

[B] Recovering Costs for a Cardinal Delay

Page 13-66, replace note 323:

Unpublished private contract.

§ 13.13 THE BURDEN OF PROOF

Page 13-67, add at end of carryover paragraph:

However, if the owner grants a time extension by a change order, there is a rebuttable presumption that the time extension is compensable. For example, in *Gottfried Corp.*, ASBCA No. 51041, 98-2 B.C.A. (CCH) ¶ 30,063 (1998), the contractor was awarded a contract for the construction of a classroom building at an Air Force base. During the course of construction, the government issued a contract modification to the contractor, extending the contract time by 16 days to deal with a structural steel design defect. Later, when the contractor sought to recover its delay costs for extended overhead, the government argued that contractor-caused delays were concurrent with the design-related delay such that the contractor was not entitled to compensation for the time extension. The Board held that the granting of a time extension created a rebuttable presumption that the delay: (1) was solely due to government-caused delay, (2) was not concurrent with contractor-caused delay, and (3) was critical (delayed the overall completion of the contract).

In *Gavosto Associates, Inc.*, PSBCA No. 4058, 01-01 B.C.A. (CCH) ¶ 31,389 (2001), Gavosto was awarded a fixed-price construction contract for a truck wash facility to be completed within 150 days. During the course of construction, the Postal Service issued numerous unilateral change orders, some of which extended the performance period, but did not include an overhead markup or extended field or home office overhead costs. Gavosto brought its claim for extended field overhead to the board, which, in finding for Gavosto, noted that when an owner directs additional work by unilateral modifications and grants time extensions for the work, it creates a rebuttable presumption that the owner is responsible for the delays, and the contractor is entitled to compensation for time-related costs.

When the claim asserted by the contractor is based upon alleged government-caused delay, the contractor has the burden of proving the extent of the delay, that the delay was proximately caused by government action, and that the delay harmed the contractor. In addition, when the contract utilizes CPM scheduling, the contractor must prove that the critical path was prolonged in order to prove a delay in project completion. Some courts have stated that a contractual requirement for a schedule is not necessary of this obligation. *Morrison Knudsen Corp. v. Fireman's*

Fund Insurance Co., 175 F.3d 1221 (10th Cir. 1999). The critical path is crucial to the calculation of delay damages because only construction work on the critical path had an impact upon the time in which the project was completed. If work on the critical path was delayed, then the eventual completion date of the project was delayed. Delay involving work not on the critical path generally had no impact on the eventual completion date for the project. *Hoffman Construction Co. v. United States*, 40 Fed. Cl. 184 (1998) *citing G.M. Shupe*, 5 Cl. Ct. at 728; *Kelso v. Kirk Bros. Mechanical Contractors, Inc.*, 16 F.3d 1173, 1177 (Fed. Cir. 1994); *J.A. Jones Construction Co.*, ENG BCA No. 6252, 97-1 B.C.A. (CCH) ¶28,918 at 144,168 (to be compensable, government-caused delay must interfere with the project's critical path and extend completion of the project).

Page 13-67, add to note 325:

1992

Page 13-67, add to note 328:

SIPCO Servs. & Marine, Inc. v. United States, 41 Fed. Cl. 196 (1998) (contractor had fulfilled its duty by informing NASA of patent ambiguity during the bidding process. At that point, the court held that the burden was shifted to NASA, as the drafter of the specifications, to correct the defect in the specifications. The court held that NASA's failure to solve the problem required it to pay the contractor the costs expended in excess of its bid).

Page 13-68, add at end of section:

In addition to delay claims, a burden of proof is imposed after actions during a contract's administrative process. Under certain circumstances, a trier of fact may find that owner approval of the contractor's schedule created a "rebuttal presumption of correctness." In *Santa Fe, Inc.*, VABCA No. 2168, 87-3 B.C.A. (CCH) ¶20,104 (1987), the Veterans Administration entered into a written agreement with the contractor to build the second phase of a hospital. Problems arose on the project that resulted in the contractor seeking additional compensation and a time extension. The contracting officer denied the contractor's claims, and the contractor appealed.

In examining the contractor's time extension claim, the Board of Contract Appeals for the Veterans Administration turned its attention to the various CPM schedules upon which the parties agreed during the course of construction. The government sought to have the Board disregard one CPM schedule in favor of a subsequent CPM schedule because certain delayed work was on the critical path in the earlier CPM schedule, but not in the later one. The Board refused to do so, stating:

> . . . [T]he adoption of an alternate method of performance in the later CPM does not, of itself, contradict the existence of a delay as shown in the preceding

> CPM. There is a rebuttable presumption of correctness attached to CPM's upon which the parties have previously mutually agreed. . . . To put it another way, in the absence of compelling evidence of actual errors in the CPMs, we will let the parties "live or die" by the CPM applicable to the relevant time frame.

Id. This presumption of correctness may extend to certain key aspects of the approved schedule, such as approved time extensions, activities that constitute the critical path, owner-related activities (such as delivery times for owner-furnished items), notice to proceed and start dates, approval periods, and other activities for which the owner or its agents have significant involvement.

The government has the burden of proving that at the time of the default termination, the contractor had no reasonable likelihood of achieving completion in accordance with the contract. *CJP Contractors v. United States*, 45 Fed. Cl. 343 (1999).

CHAPTER 14
DISPUTE RESOLUTION FORUMS

§ 14.03 ARBITRATION

Page 14-11, add at end of first paragraph:

While arbitration is generally considered to be a consensual forum, in some construction contracts one party may have the right to choose between arbitration and litigation. For example, the following language may appear in a subcontract.

> The General Contractor shall have the sole and exclusive right to determine whether any dispute or claim arising out of or relating to this Subcontract shall be submitted to arbitration or to a court of law.

Similar provisions may appear in the construction contract between an owner and general contractor where the owner retains the right to decide whether to litigate or arbitrate. In either event, arbitration may not be a matter of mutual consent as one party has the unilateral right to decide whether to arbitrate.

Courts have considered whether such provisions are enforceable. For example, in *Stevens/Leinweber/Sullens, Inc. v. Holm Development and Management, Inc.*, 795 P.2d 1308 (Ariz. Ct. App. 1990), the general contractor filed suit against the owner and architect for the unpaid balance and to foreclose on its mechanic's and materialman's liens. The owner moved to compel arbitration pursuant to a provision in the construction contract. The construction contract also contained an addendum that gave the owner the option of arbitrating or litigating any disputes, and even allowed the owner to change its choice if it was timely exercised. The trial court denied the owner's motion to compel arbitration on the grounds that: (1) the arbitration provision was unenforceable because of the lack of mutuality and (2) the unilateral arbitration provision lacked consideration. On appeal, the owner argued that since the entire contract was supported by consideration, it was sufficient to support the arbitration option. However, the court reasoned that since the owner did not promise to do anything in consideration of the rights given to it in the arbitration provisions, there was a lack of consideration to support the arbitration option. Further, since arbitration provisions are independent and separate agreements, the owner could not "borrow" consideration from the main contract to support the arbitration provision. There was no mutuality because the owner had absolute discretion to choose arbitration or litigation.

Other courts considering unilateral arbitration provisions have similarly held that arbitration agreements must be mutually binding to be enforced. *R.W. Roberts Constr. Co. v. St. John's River Water Management Dist.*, 423 So. 2d 630 (Fla. App. 1982); *Miner v. Walden*, 422 N.W.2d 335 (1979).

There is no uniformity of case law in this area. Rather than being deemed "unilateral provisions," such arbitration agreements may be considered "discretionary" and as such are valid and enforceable. For example, in *Willis Flooring, Inc. v. Howard S. Lease Constr. Co.,* 656 P.2d 1184 (Alaska 1983), the subcontract gave the general contractor the "sole option" of requiring the disputes to be arbitrated or litigated. The subcontractor argued that the arbitration provision was a unilateral option to arbitrate, that it lacked mutuality of remedy and that it would be inequitable to coerce a party to arbitrate. The court rejected these arguments, and upheld the enforceability of the arbitration provisions, reasoning that it was neither unfair nor coercive to require the subcontractor to arbitrate. Arbitration is not so clearly more or less fair than litigation that it is unconscionable to give one party the right of forum selection. *Id.* at 1185–86.

Page 14-13, add after second full paragraph:

Another form of arbitration that may have application in construction is commonly known as "baseball arbitration." It is a form of binding dispute resolution where the arbitrators' award is limited to the choice of one of the two alternative offers made by the parties—an either/or resolution. A formal arbitration clause may include the following: "The parties shall exchange with one another and submit to the arbitrator(s) their last, best offer to the other. The arbitrator(s) shall select only one or the other of such two offers that are submitted and shall award such selection." Rumbaugh & Powell, *Baseball Arbitration: Part Two,* 23 Disp. Resol. Times 18 (Oct.–Dec. 2000).

Baseball arbitration provides a mechanism or incentive for the parties to resolve the dispute in a timely manner that gives them more control over the outcome of the process and the form of the award. Good-faith attempts to be reasonable are often made by the parties to baseball arbitration, or at least the parties will attempt to avoid the appearance of being unreasonable for fear that their offer will not be selected by the arbitrator.

Baseball arbitration has the potential for being a good way to solve many types of disputes in the construction industry and to streamline the resolution of disputes. The value in baseball arbitration is that the parties may be motivated to resolve the dispute themselves when no trendsetting legal issues are at stake and business decisions can be made. For example, when the pricing of sole-source items is in dispute, it may give incentive to the seller to provide pricing information and negotiate in good faith. If the timely resolution of a dispute is crucial to the parties, an "either/or" decision may be made based upon expedited hearings. Pricing disputes related to extras, change orders, credits, and payments may be good areas for application of baseball arbitration. The resolution of multiple-item punch lists and substantial completion disputes could be done among the parties if they knew there would be an either/or decision in baseball arbitration. *Id.* However, some parties may resist the idea of such a simplistic decision-making process, which gives the appearance of being arbitrary.

§14.05 MEDIATION

Page 14-23, add after second full paragraph:

Another survey commissioned by the American Bar Association's Forum Committee on the Construction Industry was reported in Thomson, *A Disconnect of Supply and Demand: Survey of Forum Member's Mediation Preferences,* Constr. Law. 17 (Fall 2001). Three key aspects of mediation were covered by the survey. The first aspect dealt with how and on what basis mediators were selected by the respondents. Seventy-two percent of the respondents stated they chose mediators through mutual agreement of the parties, rather than through an organization that provides mediation services. The respondents ranked by importance the factors that most influenced the selection of the mediator. The top four most valued criteria were: (1) the mediator's subject matter expertise, (2) the mediator's reputation, (3) the mediator's mediation experience, and (4) the mediator's acceptability to all parties. Conventional wisdom concerning mediator selection is that the mediator's subject matter expertise is not as important as the mediator's experience in mediation. However, based upon the survey results, this often stated point is not substantiated by the reported inclination of construction attorneys. *Id.* at 18. The survey respondents preferred lawyer mediators over lay mediators except in cases where the lawyers lacked substantive knowledge or experience in the field of the dispute.

The second key aspect of the survey concerned the conduct of the mediation: pre-mediation preparation, opening remarks, and multi-party mediation. The respondents overwhelmingly preferred the submission of pre-mediation position papers over pre-mediation discussion with counsel or the parties. While it has been argued that opening statements by counsel and parties tend to alienate the parties and make settlement more difficult, the survey respondents preferred mediators who insisted upon opening statements, or gave the parties the choice to make opening remarks as opposed to mediators who insisted that the parties not make opening remarks. Concerning opening remarks, more of the respondents preferred giving the parties the choice of making opening remarks rather than requiring opening statements. *Id.* at 18–19.

The issue of multi-party mediation was of concern because of the problems with the amount of time required to mediate cases with numerous parties, especially the "waiting time" to meet one-on-one with the mediator. The survey solicited suggestions to improve the efficiency and effectiveness of the mediation with numerous parties. The majority of suggestions dealt with pre-mediation preparation such as submission of confidential settlement positions to only the mediator, pre-mediation meetings between the mediator and counsel, and pre-mediation exchange of key documents and expert reports. Another suggestion included separate mediations prior to the joint mediation meeting with all parties. A number of respondents suggested the use of multiple mediators allowed the parties to spend more time with the mediators as counsel; they believed this was more important and effective than waiting to meet individually with a mediator. *Id.*

The third aspect of the survey concerned the desired attributes of the mediators. The respondents most frequently indicated their desire for a mediator who presents the various settlement options or solutions to the parties based upon the mediator's understanding of the case. However, in actual practice, the respondents reported that they received this service from mediators only 60 percent of the time for the mediation in which they participated. The second most desired attribute in a mediator was confidentiality. The respondents preferred a mediator who discusses confidentially with each party the merits and potential outcomes of the case. Both attributes relate to the evaluative character of the mediators. The least desired attribute was a mediator who asks the parties to confidentially evaluate their own case, but does not suggest any evaluation of the merits. The respondents ranked third in order a mediator who communicates the parties' settlement positions and helps the parties avoid personality disputes. It was noted that parties often resort to mediation because the parties are unable to communicate effectively and civilly concerning the dispute and its resolution. *Id.* at 20. The results of the survey indicate that counsel prefer evaluative mediators and only secondarily emphasize the facilitative skills.

§ 14.08 DISPUTES REVIEW BOARDS

Page 14-33, add after note 96:

The Disputes Review Board (DRB) members may visit the project site to keep abreast of the construction activities and potential problems. When a dispute arises between the parties, a hearing may be requested by the parties for the DRB panel to consider the issues. The results of the hearing may be a DRB panel recommendation that includes an explanation or reasons for the recommendations. The DRB panel results may be oral or written. Oral recommendations allow the parties greater flexibility in implementing the resolution of a problem, in addition to the benefit of a third-party independent review. Many DRB procedures require a written, advisory decision or report from the DRB panel. Typically, DRB written decisions are not binding on the parties, but are often accepted by the parties to avoid an escalation of the problem or a loss of credibility or trust between the parties to the construction contract. In certain circumstance, the DRB panel's recommendations may be admissible in court or other forums for certain limited purposes. The extent of the admissibility and other uses of the DRB panel recommendations should be addressed in the construction contract specifications or separate agreement dealing with the DRB process. *Expert DRB Panel and Administration,* 6 Punchlist 1 (Nov. 2000–Jan. 2001).

Page 14-36, add the following new section:

§14.09 ADJUDICATION IN THE UNITED KINGDOM

In the United Kingdom, an alternative dispute resolution forum known as adjudication has become a prevalent method of resolving construction disputes. The goal is to provide a speedy and cost-effective method of resolving disputes on an "interim basis," allowing the decision to be "enforced pending final determination." *Macob Civil Eng'g Ltd. v. Morrison Constr. Ltd.,* BLR 156 (1999). The procedure can be invoked during the course of construction to address payment disputes, performance problems, time extensions, delays, change order valuations, and other claims related to the construction contract. An adjudicator generally must decide a dispute in less than 42 days from the time the matter is referred. In most cases, the adjudicator's decision is final and binding and may be implemented immediately. A party that is dissatisfied with the adjudicator's decision may have it reviewed by a court, but such proceedings may not be started before contract completion or termination. Bingham, *Adjudication Is King,* Building 52 (Feb. 14, 2003).

Adjudication became a statutory right in the United Kingdom in 1988 with the passage of the Housing Grants, Construction, and Regeneration Act of 1996. The specific implementation provisions and rules for adjudication were mandated by separate legislation, the Scheme for Construction Contracts (England and Wales) Regulations 1998, SI 1998 No. 649. Under the Scheme, a party to a written contract for construction, landscaping, design, or surveying may refer a contract dispute to an adjudicator for resolution by giving notice of the adjudication referral to the other party. The notice should describe the nature of the dispute, the particulars of the contract, the names and addresses of the parties involved, where and when the dispute arose, and the nature of the redress sought.

To appoint an adjudicator, a party may contact the adjudication organization named in the contract or an "Adjudicator Nominating Body" (an organization such as the Academy of Construction Adjudicator, the Chartered Institute of Arbitrators, the Centre for Dispute Resolution, or the Construction Industry Council) to make proper arrangements. The appointment of the adjudicator is to take place within seven days of the notice. The adjudicator may not be an employee of any party to the dispute, is to be impartial, and declare any interest in the matter. The same adjudicator may, with the consent of the parties, adjudicate one or more disputes under the same contract or related disputes under other contracts.

The adjudicator is to reach a decision within 28 days of the referral, and this can be extended by 14 days. The adjudicator is to ascertain the facts and the law relating to the dispute. The adjudicator may also decide the procedure to be followed. Typically, the adjudicator may request documents and written statements, meet and question the parties and their representatives, make site visits, perform tests and examinations, and appoint experts. If a party fails to comply with any of the adjudicator's requests, the adjudicator may continue with the matter, draw inferences from the failure, and make a decision on the available information. Knowles, *Adjudication,* Professional Education Seminar (2002); Black, *Dispute Review Boards in International Construction*, ABA Construction Forum (May 8, 2003).

The adjudicator's decision may direct payments on specified dates or within a time period and apportion fees and expenses of adjudication. A party may request that the adjudicator provide reasons for the decision. The statutes provide that the adjudicator's decision is to be complied with immediately upon delivery to the party and is binding until the dispute in finally determined by litigation, arbitration, or agreement. Enforcement of adjudication awards is generally obtained through an application for summary judgment to a court with requisite jurisdiction. Albertini, *Adjudication After Five Years—Need for Reform?* Constr. Law. 14 (May 2003).

One area for challenges to an adjudicator's decision often involves the adjudicator's jurisdiction to decide the matter. This may include the situation in which the contract involved is not a construction contract, when there is a defect in the adjudication notice, or when the adjudicator was appointed by the wrong body. A party may contend that the current dispute was one that was previously referred to adjudication and a decision was made, or the dispute differed significantly from the referral notice. Another jurisdictional issue is whether the matter has "crystallized," that is, developed to the stage of an actual dispute. Although a "claim" may have been made, the matter may require a reasonable period of time during which the parties consider each other's position and engage in an "open exchange" of views. For example, in *Hitec Power Protection BV v. MCI Worldcom, Inc.,* EWHC 1953 (2002), the court noted that the claim must be made and an opportunity given to the opposing party to consider the claimant's position and formulate reasoned arguments. Otherwise, there is not a dispute that can be sensibly referred to adjudication, and an adjudicator will have exceeded his jurisdiction in making an award. This may prevent an "ambush" by a claimant who seeks to rush off to adjudication, catching its opposition off-guard and unprepared. Albertini, *supra* at 16. The challenge based upon jurisdictional grounds must be made at the outset or a party may be deemed to have submitted to the adjudicator's jurisdiction. Typically, adjudicators do not have the authority to decide their own jurisdiction—it is the domain of the courts. Adjudicators may investigate their jurisdiction and prepare a finding as to whether they should continue to hear the matter. If successful, the jurisdictional challenge may allow a party to stop the adjudication from taking place or prevent the enforcement of the adjudicator's decision. *Id.*

The other major area of challenge is procedural unfairness. The adjudicator must act in a fair manner, be impartial, and act independently. For example, the adjudicator should avoid ex parte substantive communications with the parties or their counsel. *See Discain Project Services v. Opecprime*, BLR 285 (2001). Also, each party should have a reasonable opportunity to present its case. In the context of delay claims, the limited time for the adjudicator to consider the matter and render a decision may bring about challenges based upon fairness because of the need for detailed delay analysis. For example, in *Balfour Beatty Construction v.*

The Mayor and Burgesses of the London Borough of Lambeth, BLR 288 (2002), the adjudicator performed his own critical path method schedule analysis after the parties provided details of the schedule and the delays to the adjudicator. This was held to be improper because the adjudicator tried to remedy deficiencies in one party's submissions and failed to give the other party the opportunity to respond. However, in *Try Construction Ltd. v. Eton Town House Group Ltd.*, BLR 286 (2003), the parties agreed to use an independent scheduling expert to assess entitlement to time extensions. After the adjudicator's award was made, the defendants claimed the adjudicator allegedly based the decision on his own schedule analysis methodology without giving the parties the opportunity to make submissions. The court reviewed the matter, finding that the adjudicator's decision had been based upon the evidence and analysis of the expert, noting that the parties agreed to the use of an independent scheduling expert.

In the majority of cases dealing with the challenges to enforcement, the courts have upheld the adjudicators' decisions. Knowles, *New Adjudicator's Guidelines, supra.* Thus, courts have helped in make adjudication a viable means of resolving disputes.

CHAPTER 15

SURETIES AND DELAY CLAIMS

§ 15.02 PAYMENT BONDS

Page 15-5, add after carryover sentence in first carryover paragraph:

The total cost method can be used to calculate damages under the Miller Act, despite the absence of any general contractor fault. *Lighting & Power Servs. v. Roberts,* 354 F.3d 817, 2004 U.S. App. LEXIS 364 (8th Cir. 2004). Wayne M. Roberts, Inc. was the general contractor for the United States government, and Lighting & Power Services was the contractor's electrical subcontractor. The construction contract required the contractor to obtain a payment bond under the Miller Act. The renovation was scheduled for completion in 6 months, but ultimately took over 22 months. Both parties conceded that the government caused the delays. The subcontractor filed suit under the Miller Act seeking to recover the additional costs it incurred as a result of the delay. The subcontractor used the total cost method to calculate its damages and sought $110,641.00. The district court instructed the jury that in order to establish damages by the total cost method, the subcontractor had to prove that: it was impracticable for the subcontractor to prove its actual losses directly; the subcontractor's bid was reasonable; the subcontractor's actual costs were reasonable; the contractor had some responsibility in causing the subcontractor's actual losses; and the subcontractor incurred damages as a consequence. The jury returned a verdict in favor of the contractor. The subcontractor appealed.

The subcontractor argued that the Miller Act permitted the subcontractor to recover its additional costs of labor and materials regardless of the contractor's responsibility for causing those costs. The contractor responded that the instruction was required because the subcontractor chose to calculate its damages using the total cost method. The Miller Act did not require that the general contractor be at fault in order for the subcontractor to recover on the Miller Act payment bond. The Miller Act favored allowing full recovery from a general contractor regardless of fault because the contractor had privity of contract with the government and could have recovered delay damages directly from the government, while the subcontractor could not. The jury instructions did not require the jury to find that the subcontractor was not responsible for the added expenses, but instead required the jury to find that the contractor had some responsibility in causing the subcontractor's actual losses. The jury instructions did not require the subcontractor to prove that it was free from fault, and the contractor made no claim that the subcontractor caused its own delay damages. The district court erred when it required the subcontractor to prove that the contractor had some responsibility for causing the subcontractor's losses. Reversed and remanded.

Page 15-5, add to note 15:

See Recovering Delay Damages Against Payment Bonds—Part I, 21 Constr. Claims Monthly 1 (May 1999); *Recovering Delay Damages Against Payment Bonds—Part II,* 21 Constr. Claims Monthly 1 (June 1999).

Page 15-6, add after first sentence in the second full paragraph:

But under a labor and material payment bond, a subcontractor may recover only labor and material costs caused by delay. For example in *Lexicon, Inc. v. Safeco Ins. Co. of Am.*, 2006 U.S. App. LEXIS 3113 (6th Cir. Ky. Feb. 9, 2006), Lexicon (the subcontractor) was hired by Icon (the contractor) to erect and install equipment for the expansion of a stainless steel plant. The contractor provided a labor and material payment bond for which Safeco Insurance served as the surety. The subcontractor encountered numerous delays through no fault of its own, which resulted in substantial cost overruns. The subcontractor sought payment for the outstanding contract balance, extra work authorizations, and cost impacts suffered due to the action or inaction of other companies. The contractor and subcontractor settled some of the subcontractor's claims. Arguing that the labor and material payment bond required the surety to pay the subcontractor for the delay and impact damages, the subcontractor filed suit against the surety. The district court determined that any claims asserted by the subcontractor against the surety not based on labor and material costs caused by delay were not recoverable under the bond. The district court also held that the settlement agreement settled all contract-based claims and precluded the subcontractor from recovering from the surety on the bond. The district court awarded the surety summary judgment. The subcontractor appealed.

The subcontractor argued that the evidence was sufficient to establish that its claims were for extra labor and materials necessary to complete the project and were required because of delays. According to the subcontractor, it was making a claim for labor and materials under the bond. The subcontractor further asserted that the settlement agreement unambiguously reserved contract-based delay claims under the bond. The surety argued that claims for delay were not compensable under the bond. The surety asserted that impact claims were similar to breach of contract claims, which were not compensable under a payment bond. Insofar as the subcontractor asserted that it was due more than labor and material costs caused by delay, the surety argued that it was entitled to summary judgment under the bond on those claims. While the subcontractor was unable to recover for items not based on labor and material costs caused by the delay, the subcontractor pointed to several documents that provided a breakdown of the disputed costs and extras.

Those documents were sufficient to raise a jury issue that the bond covered at least a portion of the subcontractor's claims. The settlement agreement was ambiguous and did not fully settle all claims. The district court erred by entering judgment on the subcontractor's labor and material cost delay claims. The case was remanded for determination regarding the settlement agreement's proper interpretation and whether the subcontractor made a claim under the bond for labor and material costs related to delay.

Page 15-8, add after note 31 in first full paragraph:

The notice to the surety must also clarify the basis of the claim against the bond. For example, in *Missouri Department of Transportation v. Safeco Insurance Company of America*, 2002 Mo. App. LEXIS 2196 (Nov. 5, 2002), the Missouri Highway and Transportation Commission (MHTC) hired Robertson as the general contractor for road construction. PRD was the subcontractor for the base rock and paving. Robertson obtained a bond from SAFECO. Shortly after PRD signed the subcontract PRD was told that Robertson wanted PRD to begin its work on March 15, 1999. PRD verbally agreed to this starting date, and committed itself to starting the Project on March 15, expecting to be done within 80 working days, if not sooner. PRD was not able to begin its work until the end of May 1999. PRD sent a detailed letter to Robertson and SAFECO concerning "the substantial claim for unpaid moneys that we [PRD] have regarding this job [the Project]." The letter stated the principal causes for the "disruption and delay" which damaged PRD. The letter did not specifically state that PRD was making a claim against Robertson or against SAFECO under the contract bond, or that Robertson had breached the subcontract. SAFECO reviewed the letter, and determined that PRD's claims dealt with the actions of MHTC, not Robertson. SAFECO requested that PRD fill out the attached affidavit of claim form, and that SAFECO reserved all rights and defenses available to itself and Robertson at law and in equity. The affidavit averred that Robertson was indebted to PRD for labor and materials furnished to Robertson, and that notice of non-payment was sent to Robertson and SAFECO. PRD later filed suit against Robertson and SAFECO for breach of contract and for vexatious refusal to pay by SAFECO. The jury returned verdicts in favor of PRD on its claims, and awarded damages of $143,894.78 and $111,715.50 against SAFECO for vexatious refusal to pay. Robertson and SAFECO appealed.

SAFECO argued that there was insufficient evidence to show that it was unreasonable and vexatious in not making payment to PRD, as required by statute, and contended that there was a good-faith dispute as to whether or not Robertson had any liability for PRD's claims, and that the trial court should have granted judgment in favor of SAFECO or a new trial. PRD's letter did not explicitly state that PRD was making a demand under the surety bond between SAFECO and Robertson for breaches of contract by Robertson. The letter did not claim that Robertson had breached the subcontract, and stated that the principal causes for delay were a utility's failure to move the high voltage power lines, MHTC's providing an inadequate subgrade, and MHTC's persistent failure to stake and grade properly and in a timely fashion. Lacking an explicit demand on the bond in the letter, SAFECO was not unreasonable in failing to consider this to be a demand letter. SAFECO sent claim forms to PRD and began an investigation after receipt of the letter, even though the letter indicated PRD's claims were directed against MHTC, not Robertson, the principal, and that PRD was not claiming that Robertson had breached the subcontract. Hester testified that prior to PRD's filing suit, she had talked with Robertson's representatives, who had told her that PRD was blaming MHTC, and Robertson concurred that the problems were MHTC's responsibility. Hester further stated that after the lawsuit was filed, she spoke again with Robertson's representatives, who informed her that it disputed PRD's allegations, that it had breached the subcontract, and that it had a counterclaim against PRD. She declared that her investigation was limited after the lawsuit began because she could no longer contact PRD directly as it was represented by counsel, and that it was restricted to what could be obtained through the discovery process in litigation. She stated that she understood Robertson's letter of July 14, 2000 to MHTC concerning PRD's damages, written by Holt, with an affidavit from Parris, and an affidavit from Holt, was merely passing along PRD's asserted claims to MHTC. Hester believed that this letter was not an admission of legal responsibility by Robertson, and merely followed the procedure under the contract for submitting the claim of a subcontractor. She testified that she had no reason to believe that Robertson's defense lacked merit. SAFECO did not provide details in its answer to PRD's petition to explain why it was denying PRD's claim, other than on information and belief that Robertson had paid PRD all sums that were due. However, SAFECO stated in an admission that after it had received PRD's letter of December 10, 1999, Robertson had advised SAFECO that "since all of PRD's allegations related to the acts or omissions of MHTC, which in PRD's opinion caused extra costs to PRD, any claim that PRD had was against MHTC and not Robertson." This was an adequate explanation for SAFECO's denial. Robertson's defenses were not without merit. There were open questions of fact and law relating to whether or not Robertson had made unconditional promises to PRD. Parris could not remember who had made the promise that the site would be ready for PRD by March 15, 1999, Holt or Ben Robertson. Holt testified that he made no such promise. There was testimony that Robertson's representatives had expressed hope to have the site ready by March 15, without an expression of a promise. The specifications, contract, and subcontract were not always clear about which party was intended when the term "contractor" was used. There was testimony that Robertson had a superintendent on the project, though not the one initially designated as such, and not directly overseeing PRD's work. There was testimony that the subgrade material selected by MHTC was adequate and of the sort routinely used on MHTC projects. There were legal issues in dispute concerning whether or not the subcontract was a fully integrated document. SAFECO's refusal to pay PRD's claim was not unreasonable, vexatious, or recalcitrant.

§15.03 PERFORMANCE BONDS

Page 15-11, add after carryover paragraph:

However, just as a modification without notice to the surety may discharge the surety's obligations under the bond, an owner that demands the surety accelerate after rejecting the surety's offer to complete and pay liquidated damages may breach the performance bond. For example, in *St. Paul Fire & Marine Insurance Co. v. City of Green River,* 2001 U.S. App. LEXIS 6269, 6 Fed. Appx. 828 (10th Cir. 2001), the Wyoming Joint Powers Water Board constructed a 32-million-gallon-per-day water treatment plant. St. Paul Fire and Marine issued a performance bond to guarantee the contractor's performance. The construction agreement required substantial completion of the project by December 1, 1998 and included a time is of the essence clause with a liquidated damages penalty for time after the stated December 1 completion date. The contractor fell behind in construction of the plant and the board issued a notice of termination for default. This triggered the surety's obligations under the performance bond. The surety had several options in the event of contractor default, and chose to complete the project by September 24, 1999, paying the daily penalty after December 1, 1998. The board, however, informed the surety that it had stepped into the shoes of the contractor for all purposes and thus was bound by the time is of the essence clause and the December 1, 1998 completion date. The board rejected the surety's proposed performance and declared the surety to be in breach. The surety sued to be discharged from further performance obligations on the ground that the board had breached the terms of the performance bond. Both parties sought summary judgment. The trial court granted summary judgment to the city. The surety appealed.

Although the surety essentially stepped into the shoes of the contractor upon the latter's default, it did not follow that the surety was bound by the original completion date. The board would continue to be entitled to $2,500 per day liquidated damages, but the surety under the bond need only proceed with reasonable promptness. The appellate court concluded that because the surety selected one of the bond options available to it and was ready to proceed to completion, the board's rejection of its proposed performance breached the agreement. The surety's offer to substantially complete the project, by a date that the board conceded would have been acceptable, would have occasioned much greater expense. It was not unreasonable for the surety to seek compensation from the board for the cost of accelerating performance beyond the reasonable promptness standard the bond specified.

Page 15-11, add after note 55 in first full paragraph:

Liquidated damages clauses may serve to cap the takeover surety's liability for delay damages, shielding the surety against exposure for the owner's claims for consequential delay damages. Because the takeover surety stands in the place of the defaulted contractor, it assumes the rights and liabilities under the construction contract such as the liquidated damages provisions. *United States Fidelity and*

Guaranty Co. v. West Rock Development Corp., 50 F. Supp. 2d 127 (D. Conn. 1999).

Page 15-12, replace note 62 with:

Id. at 112.

Page 15-13, add at end of carryover paragraph:

Lost profits are generally not a recoverable element of damages under a performance or payment bond. In *U.W. Marx, Inc. v. Mountbatten Sur. Co.,* 3 A.D.3d 688, 770 N.Y.S.2d 777 (2004), U.W. Marx and Kleen Resources (contractor 2) formed a joint venture to bid on a contract with the Office of General Services. Their bid was accepted and, under the joint venture terms, Marx agreed to procure the bonding required by the agency for the project, and contractor 2 promised to complete all the work. As part of their agreement, Marx required contractor 2 to provide a performance bond, which it obtained from a surety. The surety's bond named the joint venture as the obligee. With respect to compensation under the terms of the joint venture agreement, Marx was to receive 8 percent of all contract payments made by the agency, and contractor 2 was to receive all funds left after paying suppliers, subcontractors, its payroll, and the 8 percent to contractor 1. Marx obtained the bonding required by the agency. Marx received 8 percent from the joint venture for a total of $646,130. However, Marx discovered that contractor 2 could not account for $658,759 that had been received under the agency's contract. Contractor 2 then defaulted on its promise to perform all the work on the project and assigned its rights in the joint venture to Marx. Marx demanded that the surety honor the performance bond. The surety refused, resulting in Marx undertaking to complete the contract for the joint venture. Subcontractors were reportedly owed funds by the joint venture. Marx assumed responsibility for those claims. Marx brought an action against the surety and successfully moved for summary judgment on the issue of liability. The lower court did not allow the claim for profit or the amount of the subcontractors' claims at the time contractor 2 defaulted. The court otherwise accepted Marx's damages and granted judgment in the amount of $620,512.49, plus post-decision interest. After trial on the issue of additional overhead expenses, the lower court awarded Marx a further judgment of $165,414, plus statutory interest. Marx and the surety appealed.

The surety argued that Marx was not entitled to the 8 percent fee prior to contractor 2's default. The appellate court found merit in the surety's assertion that contractor 1 was not entitled to an 8 percent fee following contractor 2's default. Under the terms of the joint venture agreement, contractor 1 was reimbursed for its actual costs in obtaining the original agency-required bonding. The 8 percent fee represented profit it anticipated from the business arrangement. Lost profits were generally not a recoverable element of damages under a performance or payment bond. Hence, the award of $212,767.19, representing the 8 percent fee, and the interest thereon, was reversed. The judgment was modified by reducing the amount awarded to contractor 1 by $212,767.19, together with the interest assessed on such amount, and, as modified, affirmed.

This case stated clearly that lost profits are generally not a recoverable element of damages under a performance bond. This is more important in cases where owners seek performance from contractors' sureties who have defaulted. However, it is not consistent with the notion that a contractor's surety is obligated to fulfill the contractor's agreement, which is incorporated into the performance bond where that agreement would have provided for the recovery of lost profits by an owner. 50 Construction Law Digest 35 (S. Stein, ed., Apr. 2004).

Page 15-14, add after note 79 in first full paragraph:

However, in some jurisdictions if a third party is not listed as a beneficiary under a performance bond, it cannot recover from the surety. In *Angelo Iafrate Constr., L.L.C. v. Potashnick Constr. Inc.*, 2004 U.S. App. LEXIS 10620 (8th Cir. June 1, 2004), the State Highway Commission contracted with Potashnick Construction, Inc. (contractor 1) to do grading and structural work on a highway. Travelers Casualty and Surety Co. served as surety for the contracts, signing performance and payment bonds for each of the jobs. The commission also hired Angelo Iafrate Constr. Co. (contractor 2) to do surfacing work on the same portions of highway contractor 1 was working on. Contractor 1 informed the commission it lacked the financial ability to continue. The commission issued a suspension order allowing the surety 60 days in which to prepare to resume work. The surety hired another contractor to complete the work. The work was completed and accepted. The delay in work affected contractor 2's ability to complete its work. Contractor 2 filed a complaint against the commission for delay damages. A stay order was entered and the commission negotiated an agreement with contractor 2 in which it assigned to contractor 2 its rights as obligee under the performance and payment bonds. Contractor 2 then filed a complaint against the surety. The trial court found the surety did not breach any duty under the performance bond, that the surety had no duty under the bond to disclose its knowledge of the principal's potential financial difficulties prior to default, and that once default had been declared, the surety took all reasonable steps to effect the completion of the job in a timely manner. Contractor 2 appealed.

Contractor 2 contended the surety breached its performance bond obligations both to the commission and to contractor 2 directly. The performance bond did not specify any beneficiaries, but the payment bond included as beneficiaries those parties with claims for materials, labor, and supplies provided incident to the contract work. Contractor 2 did not fall within this intended class of beneficiaries and therefore could not recover as a third-party beneficiary. Contractor 2 also argued the surety and contractor 1 owed the commission a duty to inform the state of contractor 1's financial problems when it first became aware of the problems. The surety was required to give notice only of a difficulty in its own finances that would preclude its ability to perform on the bond. Because the surety was in no financial trouble itself, it did not breach this duty of notification. Contractor 2 argued it may recover from the surety based on the surety's duty to indemnify the commission for claims brought because of any injuries or damage received on account of the operations of contractor 1. The scope of the surety's duty to indemnify did not clearly encompass contractor 2's claim against the commission for economic delay damages. The language of that section instead reflected an intent to impose a duty to indemnify claims for personal injury and property damage caused by the work performed under the contract. The surety owed a duty only to the commission to complete performance; contractor 2 was not an intended beneficiary of the contract.

Page 15-15, add after carryover paragraph:

In *Cates Construction, Inc. v. Talbot Partners,* 21 Cal. 4th 28, 980 P.2d 407, 86 Cal. Rptr. 2d 855 (1999), Talbot Partners hired Cates Construction, Inc. to build a condominium project in Malibu. The construction contract called for Cates to complete the project in eight months. The construction contract also required Cates to furnish a performance bond. Transamerica Insurance Company issued the surety bond in favor of Talbot as obligee. A fund control agreement required Cates to submit monthly applications to Talbot for reimbursement of costs incurred. The 23rd payment request was not paid because both Talbot's and Cates's records showed that Talbot had already paid several hundred thousand dollars more than the cost of work. Six months after the contract should have been completed, Talbot advised Transamerica that Cates intended to default and that Talbot already had paid everything it owed under the contract. Talbot demanded that Transamerica perform under the bond. Cates later abandoned the project. At the time of the abandonment, construction was not complete and some of the work was defective and required repair. Transamerica informed Talbot that Talbot had breached the contract by failing to make payments. Transamerica refused to intercede or arrange for performance of the contract, claiming a legitimate dispute existed between Cates and Talbot.

The court considered whether an obligee may of a construction performance bond recover in tort for a surety's breach of the covenant of good faith and fair dealing. The court concluded that the exceptional approach thus far reserved for breaches in an insurance policy should not be extended to breaches of a surety bond given to assure performance on a construction contract. Talbot argued that a performance bond was a form of insurance and that breaches in the performance bond could justify the same extraordinary remedies that are available in insurance policy cases. Transamerica argued that performance bonds were fundamentally different from insurance policies and failed to warrant similar treatment. The court held that a construction performance bond was neither an insurance policy nor a contract of adhesion. Obligees have ample power to protect their interests through negotiation, and sureties were deterred from acting unreasonably by the threat of stiff statutory and administrative sanctions and penalties, including license suspension and revocation. The court ruled that in the absence of an independent tort, punitive damages may not be awarded for breach of contract even where the defendant's conduct in breaching the contract was willful, fraudulent, or malicious. Since Talbot may not recover in tort for Transamerica's breach of the implied covenant of good faith and fair dealing, the award of punitive damages was reversed in its entirety.

Page 15-15, add after first full paragraph:

There are at least two lines of reasoning to bind the performance bond surety to the arbitration provisions in the contract signed by the surety's principal. The first rationale is that the subcontract and the arbitration clause are incorporated by reference into the performance bond. Given the proper factual circumstances, this position is accepted by many federal jurisdictions. *United States Fidelity & Guaranty Co. v. West Point Construction Co.,* 837 F.2d 1507 (11th Cir. 1988); *Exchange Mutual Insurance Co. v. Haskell Co.,* 742 F.2d 274 (6th Cir. 1984); *Compania Espanola de Petroleas, S.A. v. Nereus Shipping, U.S.A.,* 527 F.2d 966 (2d Cir. 1975), *cert. denied,* 426 U.S. 936 (1976). A different rationale may be employed when the surety steps in to complete the contract after its principal defaults. The obligee may then argue that the surety is bound by all of the terms of the contract it assumed, including the arbitration provisions. This position was accepted by courts in Pennsylvania, in *Buck Run Baptist Church v. Cumberland Surety Insurance Co.,* 355 F. Supp. 913 (E.D. Pa. 1973); and in Kentucky, in *U.S. Fidelity & Guaranty Co. v. Bangor Area Joint School Authority,* 983 S.W.2d 501 (Ky. 1998).

§15.04 SURETY'S RECOVERY OF DELAY DAMAGES

Page 15-18, add after carryover sentence in carryover paragraph:

Failure of the owner to grant a time extension when one is due also may materially affect the surety's interests, and the surety may be able to pursue a claim against the owner. For example, in *Lumbermens Mut. Cas. Co. v. United States*, 2005 U.S. Claims LEXIS 238 (Fed. Cl. 2005), the Navy awarded a contract to Landmark Construction for the repair and renovation of military family housing units. The first contractor obtained a performance bond from Lumbermens Mutual Casualty. The government and Landmark Construction agreed to a bilateral modification that added 21 additional housing units and increased the contract's scope, but did not extend the completion date. When Landmark informed the government that it would be unable to complete the project, the government terminated the contract. The government had paid Landmark nearly 40 percent of the total contract price at termination, although Landmark had completed less than 15 percent of the housing units. The surety hired a second contractor to complete the project, which was already behind schedule. The completion date under the takeover agreement was the same as required by the original contract. When the second contractor did not complete the project on the original schedule, the government withheld liquidated damages. The surety claimed that the delay was caused in part by the bilateral modification for additional housing units. The surety alleged that the government's contract mismanagement caused the surety to incur costs in completing the project. The surety sought damages under a theory of impairment of suretyship. The government filed a motion to dismiss the surety's claim.

The surety asserted that the government substantially departed from the contract to the surety's detriment. The surety contended that it did not consent to the bilateral modification's enlargement of the scope of work. The government argued that the surety's consent was not necessary. The contract should not have been modified without consideration of the surety's interests, though consideration did not necessarily include consent. The bilateral modification's addition of additional housing units with no extension of the completion date could have materially increased the risk of default assumed by the surety when it bonded the contract. Whether the burden of completing 21 extra units added to the contractor's financial difficulties was an issue of fact to be developed through discovery. The extent to which the government considered the surety's interests was another issue of fact. The surety needed to show that the government departed from the terms of its contract with Landmark. The contracting officer had a duty to weigh the government's interests against those of the surety. The contracting officer could have exercised proper discretion by balancing the government's interests in the efficient completion of the project against the surety's financial concerns. There were questions of fact as to whether the contracting officer abused his discretion in administering the original contract and whether modification of the original contract materially increased the risk assumed by the surety. The government's motion for summary judgment was denied.

Page 15-18, add after note 100 in first carryover paragraph:

Other elements may also be necessary to support the surety's claims. For example, a surety is not entitled to relief for pre-takeover agreement claims if it is not a party to a contract prior to the takeover agreement. In *Fireman's Fund Ins. Co. v. England,* 313 F.3d 1344 (Fed. Cir. 2002), the United States Navy contracted with Summit General Contracting Corp. to construct a government building. Fireman's Fund Insurance Company provided performance and payment bonds. The contractor and the surety executed a General Indemnity Agreement. The government terminated the contract for default after the contractor failed to complete the contract by the scheduled date. The government and the surety entered into a takeover agreement under which the surety agreed to complete the contract. The contractor was not a party to that contract, which did not mention the General Indemnity Agreement or its assignment of claims to the surety. The surety submitted a claim to the contracting officer seeking an equitable adjustment and/or rescission of assessed liquidated damages. Its claims related to delays in performance that the government allegedly caused, both before and after the takeover agreement. The contracting officer denied all the claims, and the surety appealed to the Board of Contract Appeals. The board dismissed the claims that arose before the takeover agreement. The board held that the surety was not a "contractor" under the Contract Disputes Act. The board also ruled that the contractor's assignment of its claims to the surety was barred by the Anti-Assignment Act. The board also rejected the surety's claim of equitable subrogation, determining the doctrine was inapplicable. The surety appealed.

On appeal, the government contended that the surety was not entitled to seek relief for pre-takeover claims under the Act since the surety was not a contractor prior to the takeover, and the contractor's assignment of its claims to the surety was invalid under the anti-assignment statutes. The surety asserted that, under the doctrine of equitable subrogation, the surety succeeded to the contractual rights of its contractor. According to the appellate court, before the assignment was made, no claims against the government arising from a default in the construction contract had arisen. The assignment of the claims in the General Indemnity Agreement was invalid. Neither equitable subrogation nor the Act provided the surety with any contractual rights prior to the takeover agreement. The statutes precluding assignment of any interest in a contract with the government rendered the contractor's assignment to the surety invalid, and the surety was not a financing institution, which was exempt from the statutory prohibitions. Also, prior to the takeover agreement, the surety was not a contractor that could pursue claims under the Act. While equitable subrogation otherwise provided an available remedy, the doctrine could not be used to expand jurisdiction under the Act. For the board to have jurisdiction under the Act, the claim must be brought by "a contractor," which was defined as "a party to a government contract," and be one "relating to a contract." The surety was not a party to any contract with the government prior to the takeover agreement it had with the government, and its pre-takeover claims did not arise under such a contract. Although the surety did enter into a takeover agreement, its claims all related to its pre-takeover agreement activities.

CHAPTER 16
AVOIDING AND MINIMIZING CLAIMS

§ 16.06 THE CONTRACT

Page 16-22, add at end of section:

Contract conditions also influence claims. Statistics show that construction is the riskiest of human work activities. The risks are generally allocated to the contracting parties in the construction project in accordance with standard terms of contract contained in standard contract forms. Experience shows that a significant number of disputes in construction arise from the application and interpretation of the special standard conditions of contract used. It must, therefore, be realized that where parties choose to superimpose a set of conditions into their contract, these conditions take effect as an agreed legal code in addition to the applicable law of their contract. Therefore, in construction there is an additional area of law to be mastered, which is continuously evolving and being refined to meet the particular type of project envisaged by the owner or promoter and his advisory team. N. Bunni, *Construction Disputes on the Eve of the New Millennium,* 65 Arbitration 308, at 310 (No. 4, Nov. 1999).

While it is true that construction disputes tend to turn on questions of fact, they do frequently give rise to two different types of law issues. The first is related to issues of general law, such as negligence and the rights and obligations of the contracting parties and others. More important, however is the second type of legal issue, which is more frequent and which arises out of the standard conditions of contract used on the project. These conditions have developed into specialized areas of the law and have crossed national frontiers to form an international legal code for construction activities. They have been very successful in coping with any conflict of laws which may have existed between the various legal systems involved. Bunni, at 312.

§ 16.09 CLAIMS ADMINISTRATION

[A] Partnering

Page 16-29, add at end of subsection:

"Alliance contracting" seeks to establish a "no blame environment." This may be accomplished by requiring the parties to waive all future claims against the other alliance members, using a single insurance program to cover all project risks, and ensuring that all participants share common stakes in the project. Common goal provisions may include building incentives into the alliance agreements-incentives for reduction in capital costs, improvement in schedule performance, or long-term performance cost saving for the project. While these types of incentives may be commonplace in construction contracts, what is unique about alliance contracting is that the incentive payouts are made to all parties regardless of which party is primarily responsible for the savings, acceleration, or long-term operating profits. In the same way, any downward adjustments are incurred by all the alliance members for delays, overruns, or long-term operating shortfalls, regardless of who is responsible. All the alliance members are motivated to focus on the success of the project as a whole, rather than the performance of an individual participant.

Alliance structure projects, once limited to energy projects, now include museums, buildings, water, wastewater, and transportation projects. The benefits of alliance contracting have included overall project cost savings (even after incentive bonuses are paid), reduced interest payments from shorter schedules, reduced management staff, smaller contingency risk allotments, and reduced legal costs. From a claims and dispute perspective, the risk sharing and "no blame" culture resulted in numerous instances where all parties have banded together to overcome problems, minimize delays, and mitigate costs, whereas under traditional contracts, one party would have been assessed liability. Grynbaum's Dispute Prevention Update, Issue 21, July 15, 2001, at 2–5.

TABLE OF CASES

References are to sections.

INDEX TO SUPPLEMENT

References are to sections.

M

N

P

R

S

T

W

REVISED INDEX TO MAIN VOLUME

References are to chapters (Ch.) and sections.

E

H

I

J

L

O

P